The Flexible Phenotype

'This text is a must for anybody who has remained curious about the ways animals, including humans, deal with their environment. It's scientifically sound and at the same time it's a gripping story.'

Dr Hans Hoppeler, *Institute of Anatomy, University of Bern, Switzerland and Editor-in-Chief of* Journal of Experimental Biology

'Written with zest, a sense of fun, and a deep love of nature, *The Flexible Phenotype* offers biologists a real synthesis of ecology, physiology, and behaviour, based on in-depth empirical research. The adventures of the subject of many of these studies, the red knot, a migrant shorebird, captivate the imagination, show us how physiology and morphology express ecology, and make this book not only an important and truly integrated study in biology, but also a pleasure to read.'

Professor Eva Jablonka, *Cohn Institute, Tel-Aviv University, Israel*

'Even amongst mammals and birds, animals are diverse. Biologists have long sought the evolutionary pressures that have led to particular body designs and lifestyles. I believe that to do so requires an approach that considers the interaction of body design and behaviour in an ecological context. Too often these components are considered in relative isolation. In contrast, this book is wonderfully broad and holistic, integrating across levels. Even though the book is written in an accessible style that will entertain anyone who is interested in nature, it is serious science.'

Professor John McNamara, *University of Bristol, UK*

'Change rules every individual's world, whether imposed by temporal variation in the environment or by movement through variable surroundings, and survival and reproductive success depend on adaptive responses to this change. Drawing on experience in field and laboratory research, and integrating modern ideas about acclimatization and the optimizing of individual behavior, physiology, and morphology, the authors have produced an accessible, highly readable, and stimulating synthesis of the flexible phenotype. Using examples drawn from their own work on migrating shorebirds, as well as myriad other organisms, the authors show how individuals respond to change by altering their structure and function through a variety of behavioral and physiological mechanisms. Long-standing traditions of research in physiological, behavioral, and evolutionary ecology are brought completely up-to-date in this timely treatment of organisms in their changeable worlds. The ways in which mechanisms of individual response promote survival and productivity are keys to understanding how organisms will adjust, or fail to do so, in the face of habitat alteration and global climate change.'

Professor Robert E. Ricklefs, *University of Missouri - St. Louis, USA*

'In their book Piersma and van Gils provide a timely summary of the reawakening in our knowledge of phenotypic flexibility in the context of comparative biology. They convincingly remind us of how it is a key component of the whole process by which an organism interacts with its environment. Written in an engaging style which draws the reader into the salient issues of the day with everyday examples this book is not only a landmark in the field, but an entertaining read as well. It will therefore appeal to readers across the spectrum, from interested amateur naturalists, via students of physiological and behavioural ecology to established professional researchers.'

Professor John Speakman, *University of Aberdeen, UK*

The Flexible Phenotype

A body-centred integration of ecology, physiology, and behaviour

Theunis Piersma & Jan A. van Gils

OXFORD

UNIVERSITY PRESS

OXFORD

UNIVERSITY PRESS

Great Clarendon Street, Oxford OX2 6DP

Oxford University Press is a department of the University of Oxford.
It furthers the University's objective of excellence in research, scholarship,
and education by publishing worldwide in

Oxford New York

Auckland Cape Town Dar es Salaam Hong Kong Karachi
Kuala Lumpur Madrid Melbourne Mexico City Nairobi
New Delhi Shanghai Taipei Toronto

With offices in

Argentina Austria Brazil Chile Czech Republic France Greece
Guatemala Hungary Italy Japan Poland Portugal Singapore
South Korea Switzerland Thailand Turkey Ukraine Vietnam

Oxford is a registered trade mark of Oxford University Press
in the UK and in certain other countries

Published in the United States
by Oxford University Press Inc., New York

© Theunis Piersma and Jan A. van Gils 2011

British Library Cataloguing in Publication Data

Data available

Library of Congress Cataloging in Publication Data

Data available

Typeset by SPI Publisher Services, Pondicherry, India
Printed in Great Britain
on acid-free paper by
CPI Antony Rowe, Chippenham, Wiltshire

ISBN 978–0–19–923372–4 (Hbk.)
       978–0–19–959724–6 (Pbk.)

10 9 8 7 6 5 4 3 2

Contents

CHAPTER 1

Introduction

Since an organism is inseparable from its environment, any person who attempts to understand an organism's distribution must keep constantly in mind that the item being studied is neither a stuffed skin, a pickled specimen, nor a dot on a map. It is not even the live organism held in the hand, caged in the laboratory, or seen in the field. It is a complex interaction between a self-sustaining physico-chemical system and the environment.

G.A. Bartholomew (1958: p. 83)

George Bartholomew was not the only one to struggle with phenotypes. We all do—whether on a personal level, because our bodies and performances change with age for better or for worse, or professionally, because we are farmers, veterinarians, nurses, medical doctors, athletes, physiotherapists, ecologists, evolutionary biologists or students of these matters. Our bodies, and those of the organisms that command our attention, never stay the same, and, as we will discover in this book, neither are they quite what they seem to be at first sight.

The migrant shorebird story

Although encompassing the whole living world, this book was inspired by, and finds a focus in, the spectacle of seasonal migration, the way that birds and other animals travel the world. Migration would appear a natural fascination for anyone with a burning desire to explore and discover. How do animals find their way? How do birds cope with the radically different conditions that they encounter on their northern breeding grounds compared with their tropical wintering areas? Why do they carry out their extensive seasonal movements in the first place? Are they as free as they appear to be?

Born and raised as we were in the water-rich Netherlands, our youthful fascination with the biology of migration would soon find a focus in waterbirds, and especially the shorebirds that bred in the meadows around our villages to then disappear for the winter, or migrated through in great numbers in spring and in autumn. In the late 1980s, as budding professional scientists, we quite naturally chose a coastal shorebird known to use contrasting migration routes, the red knot *Calidris canutus*, as the focal species in which to evaluate the energetic consequences of this fascinating intercontinental migration behaviour. In The Netherlands, red knots come in two types. One population, breeding on the tundra of north-central Siberia, passes by in autumn to spend the winter in tropical West Africa and makes a brief touchdown during the return migration in late spring. Another population, from northern Greenland and the far north-eastern Arctic of Canada, comes to us in early autumn, only to leave in spring, thus having to endure the wet and windy, and sometimes frosty, conditions typical of winters in north-western Europe.

Field studies in both West Africa and the Dutch Wadden Sea are revealing. We can look at the birds' behaviour, assess their diet, which consists mostly of molluscs, study their movements and foraging activity over the intertidal flats in relation to tide and time of day, and look at changes in body mass—and the size of body stores—in the course of the winter. However, for a full evaluation of the distinct energetic balances of the tropical- and the temperate-winterers, we had to bring birds into captivity, so that we could measure digestive efficiencies as a function of prey type, and metabolic rates as a function of thermal conditions. Rather than feeding them the small bivalves that red knots typically eat and ingest whole in the field, we fed them the only thing we could afford to give them: protein-rich fish food-pellets, prey that did not need crushing in their muscular gizzards, as was their wont. When, after some time, we compared the bodies of free-living and long-term captive red knots, we were in for the shock that eventually triggered this book: wild and captive birds could not have been more different! At similar body masses, the captive red knots were fatter, with much smaller digestive organs than their wild counterparts. Rather than the expected 8–10-g gizzard, the captives eating soft food pellets had a 1–3-g gizzard. All of a sudden, the long-term captives no longer seemed a particularly good model for birds in the field.

But of course, and this insight came obligingly fast, the drastic changes in body composition in general, and in gizzard size in particular, of birds brought from one (outdoor) environment into a very different (indoor) environment, provided us with the key to the natural ways in which red knots cope with the environmental variations encountered in the course of their seasonal migrations. Our little revelation, that organismal structure tends to change in concert with the ambient environment in fully adaptive ways, was not new. George A. Bartholomew, for one, had the insight almost half a century earlier. However, it did make it clear that, in order to understand migrant birds, we needed a much tighter integration of ecological, physiological, and behavioural approaches than was usual at the time. And there seems no reason why this should not be true for *any* other organism.

Since then, we have tried to elaborate on this insight, studying problems of behaviour and distribution from various directions, including, simultaneously, the ecological, the behavioural, and the physiological angles. To be sure, a jack of all trades runs the risk of being master of none, which may explain why, at a time when the term 'integrative' has become part of the name of highly cited journals and university departments, *truly* integrative studies of organisms, with due attention to physiology, behaviour, and ecology, and their inter-relationships, are still thin on the ground. This understanding, together with our conviction that ecological studies of organisms that ignore the key elements of their physiology, physiological studies that disregard the organisms' ecology and behavioural studies carried out in ignorance of the environmental context, might all be a waste of time, were all factors that inspired the writing of this book.

Bodies express ecology

By developing empirical arguments to describe the intimate connections between (animal) bodies and their environments, we have tried to formulate a true synthesis of physiology, behaviour, and ecology. We shall review the principles guiding current research in eco-physiology, behaviour, and ecology, and illustrate these using as wide a range of examples as possible. The results of our long-term research programme on migrant shorebirds—and the foods they eat—will provide the unifying narrative. Chapters in the first half of the book all end with sections showing relevant examples from the shorebird world. As shorebirds travel from environment to environment, the changing natures of their bodies reflect the changing selection pressures in these different environments. The book and its examples will be incremental. Each successive chapter will build on the previous ones and add another level of explanation to the fascinating complexity of integrative organismal biology, illustrating how changes in an individual's shape, size, and capacity are a direct function of the ecological demands placed upon it. In essence, the book is about the ways in which *bodies express ecology*.

Bodies can 'express ecology' by being sufficiently plastic, by taking on different structure, form or composition in different environments. Part of the phenotypic variation between organisms, especially differences between isolated populations and unrelated individuals, may be fixed and reflect differences in genetic make-up (Mayr 1963). Some of the variation develops in interaction with the particularities of the environment in which an organism finds itself. This part, indicated by the term *phenotypic plasticity*, can be further subcategorized on the basis of whether phenotypic changes are reversible and occur within a single individual, and whether the changes occur, or do not occur, in seasonally predictable, cyclical ways (Table 1). The non-reversible phenotypic variation between genetically similar organisms that originates during development, *developmental plasticity*, has attracted much empirical and theoretical attention, including the publication of several monographs (Rollo 1995, Schlichting and Pigliucci 1998, Pigliucci 2001a). In contrast, the subcategory of phenotypic plasticity that is expressed by single reproductively mature organisms throughout their life, *phenotypic flexibility*—reversible within-individual variation—has remained little explored and exploited in biology (Piersma and Drent 2003). This is surprising, because, as we shall discover, intra-individual variation most readily provides insights into the links between phenotypic design, ecological demand functions (performance) and fitness (Feder and Watt 1992).

The subtitle of our book contains the words 'ecology', 'physiology', and 'behaviour', but not the word 'evolution'. This is not because we have neglected evolution in any way. Indeed, we fully subscribe to Theodosius Dobzhansky's classic dictum that 'nothing in biology makes sense except in the light of evolution' (Dobzhansky 1973). Evolution is not in our title because it is somewhat on the periphery of what we want to reveal concerning the intimate connections between environment and phenotype. Rather than discussing evolution explicitly, such as the exciting but poorly understood relationships between phenotypic plasticity and evolutionary change or innovation (as examined in the impressive treatise by West-Eberhard 2003), we shall assign evolutionary thinking to a back-seat, as we discuss, on the basis of reversible phenotypic traits, the many ecological aspects of organismal design.

Table 1. Mutually exclusive definitions of the four historically most commonly used categories of phenotypic plasticity (after Piersma and Drent 2003; note that previous workers used less restrictive definitions). Phenotypic plasticity itself indicates the general capacity for change or transformation within genotypes in response to different environmental conditions.

Plasticity category	Phenotypic change is reversible	Variability occurs within a single individual	Phenotypic change is seasonally cyclic
Developmental plasticity	No	No	No
Polyphenism[1]	No	No	Yes
Phenotypic flexibility	Yes	Yes	No
Life-cycle staging[2]	Yes	Yes	Yes

[1] Can be regarded as a subcategory of developmental plasticity.
[2] Life-cycle staging is a subcategory of phenotypic flexibility.

Even current authors who do explicitly discuss the role of phenotypic plasticity in microevolutionary adaptation to new environments (e.g. Price *et al.* 2003, Ghalambor *et al.* 2007), while they may incorporate *developmental* plasticity into their arguments, tend not to mention the existence of the *reversible* plasticity categories (Table 1). Does this indicate ignorance or oversight, or is it the case that understanding the role of phenotypic flexibility in evolutionary change still lies beyond grasp? In the final chapter of this book we shall provide our views on this subject and conclude that 'nothing in evolution makes sense except in the light of ecology'.

What is an organism anyway?

At this point we should make clear how we look at an organism, both with respect to its 'visible' incarnations in the course of development from egg to maturity, and also the 'invisible' continuity between generations (Piersma 2002). The whole 'life-history' cycle is written in the genes, is expressed in the *phenotype* and is evolutionarily evaluated in the *demotype* (Fig. 1). The hierarchy of 'life-history structures' (Ricklefs 1991) and their transformations are inseparable from each other and exist only with reference to their environmental context. In brief, the phenotype is composed of fixed traits (e.g. sex, beak size, feather colour, skeletal dimensions), as well as many flexible morphological and behavioural traits. Single genotypes may show variation in fixed adult traits (developmental plasticity), variation that comes about during development as a consequence of variation in the environment by the action of reaction norms (Schlichting and Pigliucci 1998). In addition, there are traits that remain flexible even at maturity (phenotypic flexibility), and a subcategory of flexible traits is formed by those that show predictable seasonally cyclic changes. This latter category of phenotypic variation was named *life-cycle stages* by Jacobs and

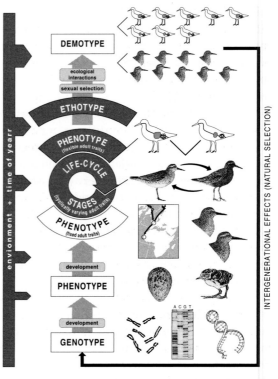

Figure 1. An organism (in this case a red knot, see Box 1) can be regarded as consisting of various intra-generational stages (indicated here by the terms genotype, phenotype, and others), in the course of which the influence of environment and time of the year increases. The 'life-history strategy' that is most successful, under the environmental conditions that an animal encounters, yields the highest numbers of offspring in the next generation (i.e. the most common demotype, the variant with the highest fitness). The sketches on the right illustrate (from bottom to top) the phenomena referred to by the genotype (chromosomes, part of a DNA sequence and the double helix), the early phenotype (egg and freshly-hatched chick), aspects of the adult phenotype that are fixed and those that are flexible (migration from Canada to Europe versus Siberia to Africa; short- versus long-billed birds, and large- versus small-gutted birds, respectively), including aspects of the phenotype that vary predominantly on a seasonal basis (life-cycle stages such as seasonal changes in plumage). Traits that can only be described in relation to environmental conditions are indicated by the term ethotype (for example, a trait like daily energy expenditure). The resulting reproductive success of such genetically instructed types, the inter-generational effect, is called the demotype (in this case illustrated by different numbers of short- and long-billed birds, or of large- and small-gutted birds).

Based on Piersma (2002).

Wingfield (2000). Life-cycle staging specifically refers to the occurrence of seasonally-structured sequences of 'unique' phenotypes with respect to state (e.g. reproductive or not; moulting or not) and appearance (e.g. nuptial plumage or not). Similarly, a subcategory of developmental plasticity called *polyphenism* refers to season-specific occurrences of particular phenotypes. The best-known examples are those butterflies in which the phenotypes of successive generations, during any given year, will depend on the season of hatching (Shapiro 1976, Brakefield and Reitsma 1991, Nijhout 1999). Both life-cycle staging and polyphenism may be under the influence of endogenous programmes, especially such as that of the circannual clock system (Gwinner 1986).

Phenotype, as classically understood, is something that one can measure in an organism independently of its environment, whereas the *ethotype* has no meaning except when set in an environmental context (Ricklefs 1991). The ethotype (derived from the now outdated label for the study of animal behaviour: ethology) encompasses the behavioural dimension of phenotypes, and includes factors such as the energy requirements of an individual as a measure of its performance in its environment. In principle, the fitness values of all these pheno- and ethotypic variants could be quantified (Nager *et al.* 2000), and in Ricklefs' (1991) terminology one would call this entity the *demotype*. The demotype, or fitness of an organism, is a function of both ecological interactions and sexual selection processes. Fitness determines which of the competing 'units of sequenced structures and transformations' (i.e. organisms) will survive in nature's never-ending struggle (Fig. 1).

Organization of the book

This book is built around a logically arranged constellation of first principles in energetics, ecology, physiology, and behaviour. Let us call them 'rules of organismal design' (including constraints on design). In Chapter 2 we start off from two relevant physical laws: (1) organisms cannot escape the first law of thermodynamics (the law of conservation of energy) and (2) they are bound by the second law of thermodynamics (the law of spontaneous 'spreading-out' of energy and matter). In order to reduce the amount of entropy in their bodies, organisms have actively to maintain a balance in terms of their gains and losses of water, nutrients and energy, and solutes (micronutrients). Depending on what needs to be kept in balance, and the size and nature of the body involved, accounting takes place over varying time-scales. In Chapter 3 we introduce the background of the cost–benefit analyses of functional capacities that recur throughout the book. Organs are always part of chains of processing units within bodies. For this reason there is no point in spending the energy and nutrients that would be required to build organs with excess capacity relative to that of the other organs in the chain. This is the principle of symmorphosis that predicts close quantitative coupling of different performance measures within single organisms. As common sense as this may seem, there are nevertheless several issues that need to be discussed. These have to do with multiple functions, reserve capacities of specific organs under extreme and dangerous conditions, and the problematic climbing of 'adaptive peaks' (Dawkins 1996). Some of the empirical arguments build around the technique of 'allometric scaling', a concept that is briefly introduced in that chapter.

In Chapter 4 we make our first excursion from physiology to ecology, moving away from physical boundary factors and design principles to the real world. In this chapter we examine the extremes of adaptiveness as we discuss peak-performance rates, also known as 'metabolic ceilings'. The idea that the working capacity of organisms would have some sort of maximum level dictated by size and other design constraints (known as maximum sustained metabolic rates), has inspired a large body of ecophysiological research, including the search for allometric constants. We shall review the field and illustrate the (conditional) presence and workings of such ceilings by examining shorebirds working hard, both under cold conditions (during breeding in the High Arctic and under harsh climatic conditions in the non-breeding season) and also during trans-oceanic non-stop migration flights. We conclude that metabolic ceilings are functional solutions to (1) the environmental conditions in which an organism lives and (2) the precise ways in which hard work precipitates death, a conclusion that reverberates through much of what we will repeatedly encounter in the rest of the book.

Expanding on our introductory Table 1, we illustrate in Chapter 5 how phenotypes, at several different levels and over several different time- and developmental scales, succeed in matching their environments. This chapter is not simply a lengthy review of odd organismal phenomena. By describing the great variation of ways in which organisms adjust to variable environmental demands, we emphasize the immense 'creativity' of the evolutionary process in response to ecological challenges. After a discussion on the possible costs of a capacity to change, we raise questions about the extent to which phenotypic adjustments are fully genetically instructed or self-organized on the basis of organism–environment feedback loops. The reversibility of some such changes is illustrated by the way in which the organs of birds are continuously adjusted to changes in energy expenditure, food quality, and other environmental factors.

The cost–benefit approach introduced in Chapter 3 is incomplete because it is limited to physical responses. After all, the behavioural dimension of phenotypes (the ethotypes) can respond even more rapidly to changing environments than bodies can. The behaviour of real animals at particular times and places is not only influenced by locally achievable energetic costs and benefits, but also by the fitness currency that the animal has been naturally selected to optimize. In Chapter 6 we discuss how currencies and design constraints influence optimal behavioural strategies, especially with respect to time and place allocations. In the process we review the use of the word 'optimal' in evolutionary ecology. Having introduced this additional layer of complexity, in Chapter 7 we focus on a particular class of behaviour: the choice of diets and feeding patches. Foraging must guarantee a steady income at the lowest possible price. We introduce the elegant theories of optimal diet choice and the use of feeding patches. Until now such theories have assumed rather constant environments and bodies. Here, based on our own work on shorebirds, we extend the story to include the dynamics of organ size, energy expenditure, and prey quality. These observations accord with new versions of foraging models that allow for flexible adjustments of food-processing capacities.

From here onwards, building on the basics laid out in the previous chapters, the book becomes increasingly 'integrative'. There is more to nature than nutrients and energy equivalents: only organisms that successfully escape predation and resist disease long enough will leave offspring. In Chapter 8 we bring these factors into context by introducing the ways in which animals cope with the threats of disease and predation. The immune system is not only amazingly complex, but it is shown to be highly flexible as well. In addition to simple behavioural measures to avoid being killed by predators, birds also appear to have physiological adjustments in store. We introduce ways in which the fields of predation, disease, and foraging can be merged under the single equation of 'common currencies'.

In Chapter 9 we come close to finding the Holy Grail of population biology by reaching an understanding of how environmental conditions and organismal strategies translate into individual fitness (the demotype), and hence into changes in the size (and the distribution) of populations. We illustrate how the relationships between variable features of the environment (i.e. prey quality) and of the organism living in that environment (i.e. gizzard size or processing capacity) combine to affect subsequent survival and population sizes of red knots. We also examine how organisms cope with various kinds of global change. Do the adjustments shown reflect changes at the level of the genome, or do they reflect phenotypic plasticity?

In Chapter 10 we try to bring everything together under an evolutionary spotlight. In his masterful book-length essay on the basics of evolutionary organismal biology, Denis Noble (2006) stopped short of using the term 'ecology' with a discussion of organs and systems, which he described as the 'orchestra'. Here we provide a venue in which that 'orchestra' can be located, a proper 'theatre' composed of the ever-stringent ecological context. Indeed, many of the riddles in understanding the distribution of red knots, other shorebirds, and many other organisms, could not have been explained without insights into their energetics (and those of their prey), organ capacities, and phenotypic flexibility. What we provide here is a discussion of the degree in which flexibility of phenotype is relevant to the evolutionary process. For example, do genes take the lead during evolutionary adaptive change, or is there merit in the suggestion that genetic changes actually follow phenotypic adjust-

ments (West-Eberhard 2003)? The writing of this final chapter on evolutionary implications was difficult. It most clearly represents this book's attempt to 'grope publicly for solutions to recalcitrant conceptual issues of the day' (Gottlieb 1992).

Scope and readership

We trust that this book will encourage a further integration of ecology, physiology, and behaviour, and will in turn foster collaborative research agendas between various kinds of organismal scientists, beyond early proposals such as those of Bill Karasov (1986). Although migrant shorebirds are the principal characters in some parts of our story, the arguments developed here are of much broader relevance. We hope that this book will be read by a wide audience of professionals (obviously including advanced students) who work with organisms, whether they have medical, veterinarian or biological backgrounds. It should attract readership among active researchers and scholars from a wide range of taxonomic specializations. Given our own background, our approaches will be most easily recognized and appreciated by those working at the interface of physiology, behavioural ecology, and evolutionary biology, but we also hope to heighten the interest of both hard-core evolutionary and molecular biologists and practising phenotype-specialists, including those in the medical and veterinary professions.

Although aimed at a diverse audience, this is still a 'scientific' book that offers in places complicated, but fundamentally instructive, drawings and diagrams, uses technical terms (that we have tried to explain as we encounter them in the course of the story) and references to the primary literature. At the same time, we have tried hard to *engage* our readers by developing narratives of excitement and discovery.

In the end, this book should appeal to all those who expound a truly integrative approach to understanding the basis of our existence on Earth, i.e. of biological systems in a global, ecological setting. We hope this will include theoretical physicist Stephen Hawking (Fig. 2), the author of the best-selling *A Brief History of Time* (1988). At a pre-flight news conference before his first experience of weightlessness, Hawking said that he wanted to encourage public interest in space exploration. With an ever-increasing risk of being wiped out on Earth, he argued that humans would need to colonize space. Apart from the *need* for large human populations and strong societies to succeed in space travel (which is perhaps

unlikely still to exist in the event of an earth-ecosystem collapse), we find it very strange to believe that humans *could* sustain themselves in space in the first place. As this book makes abundantly clear, we are connected to earth as with an umbilical cord; our functioning organism simply *needs* earth. Whether we think about the bacteria in our guts, the peace of mind that comes from strolling through earthly habitats, the way that our developing bodies need normal gravity and even pathogens (Zuk 2007), or the multifold ways in which bodies express ecology, human beings, alongside most other organisms, are a physical extension of it. Bodies *are* earth.

Acknowledgements

Our decades-long work with shorebirds was the impetus for what we have presented here, especially because we saw it as part of a much broader and as yet unfinished canvas. Over time our impatience for a unifying thesis grew and we realized that the quickest way to this end was to attempt to do it ourselves, to write a book. In an interview with Christine Beck (2004), Bart Kempenaers, then the newly appointed research director of the Max Planck Institute for Ornithology in Seewiesen, Germany, nicely expressed how we felt when he said that 'books are exciting because they go beyond pure fact, because they give shape to developing ideas, disclose backgrounds, reveal associations; books change perspectives'. However, we were somewhat naïve in thinking that our 'task would involve little more than writing down all we knew before we forgot it' (paraphrasing Dolph Schluter 2000, p. vii, replacing 'I' with 'we'). We can now certainly concur with him that 'as the book got underway the limits of our knowledge became distressingly apparent, and we now feel we learned most of the contents along the way'.

Figure 2. After a life immobilized by a debilitating neuromuscular disorder (but still able to study the bone-crushing gravity of black holes), Stephen Hawking, a theoretical physicist, enjoyed a total of four minutes of zero gravity in a modified Boeing 727 flying a series of parabolic arcs between 11 and 8 km height in April 2007. Weightlessness is achieved during the 45-degree nose dives. In Chapter 5 we discuss the effects of long-term zero gravity conditions on body function and organ sizes.

Photo freely available at www.ZeroG.com.

Box 1 Brief description of the book's empirical backbone: a molluscivore migrating bird interacting with its buried prey

Red knots (Fig. 3) are close to what an 'average bird' would probably look like with respect to mass, size and dimensions. They are slightly bigger than a blackbird *Turdus merula* (Old World, southeast Australia, New Zealand), or an American robin *Turdus migratorius* (New World), but they show much greater seasonal variation in mass and feather colour than either of these thrushes. This has to do with the very long migrations of red knots that breed on barren tundra as far north as one can get. Outside the breeding season, one finds red knots throughout the world except Antarctica, but only in marine coastal habitats, usually large wetlands with extensive intertidal foreshores (Piersma 2007). The discontinuous circumpolar breeding range of red knots incorporates the breeding areas of at least six populations (Fig. 4), populations that are sufficiently distinct morphologically to count as subspecies. These subspecies are certainly distinct when it comes to their migratory trajectories and the seasonal timing of their movements, and show little or no overlap at the final wintering destinations, with a limited overlap during south- and northward migration. Arguably the most fascinating aspect of the whole substructuring of the world's red knots is the suggestion, based on genetics, that all of the world's six or seven subspecies have diversified recently from a single founder population that survived the last glacial maximum of ca. 20 000 years ago (Buehler *et al*. 2006, Buehler and Piersma 2008).

Red knots are specialized 'molluscivores' (i.e. predators eating only shellfish), but briefly shift their diet to surface arthropods during the summer breeding season on High-Arctic tundra. Wherever people have studied their diet, red knots eat hard-shelled

Figure 3. A red knot on Banc d'Arguin, Mauritania, a wintering area in West Africa, that is ingesting a hard-shelled food item: a bivalve. The back-and-white margins of the wing covering feathers give it away as a juvenile bird, only a few months old when photographed shortly after capture in December 2004. The bird carries rings that combine the colours Red, White, and Blue, and is known to us as R6WRBB. Molecular sexing on the basis of DNA in a small blood sample (Baker *et al*. 1999) showed it to be a male. R6WRBB has been sighted many times in Mauritania for a further two years after capture.

Photo by Jan van de Kam.

molluscs, sometimes supplemented with easily accessed softer prey, such as shrimp- and crab-like organisms and even marine worms (Piersma 2007). This is not because red knots necessarily prefer hard-shelled molluscs (in fact they do not, when given the choice), but because they are specialized in finding and processing such prey, to the detriment of being able to find the actively crawling soft-bodied worms and small crustaceans on which other sandpipers specialize. In fact, one of their sensory capacities, the ability to use self-induced pressure gradients around hard objects in soft, wet sediments, has not been described

continues

Box 1 *Continued*

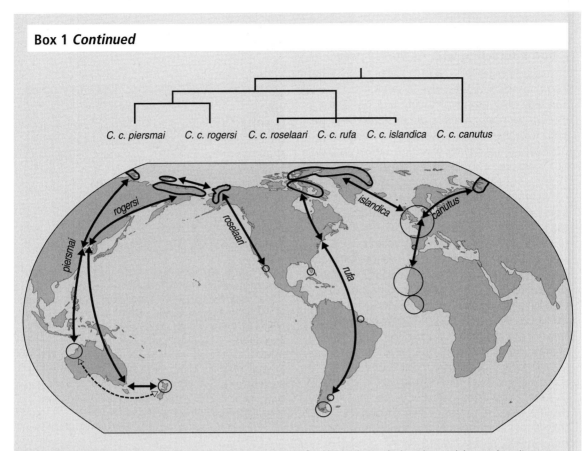

Figure 4. Global distribution of the six recognized subspecies of red knots during the breeding and the non-breeding season. Polygons in the arctic depict breeding ranges; circles depict principal wintering areas and the diameter of the circles indicate the relative number of birds using each area. Arrows depict north- and southward migration routes and use of en route sites. Projected above the contemporary distribution of red knots is a phylogram summarizing the genetic population structure as presently understood (Buehler and Baker 2005).

This figure is modified after Buehler and Piersma (2008), an update of the original synthesis by Piersma and Davidson (1992).

for any other animal. The 'remote detection' of buried hard-shelled prey is probably enabled by their bill-tip organ, the dense conglomerations of pressure sensors (Herbst corpuscles) clustered in forward-pointing 'sensory pits' in the outside surfaces of the tips of both upper and lower mandible (Piersma *et al.* 1998).

Unlike other molluscivore shorebirds, such as oystercatchers, that remove the flesh from the shell with their stout bill, red knots, in the same way as molluscivore-diving ducks, ingest their prey whole, then crush the shell in the muscular part of the stomach, the gizzard (the glandular stomach is rudimentary; Piersma *et al.* 1993b, van Gils *et al.* 2003a, 2006a). Crushed shell material is usually voided as faeces and not as pellets, which is the habit of related shorebirds, such as dunlin *Calidris alpina* and redshank *Tringa totanus*. As a consequence of the work carried out on the shell material by the gizzard and intestine, and in order, perhaps, to prevent the wear and tear inflicted by shell fragments on the sensitive intestinal wall, both gizzard and intestine are relatively heavy in shorebirds that eat hard-shelled-prey. In an allometric comparison among 41 shorebird species, the red knot came out as the one with the largest gizzard for its body mass (Battley and Piersma 2005).

Although red knots are able to crush and process reasonably heavily built molluscs, they rather prefer, as we shall see, thin-shelled prey. It also helps their food-finding if the shellfish live close to the sediment surface, or on top of it, and if they are not attached to any hard surface, such as a piece of rock, a mussel- or oyster-bank, or mangrove roots. The state of traits that red knots, indeed most predators, prefer, obviously precisely mirrors the state of traits that molluscs use to avoid being eaten. As prey they can develop heavy or extravagant armature, they can attach themselves to a hard substrate, they can sit deep in the sediment beyond the length of a probing bill, or, upon discovery by a predator, they can try to move away fast (Vermeij 1987, 1993). Red knots do eat gastropod molluscs (snails, usually protected by a heavy or profusely ornamented shell), but most of their diet consists of bivalves, and these bivalves employ all the different strategies just listed (Stanley 1970, Piersma *et al.* 1993a). All bivalves are protected by hard valves, but shallowly buried species, such as edible cockles *Cerastoderma edule*, are much more heavily armoured than deep-buried bivalves, such as the Baltic tellin *Macoma balthica* and the other relatively thin-shelled tellinids. There are no examples of intertidal bivalves exhibiting ridges or spines, although the heavily ridged venerid *Placamen gravescens* from north Australia comes close. Mussels of the family Mytilidae come in heavily armoured forms, or may secure themselves to rocks or conglomerates of conspecifics with so-called 'byssal threads'. A few bivalves of intertidal flats, such as the very thin-shelled *Siliqua pulchella* that lives in the soft muds of Australasia, are fast burrowers and swimmers and still have a chance to get away when detected by a probing shorebird.

We dedicate this book to our mentor and teacher, Rudi Drent. We are sad that he did not live to be part of the entire writing process. Sharing thoughts with him and entertaining ideas about our science were among the most formative and fantastic experiences in our professional lives. Rudi influenced this book in many more ways than by introducing us early on to the concepts of optimal working capacities, also known as 'metabolic ceilings' (Chapter 4). As an ethologist, Rudi certainly never lost sight of ecological context. As professor in animal ecology at the University of Groningen he created a great ambiance for budding ecologists. He was forever stimulating ambitious, if not daring, programmes of field research, whilst still attending to the physiological nuts and bolts that are better studied in laboratory settings. We appreciate, and cherish, his many fights for the institutional embedding of our endeavours, the very force that gave our generation a chance in science.

Yes, finishing this book has been a long journey. We would like to acknowledge the trust, patience, encouragement, help and tolerance of the people that have to live with us in their daily lives—both our colleagues at work and our family members. TP is grateful to Petra de Goeij for creating great ambiances for book writing. JAvG admires the patience of his wife, Ilse Veltman, and their children, Lieke and Jort, which they expressed so many times during the writing process. We are also grateful to the artists who helped make this book what it is. Dick Visser prepared all the new artwork and, as usual, it was both exciting and a great pleasure to work with him. We thank Barbara Jonkers for her smart cover design, Jan van de Kam for providing stunning photos, and Ysbrand Galama for the dreamcow of Chapter 2. TP thanks David Winkler and the Cornell Laboratory of Ornithology for generous hospitality during a minisabbatical in Ithaca—laying the ground for Chapter 10. We are very grateful to our contacts at Oxford University Press, especially editors Ian Sherman and Helen Eaton, for their guidance, encouragement, patience, and general good faith in the enterprise. Marie-Anne Martin provided the index, was a great text editor and a constant source of encouragement during the final phases.

Two long-time friends from Canada, Hugh Boyd and Jerry Hogan, read all chapters. Apart from giving us confidence, they kept us honest in the use of English, the terminology of proximate and ultimate causes, and helped us with context. We are very grateful for their enduring support. Bob Gill provided a refreshing review of the introductory chapter. Irene Tieleman shined her bright light on Chapter 2.

Hans Hoppeler, Ewald Weibel, Maurine Dietz and Rob Bijlsma provided helpful comments on Chapter 3. We enjoyed interacting with Tim Noakes with respect to Chapter 4, and appreciated the constructive comments by John Speakman, Bob Gill, Bob Ricklefs, Kristin Schubert (who additionally provided material for an entire paragraph), Bernd Heinrich, Irene Tieleman, Joost Tinbergen, John McNamara, Maurine Dietz, Klaas Westerterp, and Dan Mulcahy. For comments on Chapter 5 we thank Matthias Starck, Paul Brakefield, Chris Neufeld, Rick Relyea, and Massimo Pigliucci. Ola Olsson commented on Chapter 6, John McNamara on Chapter 7, Piet van den Hout on Chapter 8 and Christiaan Both helped with Chapter 9. Ritsert Jansen tinkered with the first and the last chapters, but what we may really need is a new book! Eva Jablonka, Massimo Pigliucci, and Bob Gill helped with the final chapter, as did the 2009/2010 cohort of the University of Groningen TopProgramme in Evolution and Ecology: Adriana Alzate-Vallejo, Lotte van Boheemen, Rienk Fokkema, Jordi van Gestel, Oleksandr Ivanov, Hernan Morales, Froukje Postma, Andrés Quinones, and Michiel Veldhuis. Finally, we thank those who either supplied or helped us find original photos or artwork: John Speakman, Colleen Handel and Bob Gill, Chris Neufeld, Mike Stroud, Duncan Irschick and Maria Ramos, Ola Olsson, and Alexander Badyaev.

Part I

Basics of organismal design

CHAPTER 2

Maintaining the balance of heat, water, nutrients, and energy

Dutch dreamcows do not exist

We grew up in the Dutch countryside. Land of cattle. One of us vividly remembers the cartoon in the milking shed at one of the local farms: the Dutch dreamcow, with udders at both ends to produce milk, a backside at each end to produce manure and no mouth to feed (Fig. 5). Such a cow would defy the first law of thermodynamics, the law of 'conservation of energy': a *perpetuum mobile* that could not exist. Applied to cows, the first law simply states: what comes out (nutrients and energy in the form of milk and manure), needs first to go in (nutrients and energy in the form of food). The exact form of energy is not important as it may be transformed from one (food) into another (milk, manure). This necessitates work and the energy (in the form of heat) lost along the way.

Figure 5. Cartoon of the Dutch dreamcow, a cow that produces twice the amount of milk and manure but does not have to eat. A living perpetuum mobile!

Drawing by Ysbrand Galama.

It was recognized long ago that all living organisms have to face this basic physical law. Max Rubner (1854–1932), a German physiologist, measured the heat produced by a dog placed in a metabolic chamber for 45 days! He found that this amount (17 349 kilocalories = 72 588 kJ) precisely matched the energy that was released when combusting the dog's food (17 406 kilocalories), correcting for the amount of energy retained in the dog's faeces. He concluded that the only source of heat production by his dog was the chemical energy captured in the dog's food (Chambers 1952). No one since Rubner has refuted these simple physical facts, and no one ever will.

Beside the first law that can never be broken, organisms also need to cope with the ever-present second law of thermodynamics. Any system, be it dead or alive, tends to show an increase in disorder, called entropy, over time. Eggs break rather than unbreak, coffee grows cold rather than hot, and people grow older rather than younger. In order to live, organisms need to fight against this spontaneous 'spreading-out' of energy and matter. To reduce the amount of entropy in their bodies, organisms actively maintain some sort of balance in terms of the gains and losses of heat, water, nutrients, and energy, and solutes (micronutrients). Stated simply: they have to eat (Fig. 6)! If they do not, their bodies will soon reach a state of much greater entropy, a state that we call death.

With this inevitable trend toward increased entropy, useful chemical energy eventually

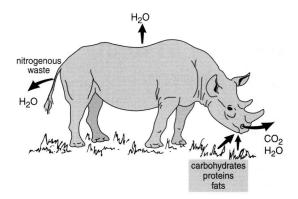

Figure 6. For all animals, this rhinoceros included, the following is true: eating is nothing more than increasing the entropy of your food, in order to prevent an increase in the entropy of yourself. By breaking down food molecules into smaller molecules, free energy is released that can then be used to drive energy-requiring reactions within the animal.

Based on Eckert *et al.* (1988).

degrades into a 'waste' form of energy: heat. Heat is considered a useless by-product, as it cannot be converted back into other forms of energy. In hot climates, heat can even be a problematic by-product, as it may raise body temperature to excessively high levels. But in cooler climates, heat is mostly a useful by-product, as it helps an animal to control its body temperature and thus the rate of chemical reactions taking place inside its body. How animals regulate their heat balance will be the first issue dealt with in this chapter.

Hot bodies in the cold

Sometime during our evolutionary history we crawled out of the water as fishlike vertebrates (Dawkins 2005, Shubin 2008). Compared with air, water has a much higher heat capacity and heat conductivity, making it difficult for our aquatically living ancestors to maintain a temperature differential between their bodies and their surroundings. Modern aquatically living mammals, such as seals, have solved this problem by depositing 'blubber', a sparsely vascularized layer of fat directly under the skin. The ability to maintain constant body temperatures was not seized upon by the land-dwelling reptiles, but once feathers and hairs had evolved, it became possible to maintain relatively constant core temperatures.

Great benefits were thus to be gained. Places where insufficient solar radiation was available to heat up a cold body could now be inhabited. In fact, even the coldest corners of the Earth became suitable habitats—think of penguins living in Antarctica or polar bears *Ursus maritimus* doing the same around the North Pole. At a biochemical level, efficient enzymes with narrow temperature-tolerance ranges could now be utilized. Furthermore, as neurological reactions run faster at higher body temperatures, information could now be processed faster as well. Very rapid responses can be vital when trying to catch a prey or escape a predator. Finally, warm blood, especially blood that is warmed up during a fever, keeps pathogenic fungi out (Robert and Casadevall 2009).

Obviously, the advantageous warm-blooded lifestyle also carries with it a large cost: much more food is needed to cover the greater heat production. We will come back to that in a moment, but first we want to explain why the terms warm- and cold-blooded are inappropriate, even though they have been in use since Aristotle (Karasov and Martínez del Rio 2007). Under hot circumstances, the blood of a 'cold-blooded' organism (e.g. a lizard on a rock) can in fact be warmer than that of the 'warm-blooded' creature sitting right next to it (e.g. a gerbil). To get rid of this inconsistent terminology, modern classifications are based on the constancy in body temperature, giving us the terms homeotherm (*homos* = same) and poikilotherm (*poikilo* = varied). Another way of classifying the thermal biology of animals is by recognizing the heat source that keeps the body warm: endotherms (*endo* = within) mostly use their own metabolism for this, while ectotherms (ecto = outside) use external heat. Most endotherms are homeotherms (e.g. the gerbil) and most ectotherms are poikilotherms (e.g. the lizard), but there are exceptions (e.g. parasites living inside a bird or a mammal are typical homeothermic ectotherms). In this chapter we shall mostly deal with homeothermic endotherms.

Let us now return to the costs of being a *homeotherm*. Much of what we now know about temperature regulation in homeotherms is due to the pioneering work of Per Scholander (Scholander *et al.* 1950a). This man, who moved to the United States after growing up and being trained as a botanist in Norway, had a lifelong fascination with the physiological responses to extreme conditions of both plants and animals. His adventurous mind, together with the need to take measurements along an 'extreme condition axis', brought him around the world at a time when not so many could do so (Scholander 1990). For his work on thermoregulation, his original plan was to build a mobile laboratory in a huge military troop glider (an engineless aircraft), which could then be brought to remote places (e.g. the inaccessible and practically unknown Prince Patrick Island in the Canadian Arctic Archipelago was on his mind). His boss thought this idea was insane, but at the same time suggested building a more permanent laboratory at an alternative location, Point Barrow on Alaska's North Slope. And this he did.

In Alaska, Scholander measured the costs of homeothermy by placing individual animals in small metabolic chambers. During trials lasting eight to twelve hours, an individual's O_2-consumption and CO_2-production were quantified across a wide range of temperatures (from the lowest Alaskan winter temperatures of –40 °C up to over +30 °C generated by an electric heater). The animals were of various kinds: snow buntings *Plectrophenax nivalis*, jays *Perisoreus canadensis*, arctic gulls *Larus hyperboreus*, weasels *Mustela rixosa*, lemmings *Dicrostonyx groenlandicus rubricatus*, ground squirrels *Citellus parryii*, arctic foxes *Alopex lagopus*, Eskimo dogs *Canis familiaris*, and polar bear cubs *Thalarctos maritimus*. With this diversity of animals locked up in out- and indoor cages, there were times when the Point Barrow laboratory must have looked like a small zoo.

Scholander and his colleagues predicted that simple Newtonian cooling laws governed the heat loss of an animal, i.e. that the heat loss (and thus heat production under a constant core temperature) would be proportional to the temperature gradient between the animal and its surroundings (Newton 1701, Mitchell 1901). They were right. Cooling down the metabolic chamber, such that the gradient between body temperature and environmental temperature doubled, did indeed double an animal's metabolic rate. However, this simple and elegant result was found only when the environmental temperature was *below* a certain value. *Above* this threshold, the so-called lower critical temperature, metabolic rate remained remarkably stable; in this region, heat as a 'waste product' of resting metabolism could be used to maintain constant body temperature. As shown later (e.g. King 1964), an upper critical temperature also exists. Above this threshold additional costs come into play in order to prevent overheating.

The lower critical temperature differed enormously between their study animals: lemmings and weasels reached this threshold between 10 and 20 °C, while arctic foxes could go down to –30 °C or more before they elevated their metabolic rate.[1] In general, Scholander *et al.* found that the lower critical temperature was lower in larger animals than in smaller

---

[1] Note that a recent study on arctic foxes found much higher values (Fuglesteg *et al.* 2006). It was argued that Scholander's foxes were not at rest during the measurements.

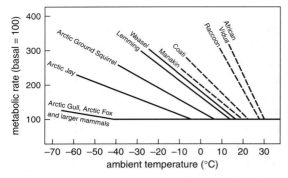

Figure 7. The metabolic response to cold in arctic (solid lines) and tropical (dashed lines) birds and mammals as measured by Scholander *et al.* (1950a). Each species' basal metabolic rate has been standardized to a value of 100 to facilitate comparison between species.

From Scholander *et al.* (1950a).

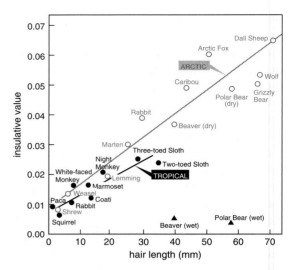

Figure 8. Hair length largely determines the insulative capacity of a mammal's pelt. Mainly for this reason, arctic mammals (open dots) are better insulated than tropical mammals (closed dots), as their hairs are usually longer. In addition, for a given hair length, tropical pelts have a slightly lower insulative value than arctic pelts. Note that immersion in water (closed triangles) reduces an animal's capacity to maintain body heat enormously.

Based on Scholander *et al.* (1950a).

animals (Fig. 7). There are two reasons for this. One reason, already recognized by Scholander *et al.* (1950b), is that larger animals have longer fur or longer feathers, which trap the heat more efficiently (Fig. 8). The other reason is that, with an increase in body mass, body surface area, through which most heat is lost, increases more slowly (mass$^{2/3}$) than does metabolic rate (mass$^{3/4}$; Schmidt-Nielsen 1984; see Box 2). Thus, at least within a given species, the larger an individual, the better it would be able to live in a cool climate, and the more difficult it would be to live in a warm climate; heat dissipation would be more difficult for larger bodies. This principle underpins a much older 'rule', named after Bergmann (1847). Bergmann's rule on the basis of surface-to-volume ratios states that body size increases when going from equator to pole. There is good evidence that this rule does indeed hold *intra*specifically, both among mammals (Ashton *et al.* 2000, Freckleton *et al.* 2003, Meiri and Dayan 2003) and birds (Ashton 2002, Meiri and Dayan 2003), though it is sometimes questioned whether heat conservation is the driving force (Ashton 2002, Ho *et al.* 2010). The existence of elephants, rhinoceros, and giraffes, all distributed in the tropics but not around

the poles, tells us that this rule does not give robust predictions *inter*specifically.

On the subject of latitudinal gradients, Scholander was interested to find out whether tropical animals, for other reasons than body size, had different lower critical temperatures from arctic animals. Thus, after spending two winters at Point Barrow (at 71° N), he and his team went to Barro Colorado Island, Panama (at 9° N), in order to take similar measurements, but this time on sloths *Choloepus hoffmanni*, monkeys *Aotus trivirgatus* and *Leontocebus geoffroyi*, raccoons *Procyon cancrivorus*, coatis *Nasua narica*, jungle rats *Proechimys semispinosus*, and two tropical bird species *Pipra mentalis* and *Nyctidromus albicollis*. The results were astonishing: tropical animals began to increase their metabolic rate when the air was cooled down just a few degrees below the ambient air temperature, i.e. in the range of 20 to 30°C! As raccoons and jungle rats are not smaller than weasels and lemmings, this difference could not be explained by body size. Scholander *et al.* skinned several of their study animals and found that a piece of tropical pelt had a lower insulative capacity than an equally sized piece of arctic pelt; this result could be partly explained by the fact that tropical fur is generally sparser and shorter (Fig. 8). As human beings, we are a good example of how the low insulative value of our 'pelt' over-rules the effect of a relatively large body size. Our lower critical temperature lies between 27 and 29 °C (Du Bois 1936, Winslow and Herrington 1949). Needless to say, this is with our clothes off. Scholander *et al.* (1950a) concluded: 'Man is indeed a tropical animal, carrying his tropical environment with him.' Note that better insulation in northern arctic/temperate species is not a general rule. For example, Tieleman *et al.* (2002) found 'desert larks' (exposed to cold nights) were better insulated than equally sized larks living in the temperate zone.

Apart from body size and the insulative value of fur, there may be another reason why arctic homeotherms can better withstand the cold than their tropical equivalents: they have a bigger 'engine', a higher basal metabolic rate (BMR). This would have the effect of lowering an animal's lower critical temperature. Is there evidence for higher BMRs in arctic mammals and birds? Scholander *et al.* (1950c) concluded that there were no general BMR differences between 'their' tropical and arctic animals, notwithstanding a few exceptions (the 'lazy-acting' sloths had relatively low metabolic rates, unsurprisingly). This was consistent with a study on five tropical bird species by Vleck and Vleck (1979). However, later studies, both on birds (Weathers 1979, Hails 1983, Kersten *et al.* 1998, Wiersma *et al.* 2007) and mammals (Lovegrove 2000), consistently found BMRs that were 20–40% lower in the tropics than in the temperate and

arctic zones. This difference is now widely accepted (Wiersma *et al.* 2007).

Although tropical nights can sometimes be quite chilling, there is of course a good reason for tropically dwelling animals not to have warm fur or a high BMR: during the heat of the day, they run the risk of being overheated. Counter-intuitively, in order to prevent overheating, metabolic rate increases above the so-called 'upper critical temperature', due to heat-dissipating mechanisms, such as sweating or panting. At a certain point (i.e. at an ambient temperature near core temperature), these measures are insufficient and core temperature will increase (Fig. 9)—a situation ultimately leading to death (i.e. a lost fight against

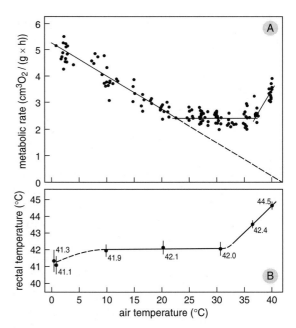

Figure 9. (A) A typical Scholander curve across its full range, extending beyond the upper critical temperature (UCT, the temperature at which metabolic rate goes up again after a plateau value), applied to data on metabolic rates in white-crowned sparrows *Zonotrichia leucophrys gambelii*. Body temperature is given in (B), and note that it increases when the ambient temperature is still below the UCT.

Modified from King (1964).

Nature's second law, occurring at core temperatures of 46–47 °C; The Guinness Book of Records reports a lethal core temperature of 46.5 °C in humans; McWhirter 1980). Scholander *et al.* did not 'heat up' their study animals enough to reach upper critical temperatures, but with experimental temperatures approaching +35 °C they must have brought them very close. In fact, the existence of the upper critical temperature was established later in studies on birds, such as the one by King (1964).

With *lower* critical temperatures being lower in arctic animals as an adaptation to the cold, we may expect *upper* critical temperatures to be higher in tropical animals as an adaptation to the heat. This has never been thoroughly surveyed along the arctic–tropics gradient, but there is some sparse evidence from a comparison between birds living in hot deserts and related species from the temperate zone. Tieleman *et al.* (2002) contrasted two temperate-living larks (skylarks *Alauda arvensis* and woodlarks *Lullula arborea*) with two desert-dwelling larks (hoopoe larks *Alaemon alaudipes* and Dunn's lark *Eremalauda dunni*), and did indeed find that the desert birds had higher upper critical temperatures (38–44 °C vs. 35–40 °C). However, this effect was not found in a general analysis containing more desert and non-desert species (Tieleman and Williams 1999). Three mechanisms may underlie this possible adaptation of tropical/desert homeotherms: not only do they produce less heat (because of a lower BMR) and are better able to get rid of this heat (because of thinner fur or sparser plumage— except for the better insulated desert larks just discussed), but they may also be able to tolerate higher core temperatures. For example, variable seedeaters *Sporophila aurita*, small passerines that live in humid lowland tropics and thus have difficulty in cooling off via evaporation, are able to tolerate core temperatures approaching 47 °C (Weathers 1997)!

So far we have seen very significant differences between arctic and tropical animals in terms of their thermoregulation. Red knots, key players in this book, breed in the High Arctic and some winter in the tropics. They thus need to tolerate both cold and heat stress within their same small bodies. No wonder that flexibility became a way of life for these impressive globetrotters. And the differences between arctic and tropical environments can get even greater when we consider more than just air temperature alone.

Thermometers do not measure feelings

'It feels cold' is an expression we often hear. It captures the notion that air temperature alone does not always describe our sense of heat loss. Strong winds on a sunny, seemingly warm day, make us reluctant to take off our sweaters and we would certainly leave them on if a cloud moved in front of the sun, taking away the pleasant warmth of the sunlight. The way to encapsulate all the factors that affect an animal's heat balance is to talk in terms of 'environmental temperature' or 'operative temperature'. In Dutch, we literally use the term *gevoelstemperatuur*, 'feelings temperature', when we speak about environmental temperature.

Thus, in addition to air temperature, environmental temperature is affected by convection through wind, different forms of radiation and conductance (Fig. 10). Furthermore, animal properties such as size, shape, and colour influence the thermoregulatory significance of given levels of wind and radiation (Porter and Gates 1969). Therefore, under specified weather conditions, environmental temperature is different for different animals! The early laboratory measurements on thermoregulation by Scholander and others had ignored these complexities, and it was only after George Bakken had published an influential theoretical paper

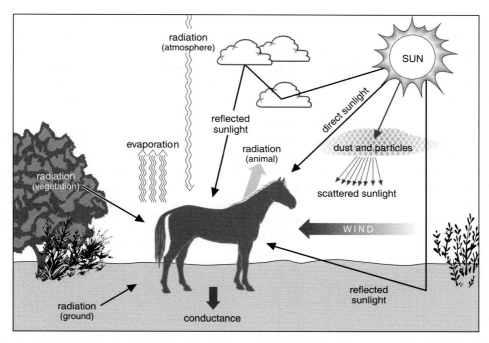

Figure 10. Overview of the ways in which an animal can lose or gain thermal and radiant energy due to the effects of weather.
Adapted from Porter and Gates (1969).

on how to 'sum up' these different factors (Bakken 1976), that biologists started to quantify environmental temperatures.

There are basically two ways in which operative temperatures can be quantified. First, all components making up operative temperatures can be measured independently, after which the Bakken equations (Bakken and Gates 1974, Bakken 1976) can be used to integrate these different measures (e.g. Gloutney and Clark 1997). Secondly, taxidermic models can be used to measure operative temperature (reviewed by Dzialowski 2005). Such copper casts mimic the morphology and absorptivity of an animal, but with these models the effect of wind on heat loss is ignored. A *heated* taxidermic mount measures so called 'standard' operative temperature, which combines all factors affecting heat flow, including wind. It does so in a direct way, by relating the rate of

heat required to keep the mount's temperature constant in a natural thermal environment to a reference laboratory environment with known wind speed (usually 1 m/s). For example, if the mount in a natural setting requires the same amount of heat per time unit as the same mount in a standard laboratory setting at –20 °C, then the standard operative temperature equals –20 °C (even though the air temperature in the natural setting may have been tens of degrees higher).

What new insights have been gained, compared to those already obtained by Scholander and his team, since biologists escaped from their confined laboratory settings and measured the actual thermal conditions experienced by animals in the wild? First, the use of heated mounts made it possible to convert the 'old' standard laboratory measurements on metabolic rate into an ecologically and behaviourally important parameter—the maintenance metabolism of free-living birds in their natural environment (Wiersma and Piersma 1994). Maintenance costs represent the sum of BMR and the extra costs of thermoregulation at environmental temperatures below the thermoneutral zone. Second, it yielded insights into the thermal properties of available microhabitats, so that we now understand how animals can reduce thermoregulatory costs via microhabitat selection (e.g. Wiersma and Piersma 1994, Cooper 1999) or prevent overheating (e.g. Walsberg 1993). It also made us realize that animals in open landscapes, where strong winds can prevail and where microhabitat selection is simply impossible, experience much greater heat losses than expected on the simple basis of air temperature alone.

This brings us back to red knots, which dwell on wide, open mudflats during most of their lives. The *islandica* subspecies overwinters in northwest Europe, notably the Wadden Sea and the British estuaries, where, during the coldest period of winter, maintenance plus activity costs frequently exceed the inferred metabolic ceiling of 4.5 BMR (Fig. 11; Wiersma and Piersma 1994; see Chapter 4). In contrast, the tropically wintering subspecies (notably *canutus* and *piersmai*) experience the other side of the same coin: on a wide, open mudflat there is no shade to hide from a burning sun! At midday in such conditions, operative temperature easily exceeds body temperature (Rogers *et al.* 2006); yet shorebirds cannot permit themselves to remain inactive during the hottest part of day, since feeding activity is tidally, rather than diurnally, regulated. We have seen that resident tropical animals tend to have lower BMRs, which prevent heat-load problems; but it is known that shorebirds have relatively high BMRs (Kersten and Piersma 1987), which make them little heat machines. It is most likely that the worst problems occur during the end of the

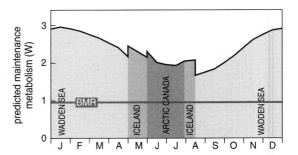

Figure 11. Variation in maintenance costs of *islandica* knots throughout their annual cycle. At their non-breeding grounds (the Dutch Wadden Sea), during the coldest period of winter, *islandica* knots frequently exceed the inferred metabolic ceiling of 4.5 BMR, assuming an average activity cost of 1.5 W.

Adapted from Wiersma and Piersma (1994).

wintering period, when fuelling demands larger organs (see Chapter 5), leading to elevated BMRs (Piersma *et al.* 1996a, 1999b), and thus, presumably, to additional heat production. Furthermore, the growing layer of fat hampers the capacity to lose all the extra heat, and, as if that were not enough, the change into an often darker breeding plumage also increases heat absorption (Battley *et al.* 2003).

How do tropically wintering shorebirds solve all these heat problems? Most significantly, they seem to reduce their BMR upon arrival in their warm winter quarters (Kersten *et al.* 1998), probably through the shrinkage of several visceral organs (Piersma *et al.* 1996a, Guglielmo and Williams 2003, Vézina *et al.* 2006). Their BMRs are 30% lower than those of comparable shorebirds wintering in the temperate zone (Kersten *et al.*1998). Besides this, they have several other behavioural options available. For example, by increasing the blood flow through their legs (giving them 'hot legs'; Fig. 12), knots standing on dry sand can lose 16% of their metabolic heat production at 34°C air temperature (Wiersma *et al.* 1993). As the thermal conductivity of water is 25 times that of air, knots can potentially lose a lot of heat when standing in water (Battley *et al.* 2003). Furthermore, by raising their back

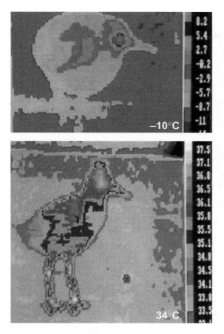

Figure 12. Infrared photos of red knots at two ambient temperatures. The temperature scale given on the right is unique for each photo. By increasing the blood flow through their legs, and by raising the feathers on their back, knots can increase their heat loss at high ambient temperatures such as 34 °C.

See Wiersma *et al.* (1993).

feathers (ptiloerection), they reduce heat gain and increase convective and evaporative cooling (Fig. 12; Battley *et al.* 2003). One other avenue for evaporative cooling (losing heat by bringing body water from a liquid into a gaseous state) is to speed up respiration. However, the resultant panting may solve a heat-load problem, only to lead to another difficulty. How to maintain a water balance when using so much (osmotically free) water for evaporative cooling, whilst being surrounded by salty seawater?

Balancing water

Evaporative water loss is an inevitable consequence of thermodynamics' second law: water molecules, nicely ordered in an animal's body,

tend to 'seek' randomness by evaporation. It is the faster-moving molecules that are able to do this, i.e. those which have the most kinetic energy. Those molecules which are unable to escape have a lower average kinetic energy, and thus the overall temperature of the remaining liquid water decreases. This is the simple physical principle of excessive heat loss via evaporative cooling. There are two main avenues along which water can evaporate from an animal's body: via the skin (cutaneous evaporative water loss, CEWL) and via the lungs (respiratory evaporative water loss, REWL). Mammals have evolved a special way to speed up their cutaneous water loss: their skin is full of sweat glands. Birds lack sweat glands (Calder and King 1974), but some of them living in extremely hot environments have found a remarkable alternative. Turkey vultures *Cathartes aura*, black vultures *Coragyps atratus*, and wood storks *Mycteria americana* all excrete onto their legs when exposed to increasing environmental temperatures, a phenomenon known as urohidrosis (Kahl 1963, Hatch 1970, Arad *et al.* 1989). This excretion takes place at a high rate (approximately every 5 minutes), with each 'dropping' containing about 5 ml of liquid—a stinky, but effective, way to keep cool!

In addition to evaporation, water leaves a body by excretion of (normal) faeces and urine as a means of eliminating indigestible matter and toxic products of metabolism (and through the occasional production of eggs or milk). With 99% of all molecules in a body being water molecules (Robbins 2001), there is a clear need to compensate for all this water loss. Part of this compensation comes naturally, as water is a by-product of metabolism (so-called oxidative water). Furthermore, food contains a variable amount of water that can also be utilized. Finally, all remaining water deficits can be made up by drinking.

Until recently, and still echoed in some scholarly textbooks, it was thought that birds living in hot and dry places showed no specific physiological adaptations to reduce water loss (e.g. Maclean 1996, Hill *et al.* 2004). Mammals were known to have specific water-loss adaptations, such as the unusual ability to concentrate urine found in desert-dwelling kangaroo rats *Dipodomys merriami* (Walsberg 2000), but birds were thought to prevent dehydration problems simply by their behaviour (e.g. hiding in the shade, or flying to a water pool when thirsty) and not by physiological means. This idea probably stems from the work of George Bartholomew and colleagues, who concluded, on the basis of almost a decade of work, that desert birds lack any physiological specialization (Bartholomew and Cade 1963, Dawson and Bartholomew 1968). As Bartholomew did his work in an area that is relatively young on an evolutionary time-scale (the deserts of the south-western United States), the generality of this conclusion was questioned by Joe Williams and Irene Tieleman (2005). In a collaboration with Haugen and Wertz, they carried out their work on Old World larks and showed that the upper skin layer in larks living in the hyper-arid deserts of Arabia has a different lipid structure from that of related larks living in The Netherlands, and that this diminishes the rate of cutaneous water loss in the Arabian larks (Haugen *et al.* 2003a). They further showed that one of these Arabian desert larks, the hoopoe lark, could flexibly increase skin resistance to water vapour by changing the lipid structure of its skin in response to increasing temperatures (Haugen *et al.* 2003b). The other avenue of evaporative water loss, via respiration, was also found to be lower in desert larks. Compared to Dutch skylarks and woodlarks, the Arabian desert larks had lower BMRs (Tieleman *et al.* 2002), a result consistent with the earlier mentioned general trend of lower BMR in the warm tropics. A reduced BMR reduces evaporative water loss in two ways. First, less heat is produced and thus

there is less need for evaporative cooling. Second, as ventilation frequency is reduced, less water is able to escape via warm and water-saturated breath. So Bartholomew's conclusions were incorrect, and ecophysiological textbooks will need to update their section on desert birds.

Pigeon racers know it too: the first thing avian migrants do upon arrival is to drink heavily to quench their thirst (Biebach 1990, Pennycuick *et al.* 1996, Schwilch *et al.* 2002). Being aloft for a few days apparently places heavy demands on a bird's water economy, which is not surprising, as flying and drinking are simply incompatible. In addition, metabolic water production is relatively low during flight, despite an obviously high metabolic rate. This is because the main fuel, fat, has a low water yield per unit energy (Jenni and Jenni-Eiermann 1998). On top of this comes water loss through respiratory evaporation, and this amount can be considerable, as respiration rates are 'sky high' during flight (Carmi *et al.* 1992, Klaassen 1995).

There are, therefore, both behavioural and physiological solutions to all these potential dehydration problems in long-distance migrants. The simplest solution is not to fly during the heat of the day in order to prevent extraordinary water losses. There are many records of birds travelling by night and resting by day, notably among trans-Sahara migrants (Biebach 1990, Biebach *et al.* 2000, Schmaljohann *et al.* 2007). Another option may be to make regular 'drink stops'. Indeed, large waterfowl do make frequent stops of up to a couple of hours when travelling long distances (e.g. Green *et al.* 2002). These breaks are too short to replenish energy stores, which require a period of weeks in these species (Lindström 2003), but quite long enough to 'refill' with enough water. Flying at high altitudes is another way to reduce respiratory water loss. Although the air gets thinner and thus more air needs to pass through the lungs, the reduced air temperature also reduces the temperature of the exhaled air and thereby the amount of saturated water that it can contain (Carmi *et al.* 1992, Klaassen 1995, Schmaljohann *et al.* 2008, 2009). On a physiological level, birds may use protein rather than fat as their source of energy. Per unit distance flown, the amount of water produced when burning protein is six times that of burning fat (Jenni and Jenni-Eiermann 1998)! Although this mechanism has been found in dehydrating humans (Bilz *et al.* 1999), so far it has not been explicitly investigated in migratory birds. However, it is well known that migrant birds catalyse protein during their flights (e.g. Piersma and Jukema 1990) and the striking observation of migrant birds found in the Sahara with good fat stores but with low breast muscle scores (L. Jenni unpublished data as cited in Klaassen 2004) points in the right direction.

In spite of all these water problems that migration may entail, the idea that dehydration risk may govern migratory behaviour is not widely accepted. Klaassen (2004) suspects that this is because lowered body water contents have only rarely been found in long-distance trans-desert migrants (Biebach 1990, Landys *et al.* 2000). However, as noted by Klaassen, body water content remains remarkably stable in animals dying from water restriction (Chew 1951, 1961, Dawson *et al.* 1979), and it is thus a poor indicator of dehydration stress.

Indirectly, the (un)availability of water may affect migratory performance during the fuelling phase. Blackcaps *Sylvia atricapilla* stopping-over in dry fruit plantations in southern Israel during southward migration gain body mass faster when provided with extra fresh water (Sapir *et al.* 2004, Tsurim *et al.* 2008). The extra water available enables them to take up higher quantities of fat-rich but water-poor (35%) fruits. During water-restricted conditions, the blackcaps stop feeding on fruits and switch to energy-poor but water-rich (62%) insects. This shows that food may sometimes be too dry—but, as we shall see next, food may also be too wet!

When red knots feed on shellfish under temperate winter conditions, they ingest 27 times the amount of water per day than is predicted on the basis of their body mass alone (Visser *et al.* 2000). They achieve this notable water uptake for two reasons. First, shellfish contain little energy relative to the amount of water (0.3 kJ per g wet mass, while, for example, the knots' insectivorous summer diet contains 4.9 kJ per g wet mass). Second, the knots' energy demands during winter are relatively high, for reasons explained in the previous section. Thus, it may seem that, whenever enough harvestable molluscs are available, red knots should never run out of water. There is one problem, however. All this ingested water is too salty to become part of the bird's body fluids or to be used for evaporation. Red knots, together with other marine molluscivores, need a highly efficient salt-excretion system (Peaker and Linzell 1975, Hughes *et al.* 1987). And yes indeed, they do have one: corrected for body mass differences, their nasal salt glands were the largest in a comparison with 21 other Charadriiformes species (Staaland 1967; Fig. 13). In spite of this, it may be that the constraint on food intake and the ability to evaporate osmotically free water lies in the bird's capacity to excrete salt. Indeed, knots reduced their food intake when the water provided was experimentally changed from fresh to salt (Klaassen and Ens 1990). Nevertheless, over a time period of just a few days, the knots adjusted to the salty water and increased food consumption again, suggesting that they were able to increase the capacity of their salt glands. This flexible adjustment in the capacity and size of the nasal glands has also been found in several other avian species (Peaker and Linzell 1975).

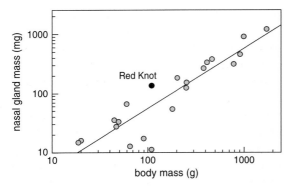

Figure 13. Scaled to body mass, nasal salt glands are largest in red knots in a comparison with 21 other Charadriiformes species.

Based on data presented by Staaland (1967).

Elements of a body

It is obvious that shellfish-eating knots ingest more salt than they need, but, on the other hand, they cannot live without a certain amount of it. The sodium in salt is needed for the regulation of body fluid volume and osmolarity, acid–base balance and tissue pH, muscle contraction and nerve impulse transmission (Robbins 2001). As well as sodium, there are many more elements that need to be kept in balance in an animal's body, since deficiencies and imbalances of minerals are considered to be an important determinant of animal condition, fertility, productivity, and mortality (Underwood 1977). The field that studies the balance between the elemental make-up of animals and their food is called 'ecological stoichiometry', an area of research that has flourished over the last decade (Sterner and Elser 2002, Elser and Hamilton 2007). One of the best-known applications of stoichiometric principles to ecology is the Redfield ratio, named after Alfred C. Redfield, an oceanographer from Harvard and the Woods Hole Oceanographic Institute. He discovered a remarkably constant ratio between carbon (C), nitrogen (N), and phosphorus (P) (106:16:1), both in the world's oceans and in the phytoplankton living in them (Redfield 1934), and

which he explained by the continuous degradation of phytoplankton keeping this ratio in the water column constant (Redfield 1958). In more recent times, larger datasets and more precise measurements have yielded some small modifications here and there, but overall they still support the generality of the magic ratio 106:16:1 in the offshore ocean (Hoppema and Goeyens 1999).

Unlike the elemental equality between the sea and its phytoplankton, most 'higher' organisms face a dramatic difference between the elemental composition of their food and their own bodies. Take termites, for example, whose bodies have a C:N ratio of approximately 5, but consume dead wood with ratios between 300 and 1000 (Higashi *et al.* 1992). That represents a lot of wood that needs to be chewed before a termite satisfies its nitrogen needs! Since photoautotrophs are almost always made up of much more carbon than their herbivorous consumers, a surplus of carbon is a general phenomenon in herbivores. With the initial focus in ecological stoichiometry on small ectothermic organisms, notably the freshwater cladoceran *Daphnia* (Sterner and Hessen 1994, Urabe *et al.* 1997), stoichiometrists mostly considered the surplus of C as a problem rather than a benefit, as it reduces the efficiency of growth in such organisms (Anderson *et al.* 2005, Hessen and Anderson 2008). However, this point of view changed drastically when they turned their attention to the endotherms.

Marcel Klaassen and Bart Nolet (2008) hypothesized that endothermy is the perfect solution to excess carbon: just by 'burning it off' endothermic animals get rid of their surplus carbon, while at the same time taking advantage of the heat-release to keep their bodies warm. They tested this idea by allometrically comparing nitrogen requirements relative to energy requirements in birds, eutherian mammals, marsupials, and reptiles. Birds and mammals did indeed have relatively low nitrogen requirements compared to reptiles, which explains why they are able to tolerate lower quality food (i.e. food with higher C:N ratios). In fact, even among birds and mammals, the phylum with the highest energy requirements (birds) had the lowest (relative) nitrogen requirements. This way of thinking sheds new light on the evolution of endothermy. As mentioned earlier in this chapter, one of the advantages of being homeothermic is the ability to capture cold-blooded, and thus slower, prey. It is thus generally thought that homeothermy originated in small predatory carnivores (Bennett and Ruben 1979, Hayes and Garland 1995). In contrast, the stoichiometric hypothesis of Klaassen and Nolet opens up the possibility that endothermy began in herbivores eating too much carbon to meet their nitrogen requirements. As is often the case with ideas about evolutionary origins, the hard-to-prove facts lie hidden in the fossil records, and we have not yet heard the last word on this subject (Lane 2009b)!

'Burning it off' may be one solution to the problem of excess carbon, but growing very large may be yet another way to deal with it. Long ago, in the Jurassic era, typical plants were conifers, cycads, and ferns, which have inherently low nitrogen content. The higher carbon dioxide concentrations in the air during that period made things worse, suppressing the nitrogen levels in those plants even further. Jeremy Midgley and colleagues (2002) postulated the idea that sauropod dinosaurs could only live on such low quality food by growing very large. With an increase in body size, relative metabolic rate and relative growth rate (both expressed per unit body mass) decline. The lower the metabolic rate and growth rate, the less protein, DNA and RNA needs to be produced, which tempers the nitrogen needs and thus allows a low-quality diet. Midgley *et al.* suggest that juvenile sauro-

pods must have been partly carnivorous in order to grow, a pattern still seen in juvenile marine iguanas *Amblyrhynchus cristatus* (White 1993). However, if growing larger has been the evolutionary solution that overcame one dietary problem in sauropods, it may have led to yet another dietary problem in its place. Larger animals need relatively bigger bones (Schmidt-Nielsen 1984; and already recognized by Galileo in 1637). Of all the organs in a vertebrate, bones contain the highest levels of phosphorus (Elser *et al.* 1996; Fig. 14), as they are made up of hydroxyapatite, a phosphorus mineral. Adult sauropods were thus likely to have high P-demands (Lane 2009b). Being giants, they may have fulfilled these needs by consuming the leaves of large and fast-growing trees, which contain relatively high fractions of phosphorus. Although these can only be mere speculations, it is clear that living in a true Jurassic Park was not at all easy!

The stoichiometry of bones brings us neatly back to red knots. Apart from phosphorus, bones also contain a lot of calcium. When reproducing, most arthropod-eating birds need

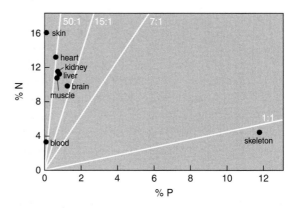

Figure 14. Stoichiometry of organs: both blood and, especially, the skin are very protein-rich and thus have high N:P ratios. By contrast, bones are rich in phosphorus, due to the deposition of apatite (a group of phosphate minerals), and thus have a low N:P ratio. Percentages refer to dry masses, while the white lines refer to atomic ratios.

From Elser *et al.* (1996).

extra calcium in order to produce strong egg shells, and many obtain this by supplementing their insectivorous diet with molluscs, e.g. landsnails in the case of great tits (Graveland *et al.* 1994). On their tundra breeding grounds, red knots are specialist insectivores too, and female knots may thus also face calcium deficiencies prior to egg-laying. As knots daily ingest huge amounts of calcareous shell material when not on the tundra (which is during most of their life cycle), one would expect them to store calcium in their bones before departure to the breeding grounds—and indeed they do. However, the highest rates of calcium accumulation in their bones occur after arrival at the breeding grounds (Fig. 15), presumably from ingested teeth and bones extracted from lemming carcasses (MacLean 1974). The most likely functional explanation for not storing all the required calcium at the stopover is that lighter bodies make for cheaper flying. If the remaining calcium can be collected upon arrival, then that is what should be done. With an almost 50% increase in skeleton mass, knots

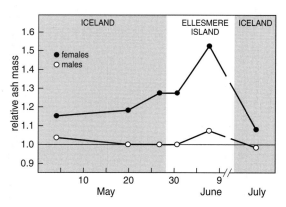

Figure 15. Changes in ash mass of the skeleton of red knots during their stopover in Iceland and while at their Canadian breeding grounds at Ellesmere Island. The ash of skeleton contains 25–30% calcium (Grimshaw *et al.* 1958), and thus these changes also reflect changes in a knot's calcium stores. Ash masses are scaled relative to the average ash mass of males in Iceland.

From Piersma *et al.* (1996b).

hold the record in skeletal calcium dynamics (Piersma *et al.* 1996b). But such an increase is still small compared to their real asset: red knots, together with bar-tailed godwits and other long-distance migrant shorebirds, can store fuel in the form of fat in such amounts that upon take-off on long flights half of their body can be pure lipid (e.g. Piersma and Gill 1998). At other times, e.g. upon arrival or when close to starvation, their bodies contain very little fat. The amazing dynamics of fat in shorebird fat stores are a consequence of variable energy balance states.

Birds are not airplanes

Red knots ready to take off for their 2800-km long non-stop journey from Selvogur, southwest Iceland, to their Canadian/Greenlandic breeding grounds carry about 80 g of fat. At the end of the flight, there is 15–40 g of fat left (Piersma *et al.* 1999b, Morrison *et al.* 2005). These knots build up and break down this amount of fat four times a year (fuelling four non-stop flights), a seasonal routine that is maintained throughout an average life of 5–10 years. Fat is by far the most cost-effective way to store chemical energy. Its energy density amounts to 38 J/g, compared with approximately 5 J/g for wet protein or 4 J/g for wet carbohydrates. The main reason for this difference is that fat can be stored 'dry', whereas stored protein and glycogen typically contain >70% water. Even without water, fat (40 J/g) is more than twice as energy-rich as protein (18 J/g) or carbohydrates (18 J/g; Jenni and Jenni-Eiermann 1998). Furthermore, the energetic maintenance cost for adipose tissue (the tissue holding body fat) is at least an order of magnitude lower than that for other tissues (Scott and Evans 1992). It is, therefore, not surprising that fat was long thought to be the only source of energy during migratory flights (Odum *et al.* 1964). Birds were thought of as airplanes, filling up and emptying their fuel tanks, while holding structural mass constant.

Gradually this 'aircraft-refuelling paradigm' was abandoned. Studies accumulated evidence that fuelling birds also increased the amount of protein tissue, e.g. in the form of larger flight muscles or a bigger heart (in order to gain power; e.g. Fry *et al.* 1972, McLandress and Raveling 1981, Piersma *et al.* 1999b), while simultaneously reducing the mass of their nutritional organs before take-off (so as to fly with a total mass as low as possible; e.g. Piersma 1990, 1998, Piersma and Gill 1998). In fact, protein may be used as actual fuel: the central nervous system is unable to oxidize fatty acids and thus needs glucose as an energy source, which is derived from amino acids (Jenni and Jenni-Eiermann 1998). In addition, the burning of fat itself requires some protein because intermediates in the citric acid cycle need to be replenished on a continuous basis in the form of amino acids (Jenni and Jenni-Eiermann 1998). Furthermore, as burning protein yields so much more metabolic water than burning fat, a fuelling migrant storing extra protein may prevent dehydration during the flight. Finally, muscle tissue undergoes continuous damage during strenuous exercise and needs to be repaired, probably immediately after ending a long-distance flight (Guglielmo *et al.* 2001).

A bird in flight burns both fat and protein (Jenni-Eiermann *et al.* 2002), but the ratios of fat catabolism and protein catabolism may change with depletion of the stores. For example, when fat stores become depleted the body needs to rely on protein as the prime source of energy (van der Meer and Piersma 1994, Jenni-Eiermann *et al.* 2002). Such prolonged fasting periods occur, for example, in male emperor penguins *Aptenodytes forsteri*, which do not eat for as long as 115 days while incubating the eggs at colonies far from the sea in Antarctic winter conditions, losing 40% of their initial body mass (Groscolas 1986). Yvon Le Maho and his team studied the metabolism and chemical body composition changes during long-term fasting in these penguins (Robin *et al.* 1988) and in other birds (Le Maho *et al.* 1981, Cherel *et al.* 1988), and concluded that three distinctive fasting phases can generally be recognized (Fig. 16). During the short phase I, characterized by a rather rapid body mass decline, the animal adapts to long-term fasting by decreasing its protein catabolism and mobilizing its lipid stores. The loss of body mass is tempered during phase II, when the animal has fully switched to relying on its fat stores. During the last phase, phase III, fat stores have run out and the animal starts to burn structural tissue, thus catabolising mostly protein again and losing body mass at a high rate. As the animal is now burning its 'true self', it will face death unless it soon finds food.

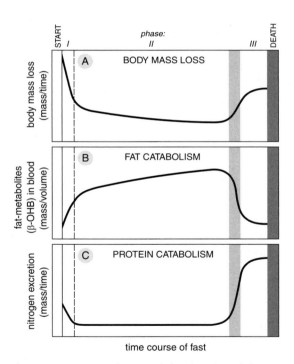

Figure 16. Conceptualization of the physiological changes that occur in the course of prolonged fasting: (A) changes in body mass, and shifts in the metabolic dependence on (B) fat and (C) protein.

From van der Meer and Piersma (1994).

Shorebird insurance strategies

Animals are able to anticipate a forthcoming fasting period or a period of high endurance expenditure by storing extra fat mass, thereby extending the 'economic' phase II of fasting. This is most obvious in migrant birds fuelling for their long-distance journeys. In a non-migratory context, birds and mammals can also lay down extra fat stores, which enable them to overcome a period of probable food scarcity. Take, for example, red knots during winter in the Dutch Wadden Sea. Not only are the environmental conditions harsh (low ambient temperatures), they are also unpredictable. Mudflats can be ice-covered for periods of more than a week, and even low tides can be too high for enough mudflats to become exposed (due to storms). Red knots wintering in the tropics face much better and much more predictable conditions: there is never frost and strong onshore winds are uncommon. Moreover, because most tropical wintering sites are located so close to the open ocean, tidal height is hardly ever affected by wind (Wolff and Smit 1990). Indeed, these differences in environmental predictability seem to be reflected in the midwinter body mass patterns of Wadden Sea wintering and tropically wintering red knots, respectively. The body mass of knots wintering in the Wadden Sea increases from 130 to 160 g during winter, while it remains low at 120–125 g in knots wintering in West Africa (Fig. 17).

In order to put these suggestive descriptional data to a true test, we performed an indoor experiment under controlled conditions. Captive knots (of the *islandica* subspecies wintering in The Netherlands) were either offered food on a continuous basis, or were given food at random moments in time. This regime was maintained for about ten weeks with clear-cut results: knots that always had food available were always about five grams

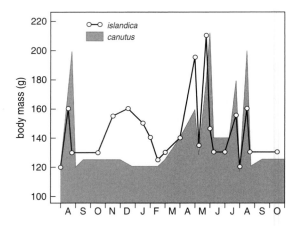

Figure 17. Summary of the annual cycle in body mass of a typical *islandica* knot wintering in northwest Europe and a typical *canutus* knot wintering in West Africa. Both subspecies fuel up four times a year, each time covering a long non-stop flight of 1850–5100 km. Additionally, *islandica* deposits extra mass during midwinter.

From Piersma and Davidson (1992).

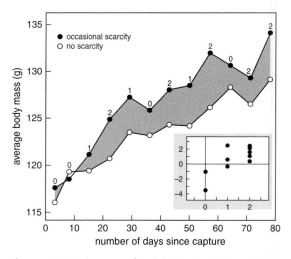

Figure 18. Body mass of red knots in relation to their experience of food scarcity. Four flocks of seven birds occasionally had no food for 1 or 2 days (the number above the closed dots indicating the number of days without food in the preceding week). Three flocks of seven birds (open dots) always had food. The inset shows the difference in weekly change in body mass between starved and fed birds as a function of the number of days without food in the preceding week.

Based on T. Piersma *et al.* unpubl. data.

lighter than the birds whose food was taken away at stochastic moments (Fig. 18). It is likely that this difference in body mass mainly reflects a difference in the amount of fat stored, but it is evident that birds can indeed insure themselves against elevated levels of starvation risk (see also Piersma and Jukema 2002, MacLeod *et al.* 2005).

Dying strategically

With the end of this chapter comes a discussion on the end of a life. You have just been reading that shorebirds wintering under harsh and cold conditions have more fat stores to deplete before they die than conspecifics wintering in the more predictable and mild tropics. But there is a snag in this argument: analysis of many hundreds of carcasses of starved oystercatchers and red knots has revealed that dying in milder climates occurs with lower lean body masses than dying under harsher wintering conditions (Piersma *et al.* 1994a). Red knots starving in a typical west European winter had 25 g more fat-free mass left than conspecifics that died in West Africa (Fig. 19). The same principle applied to oyster-catchers: individuals wintering in Scotland died at a 30–50 g greater body mass (with almost no fat left) than individuals wintering in the milder Wadden Sea. How can we explain these remarkable observations?

Think of a bird's body as consisting of an essential part, without which it cannot survive, and a remaining part consisting of reserves and stores. The essential part cannot be catabolized, since if this were to happen, the animal would die immediately because of insufficient energy output. Instead, the energy output comes from the reserves, i.e. the non-essential part, and needs to supply the entire body, essential and non-essential components included. A bird dies of starvation when the energy supply is less than the energy demand.

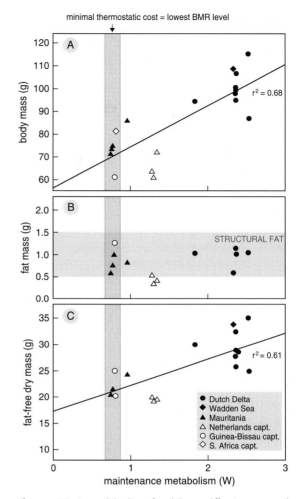

Figure 19. Starved bodies of red knots differ in mass and composition depending on where they died. (A) body mass, (B) extracted fat mass, and (C) fat-free dry mass are plotted as a function of the maintenance costs in the 5-day period preceding death.

From Piersma *et al.* (1994a).

Because the demands for a given body mass are higher under a high working level than under a low working level (i.e. because of differences in maintenance costs), the point of death occurs at a higher body mass when the animal is working harder (Fig. 20).

This higher body mass at death, when working hard, offsets the life-prolonging effect of midwinter fattening in temperate wintering

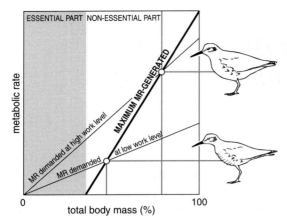

Figure 20. Why work level may determine body mass at death: both the energy production (thick line) and the energy demand (thin lines; two work levels) vary as a function of total body mass. However, because only the fat-free component of a body can produce energy, while it is the entire body that demands energy, hard-working birds die (death occurs when production equals demand) with greater body mass.

From Piersma *et al.* (1994a).

areas. Based on the wintering body masses of non-fasting individuals, it has been calculated that red knots can go without food for 4–5 days, irrespective of where they live. The average knot wintering in the Dutch Wadden Sea has about 20 g of fat and 25 g of protein stored, while a knot wintering in West Africa has only 3 g of fat but 55 g of protein to catabolize before it is 'burned out'. This point shows that a body echoes an individual's ecology. As we will point out again and again, bodies are by no means static machines carrying a single-sized fuel tank, running on a single type of fuel. On the contrary, their fuel tank can shrink and expand, it can use multiple types of fuel, and the size and capacities of the engines can be adjusted. It is indeed fair to say that 'birds are not airplanes'.

Synopsis

Animals face two relevant physical laws: (1) they cannot escape the first law of thermodynamics (the law of conservation of energy), and (2) they are bound by the second law of thermodynamics (the law of spontaneous 'spreading-out' of energy and matter). In order to reduce the amount of entropy in their bodies, organisms have actively to maintain a balance in terms of the gains and losses of heat, water, nutrients, and energy. We have seen various traits that organisms have evolved to keep these budgets in balance across a wide range of environmental challenges. In the next chapter we will consider how these traits are themselves balanced against each other in order to prevent excess capacity.

CHAPTER 3

Symmorphosis: principle and limitations of economic design

A well-trained man, a frog, and a hummingbird

'If a well-trained man runs, the mitochondria of his muscles can consume over five liters of oxygen every minute. To maintain oxidative metabolism at such a level, a continuous flow of oxygen through a series of connected compartments has to be maintained: from the pool in environmental air, oxygen is carried into the lung by inspiration, is transferred to the red blood cells, moved into the tissues by circulation, and finally reaches the cells and their mitochondria by diffusion. Conversely, the carbon dioxide produced during the metabolism of organic compounds needs to be moved along much the same pathway and by similar mechanisms, only in the reverse direction.' So began the paper in which Richard Taylor and Ewald Weibel (1981) articulated the principle of 'symmorphosis' for the first time. Symmorphosis represented the core idea germinating from a meeting of two minds who both firmly believed that animals 'are built reasonably' (Weibel and Bolis 1998). Taylor worked at Harvard University on the efficiency of animal locomotion, Weibel at the University of Berne in Switzerland on the design of well-functioning lungs. Jointly they set up a research programme to find out whether lungs were adjusted to the maximal rate of oxygen consumption in running animals of different size (Weibel 2000). Although the answer would eventually be negative with respect to the uptake of oxygen in the lungs of mammals,

most other features of the mammalian respiratory system indicated that the capacity of one part of the system was closely matched to the capacity of other parts (Weibel *et al.* 1991, 1998). Animals are not built in wasteful ways, animals are 'economically designed'.

One organismal design feature must surely have been key to this insight: the realization that most organ systems are arranged in series, rather than in parallel. The respiratory chain that leads up to muscular work is clearly serial. First, oxygen from the air is extracted by the lungs and captured by the blood. The oxygen is then transported by the blood that is pumped by the heart and delivered to the cells where it is 'burned' in the mitochondria to generate ATP to make the muscle work. In a summary address in 1985, Ewald Weibel clarified this point with a cartoon of a frog (Fig. 21). The cartoon illustrates the sequence from fuel to force, also incorporating aspects of the nervous control of muscles, for the sake of physiological completeness. In a summary of the amazing feats of aerodynamic performance by hummingbirds, Raul Suarez (1998) presented a somewhat evolved 'frog-cartoon', a hovering hummingbird (Fig. 21). His account also focused on the respiratory chain, in this case to show how hummingbirds are able to generate the highest known mass-specific metabolic rates among vertebrates (Suarez 1992). Hummingbirds can do this because they show: (1) high oxygen-diffusing capacity in the lungs, (2) high pumping performance by the heart,

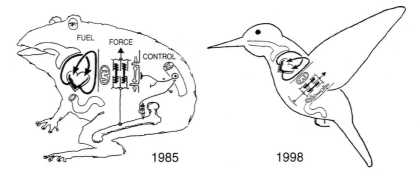

Figure 21. Schematic diagrams of a sitting frog and a flying hummingbird to show how organ systems, such as those for respiratory performance, are arranged in sequence. The frog is directly from Weibel (1985), who used it to show how the heart is coupled to the lungs, the blood stream to the heart, the mitochondria to the bloodstream, and the muscular sarcomeres to the mitochondria. He also indicated how muscle function is controlled by the nervous wiring from spine to bone and muscle. The hummingbird on the right is from Suarez (1998). It provides a visual template to make the point that hovering is made possible by large pectoral muscles, powered by abundant mitochondria to which oxygen from the lungs and metabolic fuels from the gut must be transported at high rates.

(3) high rates of oxygen delivery to muscle fibres due to dense capillary networks, (4) high intracellular densities of mitochondria that also show (5) internal surface enlargements to boost ATP production, and (6) high concentrations of enzymes involved in the energy metabolism of the muscles. Thus, hummingbirds are specialized in burning fuel fast; of course, they can only do so if there are enough substrates to burn. Note that in Suarez' version of the cartoon, that small piece of intestine is a little more visible than in Weibel's version (Fig. 21). In this way Suarez emphasized the importance of fuel sequestration in the gut, the parallel organ system with probably the tightest coupling to the respiratory system.

Economy of design

In an age when problematic arguments against evolution by proponents of intelligent design have not quite been put to bed (e.g. Coyne 2009a), it may seem hazardous to use the word 'design'. Yet, because design arguments are so helpful in appreciating the workings of evolution, this is precisely what we plan to do. After

all, 'evolution is a design process that, as in the first chapter of the Book of Genesis, makes something and then looks at it to see if it's any good' (Vogel 2003, p. 497). What we discuss in this book is what Richard Dawkins (1995), with his usual flair for understatement, has called 'God's utility functions'. What is maximized in evolution (i.e. utility in economics) is long-term reproductive success (fitness). In a world of limited resources and rampant competition, a good way to achieve this is the wise use of energy and supplies. This includes, of course, the evolutionary engineering of bodies. So, indeed, 'engineering' is another word that we will use. In fact, according to Daniel Dennett (1995), the marriage of biology and engineering is *the* central feature of the Darwinian revolution.

Enter industrialist Henry Ford (1863–1947), prolific inventor, father of mass production and assembly lines, and, of course, maker of the model T-Ford, the low-cost, mass-produced vehicle that made cars a commodity. Henry Ford wrote up his 'life and works' together with Samuel Crowther (1922), and this book makes abundantly clear how 'economic' Ford was with respect to almost everything. In regard to the economy of production and the product itself, he had this to say: 'the less complex an article, the easier it is to make, the cheaper it may be sold, and therefore the greater number may be sold'. In many ways, the concept of greatest numbers sold parallels the Darwinian idea that only lineages with more successful offspring than competing lineages will prevail. And small cost differences do seem to make the difference. An appealing example was provided by Daniel Dykhuizen (1978) who compared the growth of two strains of the common intestinal bacterium *Escherichia coli*, a wild type and a mutant unable to synthesize the amino acid tryptophan. In chemostats (vessels with a constant supply of nutrients), the two strains were brought together. Without a tryptophan supply in the chemostat's solution, the mutants would die out rapidly. However, in a solution with tryptophan, the mutant outcompeted the wild type. The real surprise was that, by not having to incur the biosynthetic costs of tryptophan, the winning mutant only saved a tiny fraction, an estimated 0.01%, on the wild-type's energy budget. In another—mammalian—example, by not developing and maintaining a useless visual system in their pitch dark world, mole rats appeared to free up 2% of their energy budgets (Cooper *et al.* 1993). With a 200 times greater saving than the mutant gut-bacteria, one can imagine how quickly mole rats went blind after they became permanent underground citizens.

However, completely simple, one-dimensional examples of the economic imperative in biological design are not so easy to find (although several, more complicated, examples will be worked up in the rest of this chapter). In fact, that other powerful Darwinian mechanism, sexual selection, or the acquisition of good sexual partners on the mating market (Cronin 1991, Ridley 1994, Miller 2001), will generate the opposite of thrifty designs in nature; it leads to a fair amount of

extravagance (Briffa and Sneddon 2007, but see Oufiero and Garland 2007). In today's world of evolved economic competition, the sex appeal of cars is certainly a much greater selling factor than it was in the days of Henry Ford. Nevertheless, a vindicated simple example of energy minimization in nature (indeed, it was published in the journal *Nature*), was provided by Richard Taylor, one of the fathers of the symmorphosis principle (Hoyt and Taylor 1981). When a horse selected her own speeds, walking, trotting or galloping, at each of the three gaits she chose a narrow range of non-overlapping speeds (Fig. 22). The same horse was then run on a treadmill where she could be coaxed into using other than her preferred speeds. With a mask on, her oxygen consumption rates were measured. The measurements (Fig. 22) clearly demonstrated that whether she walked or trotted or galloped freely, the horse always chose the speeds where her oxygen consumption rates (i.e. energy expenditure) were lowest for that particular gait.

Let us now go back to Henry Ford one more time. According to an anecdote expounded by Nicholas Humphrey (1983), Ford commissioned a survey of car scrapyards to see whether there were Model T parts that never failed. It turned out that almost anything could break: axles, brakes, pistons, steering wheels, whatever you will. There was one exception: pins called 'kingpins' invariably had years of functional life left in them. 'With ruthless logic' (Humphrey 1983), Ford concluded that the kingpins of Model T were too good for their job and should be made more cheaply. If he had wanted to build a Mercedes, the pins would have been fine. He would, however, have been obliged to improve all other parts of the car and make it much more expensive. The point is that a 'design had to balance. Men die because a part gives out. Machines wreck themselves because some parts are weaker than others. Therefore, a part of the problem in designing a universal car was to have as nearly

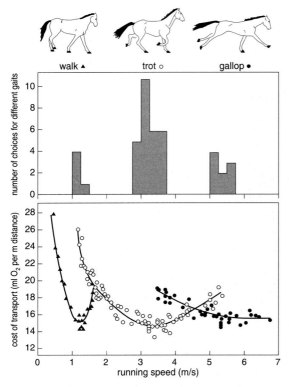

Figure 22. Horses show three distinct gaits (walk, trot, or gallop), each of which is carried out over a limited range of preferred speeds (top panel). When horses were forced to move at a range of speeds on a treadmill, the preferred running speeds at each of the three gaits corresponded to where they achieved minimal costs of transport (lower panel).

Compiled from Hoyt and Taylor (1981).

as possible all parts of equal strength considering their purpose' (Ford and Crowther 1922). The message is loud and clear: economic designs avoid unnecessary excess, a kind of symmorphic logic *avant* Taylor and Weibel (1981).

Symmorphosis: the principle and the test

The term symmorphosis comes from Greek, as technical biological terms tend to, with 'morphosis' meaning 'formation' and 'symmorphosis' literally meaning 'balanced formation' (think of symmetry). In 1981, Taylor and Weibel provided the following definition: 'state of structural design commensurate to functional needs resulting from regulated morphogenesis, whereby the formation of structural elements is regulated to satisfy but not exceed the requirements of the functional system'. Symmorphosis thus predicts that all structural elements of a body, or at the least its subsystems, are fine-tuned to each other and to overall functional demand (Weibel 2000). Because a serial system is as strong as the weakest link, any element in the chain that would be stronger than the weakest would be wasteful.

Most research efforts to develop empirical tests of the symmorphosis principle have been devoted to the respiratory chain as the model system (Fig. 23). Such tests require a comparison of performance levels of different functional elements in the respiratory chain with overall respiratory performance. However, comparisons at just any level of oxygen consumption, carried out within individuals or groups, will be uninformative. This is because, at submaximal performance levels, all functioning elements of the respiratory chain *have* to be in balance: there are no escape pathways for oxygen once it is taken up by the bloodstream. The level of performance of the different respiratory elements will therefore be determined by the demand for oxygen at the receiving end, the mitochondria (di Pamprero 1985, Lindstedt and Conley 2001, Burness 2002, Darveau *et al.* 2002). Only at truly maximal performance levels by the entire organism, i.e. when animals work so hard that they reach maximal oxygen consumption rates (Fig. 24), and when 90% of both the blood and the oxygen in the bloodstream are delivered to the muscles (Figs. 23 and 24), is it informative to determine if working capacities of elements in the chain leading up to muscular work (the supply side) match oxygen consumption rates (the demand side). Keeping in mind the first law of thermodynamics (Chapter 2), working capacities of elements of the respiratory chain cannot be lower than those

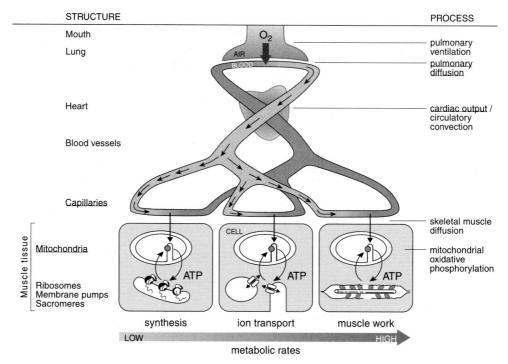

Figure 23. Schematic representation of the cascade of structural elements and functional steps in a mammalian respiratory chain, from the flow of oxygen into the lungs to the supply of ATP to fuel action by the contractile elements of the muscles, the sarcomeres. On the left of the diagram, some of the structural elements are mentioned; and on the right, some of the processes involved in getting oxygen to the mitochondria for ATP generation. The underlined structures and processes were examined in the greatest dedicated empirical test of the symmorphosis principle so far, and are discussed in detail in the text. At the bottom end of the scheme it is shown how, at rest, during low active metabolic rates, energy demands are mainly due to the maintenance processes in the tissues; for example, protein synthesis by ribosomes (left) or ion transport by membrane pumps (centre). With increasing levels of exercise, muscle work (right) takes over the respiratory cascade from lung to capillaries and mitochondria, and the ease with which oxygen can be transported through the cascade limits the maximal oxygen consumption rate.

Compiled from Lindstedt and Jones (1987), Weibel *et al.* (1991), and Weibel (2000, 2002).

achieved by the whole organism. What is critical in tests of the symmorphosis principle is to see whether the working capacities of any of the elements may be greater than necessary.

The research programme that Taylor and Weibel and their colleagues set up to test symmorphosis contained two aspects. First of all they went to East Africa to enlarge the range of mammals in which peak aerobic performance levels had been measured. Animals that varied from mice, mongooses, wildebeest, lions, to eland (the largest antelope), steers, and horses (spanning four orders of magnitude in body mass) were exercised in what was no doubt an equally large variety of treadmills.[1] They then sacrificed the experimental animals. Using

---

[1] As recounted by friend and colleague Steven Vogel (2001) 'in this business, data do not come easily. Determining condition is tricky, and providing motivation is worse; animals don't have the investigator's sense of the importance of working as hard as possible. On one occasion, a cheetah expressed its antipathy toward task and taskmaster by jumping off the treadmill and doing Dick [Taylor] quite a lot of damage.'

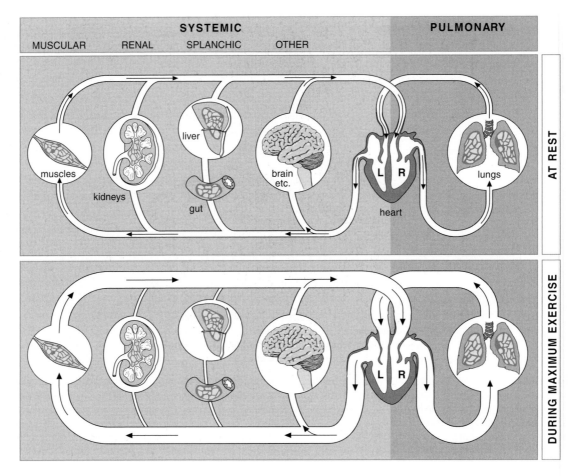

Figure 24. Distribution of blood flow among the different circulatory subsystems in an adult human at rest (top) and during maximum exercise (bottom). The width of the blood vessels indicate, in linear ways, the relative blood flows (i.e. 6 l/min through the pulmonary system at rest, and 25 l/min during maximum exercise). Although the lung receives 100% of the blood flow in both states, muscle blood-flow is 20-fold higher during intense exercise (22 l/min) than during rest (1.2 l/min), at the expense of blood flow to the kidneys and the liver. Blood flow through the rest of the body (especially the brain) is defended during exercise.

Inspired by, and based on, Weibel (2000).

specialist histological and morphological expertise from a wide range of laboratories, they determined (from the end to the beginning of the chain, see Fig. 23): (1) the functional capacity of muscle mitochondria to generate ATP (indexed by mitochondrial volume; there was no evidence for species differences in mitochondrial characteristics), (2) the capacity of the blood circulation to deliver oxygen to the muscle cells (indexed by capillary volume),

(3) the capacity of the bloodstream to transport oxygen bound to haemoglobin (indexed by the red blood cells in circulation), and (4) the capacity of the lungs to capture oxygen from inhaled air (indexed by the oxygen diffusing capacity) (Weibel *et al.* 1991). Covering the great range of body masses, the mammal data showed that the slopes of the allometric regressions of mitochondrial and capillary volumes were similar to the slope of maximal oxygen consumption

on body mass (Fig. 25). These two elements of the respiratory chain apparently were in fine balance with maximal performance on the treadmills. However, the slope of 1.08 of pulmonary oxygen diffusing capacity on body mass was significantly higher than the slope of 0.86 for maximal oxygen consumption on body mass, suggesting overcapacity of the lungs, especially for larger mammals.

These findings were confirmed in a second type of what Weibel and colleagues have called an *adaptive* rather than an *allometric* comparison. Keeping body mass approximately constant, they went on to compare the same variables of maximal aerobic performance and respiratory chain elements between athletic bodies and less athletic forms. On the premise that athletic types would show higher maximal performance levels than non-athletic types (which they did, see the data points for the East African mammalian athletes in Fig. 25), one would again expect the capacity of elements of the respiratory chain to match maximal oxygen performance. Comparing dog (athletic) with goat (non-athletic), human athlete with couch potato, pony with calf, and horse with steer (Table 2), the Weibel–Taylor team was able to confirm the findings from the allometric comparison. Mitochondrial, capillary, and circulating red blood cell volumes were close matches with maximal oxygen-consumption rates, but the dog, the athlete, the pony, and the horse had smaller oxygen-diffusing capacities relative to their maximal oxygen-consumption rates than their non-athletic counterparts, so that the ratios were significantly smaller than 1 (Table 2). Assuming that human athletes start off at the same level as their sedentary counterparts, the athletes must have upregulated all measured parts of the respiratory chain except the diffusing capacity of the lungs.

Why should lungs, in humans and other mammals, have excess capacity? Weibel (2000) presents two possible explanations. First, he considers it possible that lungs in general are

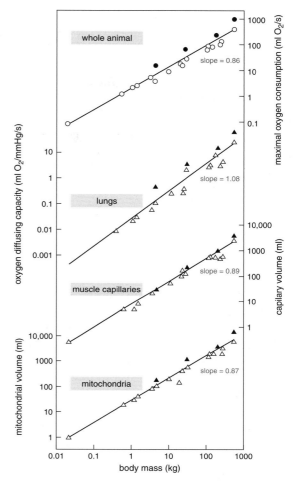

Figure 25. Double-logarithmic plots of the critical peak-performance measure (maximal oxygen consumption rate) and various elements, both structural and functional, in the respiratory chain (see Fig. 23) against body mass of mammals that vary in size by four orders of magnitude. Particularly 'athletic' species are indicated by the filled symbols. The smallest species was the mouse *Mus musculus*, the second smallest the dwarf mongoose *Melogale pervula*, and the two largest, the horse *Equus ferus caballus* and the steer *Bos primigenius*.

Compiled from Weibel (2000).

prepared for lower than normal partial oxygen pressures, such as those found at high altitudes and in underground burrows. In that case, excess lung capacity would give mammals a greater freedom of choice with respect to habitat use. This would be important as, and here

Table 2. A test of symmorphosis in the respiratory chain of mammals based on comparisons between the ratios of several down-chain respiratory variables (either morphological or physiological, see Fig. 23) and maximal oxygen consumption of four pairs of athletic versus non-athletic forms of similar body mass. A ratio of 1 is expected if the down-chain respiratory variables match maximal oxygen consumption in similar ways in athletic and non-athletic forms. Underlined ratios differ significantly from 1. Simplified from Weibel *et al.* (1991) and Weibel (2000).

Athletic/Non-athletic form	Mitochondria	Blood circulation	Heart	Lungs
	Mitochondrial volume/maximal oxygen consumption	Capillary volume/ maximal oxygen consumption	Red blood cells in circulation/ maximal oxygen in circulation	Oxygen-diffusing capacity/maximal oxygen consumption
25–30 kg dog/goat	1.20	1.26	1.08	0.61
70–80 kg athlete/sedentary person	1.03	0.77	0.96	0.67
150 kg pony/calf	0.90	0.89	0.79	0.65
450 kg horse/steer	1.00	0.82	1.00	0.76

is the second explanation, lungs have 'very limited malleability of structure'. Unlike the heart and the mitochondria that can increase their volume with training, to generate greater oxygen-diffusing capacity, alveolar surface areas can no longer increase in mature mammals. Although in humans a modest increase of lung performance is possible with training (Hochachka and Beatty 2003), attempts to train guinea pigs *Cavia porcellus* from the time of weaning were not successful in increasing lung-diffusing capacity (Hoppeler *et al.* 1995). One could say that lungs are constructed in a somewhat wasteful (and permanent) way so that their capacity is on the safe side.

Safety factors

That the lungs of large mammals have approximately double the capacity than required, even at high oxygen-consumption rates, is

good news for human patients who lose part of a lung, or even a whole lung. From this perspective, the argument of supposed wastefulness is turned on its head. Perhaps, in some cases, excess capacities are not just wasteful, but safe (Mauroy *et al.* 2004).

A celebrated example of safety factors is their relevance to the construction of elevator cables, especially in fast passenger lifts. Jared Diamond (1993) begins a wonderfully written review of safety factors with a very frightening account of what it would be like to find yourself in a crowded skyskraper elevator when the cables break. 'Fortunately, it's rare in practice that elevator cables snap. The reason is that they're designed by engineers to support a strain up to 11 times the maximum strain that they would face when lifting their advertised maximum load at their designed speed.' However, Diamond continues, 'you may not wish to learn…that safety factors of steel bridges and buildings are only about 2'.[2] Safety

[2] At the very time this part of the text was formulated, a bridge across the Mississippi, part of interstate 35W connecting Minneapolis with the sister city of St. Paul, collapsed during the evening rush hour, leaving many people dead and injured. Such accidents are not unique and led Gordon (1978) to make the comment that 'engineering amounts to applied theology'.

factors, then, are estimates of the magnitude of the load on a device that just causes failure, divided by the maximum load the same device is expected to experience under normal conditions (Vogel 2003). Safety factors of fast passenger lifts are among the highest reported for man-engineered structures (Fig. 26). Human teeth, with a safety factor of 15, are the champions among biological structures. This can only be regarded as a good thing, as, once broken, teeth could not be replaced in a pre-dentist world.

A capacity for compensation when things go wrong may explain why paired organs, such as lungs or kidneys, usually have smaller safety factors than unpaired organs, such as the pancreas (Fig. 26). It provides the logic of why a healthy person can afford to donate a kidney to a grateful transplant patient, and it can also explain why pancreatic cancer is usually only

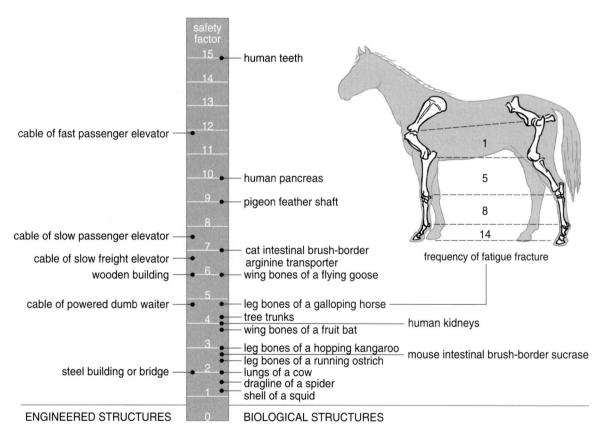

Figure 26. A review of 'safety factors' in structures built by humans and in a great variety of biological structures. Despite an average safety factor of 4.8 for bending loads, leg bones of galloping horses do sometimes break from material fatigue, and the incidence of such fractures increases towards the periphery. Safety factors are from Diamond (1998) and Vogel (2003); the diagram with the leg-bone fractures is from Alexander (1998) compiled from data in Currey (1984). Note that safety factors, such as those presented here, can also reflect a close match of system design to unrecognized and unscrutinized performance requirements (Salvador and Savageau 2003). For those of you who are curious about 'dumb waiters', these are the miniature lifts by which hotels transport room-service meals from the kitchen in the basement to the rooms at higher floors (Diamond 2002).

Table 3. Nine considerations that would either increase (+) or decrease (−) the setting of safety factors in design processes. Modified from Diamond (2002).

Consideration	Increase or decrease of setpoint value
Coefficient of variation of load	+
Coefficient of variation of capacity	+
Deterioration of capacity with time	+
Cost of failure	+
Cost of initial construction	−
Cost of maintenance	−
Cost of operation	−
Opportunity cost of occupied space	−
Cost of rebuilding	−

detected when pancreatic enzyme outputs have dropped to only 10% of normal peak values (DiMagno *et al.* 1973, Diamond 2002). That the wing bones of birds have a safety factor of approximately 6, whereas their feather shafts have a safety factor of 9 (Fig. 26), probably has to do with the higher variability of the overload on structures, such as flight feathers, when compared with wing bones (Corning and Biewener 1998). It is not difficult to imagine that the wing bones of fast-flying animals, such as birds and bats, face greater overloads (e.g. when hitting or avoiding obstacles in mid-air) than the leg bones of hopping kangaroos, which in turn may face greater stresses than the leg bones of ostriches running on flat ground. These examples clearly suggest some of the general engineering considerations summarized in Table 3: with an increase in the coefficients of variation of load or capacity, and with an increase in the costs of failure (high for fast elevators, lower for slow elevators and negligible for dumb waiters), safety factors should go up. This is also true if the capacity of structures to withstand overloads deteriorates with time; an example would be the keratin

used by birds to build their feathers (Bonser 1996, Weber *et al.* 2005). While wear and tear on feathers may well explain their relatively high safety factor, it certainly explains their regular replacement during moult.

At the low end of safety factors higher than one (remember that 1 indicates symmorphosis as originally understood) are structures such as the spider's dragline. Draglines just have to carry the spider's own weight, and some of its prey. In cases when local failure actually gives a fitness advantage, safety factors can thus better *be lower than 1* and do not quite deserve the term any more (Vogel 2003). In the summer of 1999, during an icebreaker-based expedition organized by the Swedish Polar Secretariat, we visited a series of tundra sites across the Canadian Arctic. We were interested in the spatial variation of predation pressure by shorebirds on insects. Research partner Jens Rydell then proposed to count, as a measure of predation pressure, the number of legs remaining on mature craneflies (Tipulidae; they start off with six). With too few craneflies captured per site, the plan did not quite work, but the story makes the point that giving up a leg grabbed by a bird can be a good thing for a cranefly, if it prevents it from being eaten. There are other examples where places of weak construction in a structure pay off: self-amputations occur in spiders (eight legs to lose) and lizards (one tail to spare, but the tail will grow back). Nicely obeying the evolutionary engineering principles advocated in this book, lacertid lizards *Podarcis* spp. that live in places with high predation pressure, release their tails more readily when receiving the proper prompting than similar lizards that have less to fear (Cooper *et al.* 2004). Birds adaptively drop their tail feathers when pursued or grabbed by raptors for exactly the same reason (Lindström and Nilsson 1988, de Nie 2002, Møller *et al.* 2006). Trees may survive a storm if some weaker branches snap and streamline the remaining structure (Heinrich 1997). At this point you may want to go to the garden shed to verify whether those long-legged spiders, the harvestmen (order Opiliones), do indeed move around well with only 5, 6 or 7 legs!

For spiders, the combined costs of losing a leg and rebuilding it, or losing a silky dragline and renewing it, are relatively cheap. As listed in Table 3, with increasing costs of construction, maintenance, and operation, safety factors should go down. Cost considerations may explain why, despite safety factors close to 5, race horses nevertheless break leg bones, and, particularly, why the frequencies of fatigue fractures during or prior to falling increase in the bones at the extremities (Fig. 26); these bones are made of the same material as those closer to the core of the horse. The costs of transport during running (which, as we have seen, Hoyt and Taylor's horse wanted to minimize, Fig. 22), will go up a great deal if the lower leg bones are given extra strength and weight (Alexander 1984, 1998). After all, these parts have to be powered with the greatest momentum. Incidentally, the lack of weight in the extremities explains why long- and thin-legged birds, like shorebirds, have relatively low costs when they run (Bruinzeel *et al.* 1999).

This brings us back to previous considerations about weaker than average links in serial constructions such as the respiratory chain, or indeed, the legs of a horse. As explained by Alexander (1997, 1998), if elements in a chain differ in the coefficients of variation with respect to load, capacity, and/or cost (Table 3), the performance of the system is optimized by devoting resources to extra capacity for the more variable or the cheaper element. Quite literally, in a chain made partly of expensive gold links and partly of cheap iron links, making the gold links very slightly weaker and the iron links much stronger would result in a stronger chain for the same price. This is somewhat counter-intuitive, but means, as pointed out by Alexander (1998), that because the

metabolic rate of lungs is smaller than the metabolic rate of the skeletal musculature, the oxygen-diffusing surface of the lungs is the relatively cheap link in the respiratory chain. Therefore, he predicted, lungs would be designed for higher capacities than the relatively expensive mitochondria!

Multiple design criteria

Since their grand summary in 1991 of tests of the symmorphosis principle, by detailed studies on serial capacities in the respiratory chains of mammals, and in the face of serious constructive criticism (e.g. Garland and Huey 1987, Lindstedt and Jones 1987, Dudley and Gans 1991, Garland 1998, Gordon 1998, Bacigalupe and Bozinovic 2002), Weibel and associates have regularly updated their analyses (Weibel *et al.* 1992, 2004, Weibel and Hoppeler 2005). They clearly stand by their conclusion that respiratory chain elements such as the vascular supply network and the energy needs of the cells active during maximal work are quantitatively coupled to maximal metabolic rates; these elements thus have safety factors close to 1. Although symmorphosis is no longer a hotly debated issue (Suarez and Darveau 2005), the previous discussion illustrates that the principle of elegant economic design needs qualification.

Let us get back to lungs. Although we have not acknowledged this yet, they do more than catch oxygen for the circulation of blood and eventual delivery to the mitochondrial powerhouses. Lungs also service the body by removing carbon dioxide, one of the products of metabolism, from the bloodstream. The removal of carbon dioxide and the regulation of blood pH (a delicate function of carbon dioxide level in the body circulation) may put different design constraints on the lungs (Lindstedt and Jones 1987, Dudley and Gans

Figure 27. The behavioural and morphological trick of the peacock, a butterfly that goes under the Latin name *Inachis io*, to combine multiple design criteria. For camouflage they keep the wings folded and only show their drab undersides; for advertisement they show the full glory of their colourful upper wings with eyespots.

Photos by Harm Alberts (from www.waarneming.nl).

1991). Water loss via respiration and the associated maintenance of heat balance may also impose demands on the respiratory system. Dudley and Gans (1991) point out that this may be especially true for the ungulates of the tropical open savannahs that were so prominent in the empirical verification of the symmorphosis principle (Weibel *et al.* 1991).

In some cases it is possible to do justice to multiple natural selection criteria in a single design (Fig. 27). The shape, surface area, and strength of butterfly wings are likely to be adaptations primarily to the aerodynamic demands placed on them. The colours of wings are subject to a different selection regime. Somehow, butterflies have to find a balance between the opposing forces of crypsis and sexual advertisement, and also take into account thermoregulation (Cott 1957, Kingsolver 1988, Dudley and Gans 1991, Houston *et al.* 2007). The peacock, and many other butterflies, have resolved this engineering conflict by putting their colourful sexual advertisement on the upper wings, and the cryptic coloration on the under wings (Fig. 27). Simply by closing its wings, a peacock turns from a strikingly visible object into a cryptic form that is hard to find. Behavioural solutions like this may be rare (but note that cuttlefish, fish, and chame-

leons can show rapid adaptive colour change, e.g. Stuart-Fox *et al.* 2008). Accordingly, some of the critics argue that the best we can expect from the form and function of most natural structures is that they represent 'continua of imperfection' (Dudley and Gans 1991).

One more problem: the climbing of adaptive peaks

A final criticism of the symmorphosis principle, related to the previous ones, is that natural selection will not necessarily lead to optimal design (Gould and Lewontin 1979, reiterated by Lindstedt and Jones 1987, Dudley and Gans 1991) because selection pressures will change with time, i.e. that we are studying 'moving targets' (Gordon 1998). For these reasons, the structure and function of organisms, and the parts thereof, would have difficulty in climbing any adaptive peak (Dawkins 1996). This may all be true, but such criticisms give us no reason to abolish the helpful notion of symmorphism. Quite the opposite: the various caveats mentioned suggest additional layers of adaptation and make our lives as biologists so much more interesting!

For an example, let us get back to race horses, to the appropriately named 'thoroughbreds'. For thousands of years, and especially over the last few centuries, these horses have been bred for racing performance. Selective breeding confers a consistent and intense directional selection pressure over many generations (Gaffney and Cunningham 1988, Bailey 1998). Despite the best efforts of horse breeders in over 40 countries that have more than half a million thoroughbreds to work with, racing performance, the type of hard muscle work that is highly correlated with peak aerobic power (Jones 1998a), has levelled off over the last century (Fig. 28). Yet Jones (1998a, 1998b) reports that selective breeding has indeed pushed up the performance of various elements of the equine respiratory system. Most interestingly, even the oxygen-diffusing power of the lungs of these thoroughbreds, an element that earlier was reported to have excess capacity, has reached symmorphosis with the rest of the respiratory chain. Artificial selection has done the job it set out to do.

At this point sports enthusiasts will think of similar patterns of levelling-off in athletic performances by humans. Rates of decrease in the men's world record running times began to slow down after about 1960 (Nevill and Whyte 2005, Noakes 2006, Denny 2008). The women's world record progression showed dramatic improvements between 1965 and 1980, but has also levelled off over the past 25 years. These data suggest that runners today are at the limits of achievable biological capacity (Denny 2008). This, of course, is not because human runners as a group have been selectively bred in the way that thoroughbreds have been. Training schemes have improved. Also, champion athletes originate from an ever wider range of human populations. The élite marathon runners of today usually come from high-altitude populations, especially in East Africa,

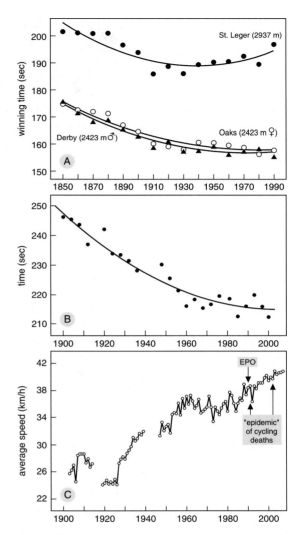

Figure 28. Limits to running performance improvements of race horses (A) and human athletes (B), and cycling speeds of human athletes during the Tour de France (C). (A) Changes in the winning times over the 2400-m long Derby and the 3-km St. Leger races in the UK (from Jones 1998a); (B) long-term changes in the record times for the Olympic 1500 m for men (from Vogel 2001); and (C) average racing speeds in the Tour de France. Note that the drug erythropoietin (EPO) was first introduced in the late-1980s and may have increased the mortality level since then (from Noakes 2007).

people that are adapted to live under relatively low partial oxygen pressures and may show 50% greater oxygen-diffusion capacity than lowlanders (Hochachka *et al.* 2002; incidentally, athletes have also become relatively taller and more slender, Charles and Bejan 2009). This hints at something we have just learned: that directional selection on the performance of the respiratory chain increases aerobic performance. In any case, now that the limits to performance have almost been reached in the world of athletes, future improvements in world records will require the use of more effective performance-enhancing drugs or the introduction of gene doping. This may already have happened in the Tour de France, the classic cycling tournament (Noakes 2007). A plateau level that had been maintained for three decades was exceeded when the new drug erythropoietin (EPO, a red blood cell booster) was first introduced in the late 1980s (Fig. 28C). In biology, 'metabolic ceilings' really may exist. We will develop the subject in the next chapter.

In addition to oxygen, fires need fuel too

As Max Kleiber (1961) implied with the title of his animal energetics classic *The Fire of Life*, oxygen needs a substrate to burn. So far, the discussion has focused on the respiratory chain: how animals make sure that there is enough oxygen for the mitochondria to oxidize fuels to generate ATP and thus power body work. Unlike oxygen, fuels can be stored in the body, and this complicates the one-to-one relationship between supply and demand, a simplification that made the respiratory chain such an attractive system to study (Taylor *et al.* 1996, Weibel *et al.* 1996). Just like the respiratory chain, the food-to-fuel-pathway (which includes ingestion, digestion, absorption, and assimilation), exhibits serial organization (Fig. 29). Food-processing chains

are interesting for their own reasons, and it is worth reminding ourselves that when it comes to whole body performance, the oxygen and fuel transport chains have to complement each other.

To the best of our knowledge, nobody has carried out formal analyses of symmorphy across all the successive elements in the food processing chain (Fig. 29). Is the level at which

Figure 29. The 'barrel model' of an organism's energy balance. Input constraints (maximum rates of foraging, digestion, and absorption) are engaged in series; different outputs (heat production, mechanical work, tissue growth) are parallel and independently controlled. If the sum of output rates does not match the input, the balance is buffered by the storage capacity of the system. Note that the first spigot always 'leaks' (basal heat loss of the organism). From Weiner (1992).

fuel becomes available to the body tissues limited by rates of prey capture, maximum food-processing, by rates of digestion in the stomach and intestine, by rates of absorption in the intestine or by rates of assimilation and biochemical adjustment in the liver? Clearly, an animal built with small safety factors for maximum food-intake rates would have to live in very predictable worlds with very predictable energy expenditures. In practice, instantaneous food-intake rates are usually at least one order of magnitude greater than long-term food-intake rates (e.g. Piersma *et al.* 1995b, van Gils and Piersma 2004, Jeschke 2007), so 10 may serve as a preliminary minimum estimate of the safety factor for food-intake rates. Skipping the second funnel in Weiner's diagram (Fig. 29) for lack of data on digestive rates, we come to the third, the one symbolizing absorption rates. In this case, we can count ourselves lucky, as in the 1990s a whole body of work was carried out by Jared Diamond and his research group on rates of intestinal assimilation in house mice (Toloza *et al.* 1991, Diamond and Hammond 1992, Hammond and Diamond 1992, Hammond *et al.* 1994, 1996a, Weiss *et al.* 1998, O'Connor and Diamond 1999, O'Connor *et al.* 1999, Lam *et al.* 2002; summary in Diamond 2002). Their approach differed from that to respiratory chains, as they did not look at maximal performance over a range of species. Instead, they examined capacities within a single species manipulated to function at different levels of energy expenditure. In small (female) mammals one can do this with ambient temperature and the presence or absence of lactation for different numbers of pups. This is a valid approach for the food-processing chain because, unlike oxygen that cannot escape once it has entered the bloodstream, nutrients do have 'escape pathways' after being absorbed by the intestine. They do not have to be burned at once, but can be stored. However, Diamond and his associates faced the complication that different nutritional components (sugars, carbohydrates, lipids, etc.) are passively or actively absorbed in a variety of ways. They had to develop different assays for different nutrients.

The evidence accumulated by the Diamond group is consistent with 'reasonably designed' nutrient-processing systems. With manipulated variations in food-intake rates, the experimental mice showed parallel variations in the size of the main digestive organ, the small intestine (Fig. 30A). Diamond (2002) suggests that, with increasing food-intake rates, the excess capacity of the intestine gradually diminishes because 'a mouse cannot increase its intestinal mass indefinitely'. This becomes clear when the measured index of absorption rates of one of the nutrients, glucose, is plotted against glucose intake rates (Fig. 30B). With an increase in nutritional demands because of lactation in the cold, the safety factor, that under relaxed conditions is close to 3 (see Fig. 26), levels off at 1.

As we shall see in the next chapter, the question whether the respiratory chain and the food-processing chain show any degree of symmorphy cannot be separated from questions about the biological engineering constraints to maximal and long-term aerobic performance, on the design of 'metabolic ceilings', so to speak. Before we turn our attention to this question, let us briefly examine symmorphosis in the food-processing systems of our prime witnesses in this book: the shorebirds and, especially, red knots.

Testing symmorphosis in shorebird food-processing systems

Diamond (2002) made a plot of small-intestine mass against food-intake rate in mice experiencing different levels of energy expenditure (Fig. 30), in order to test whether indices of capacities of the foraging- and the absorption-funnels (Fig. 29)

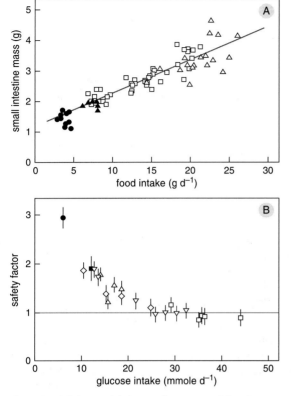

Figure 30. (A) Mass of the small intestine of female mice increases with increasing food intake during lactation and at low ambient temperatures. Virgin mice at temperatures of 23 °C and 5 °C are indicated by filled circles and triangles, respectively; lactating mice at 23 °C and 5 °C are indicated by open squares and triangles, respectively. (B) The safety factor for the small intestinal brush border glucose transporter SGLT1 of female mice goes down with glucose intake rates. Data are from the same experiments as in (A) plus experiments involving increased pup mass. Virgin mice at 23 °C and 5 °C are indicated by filled circles and squares, respectively; lactating mice are indicated with open symbols, for 23 °C (downpointing triangle, diamond) and 5 °C (squares) environments, and at 23 °C with increased pup mass (upward pointing triangle).

Data are from Hammond and Diamond (1992, 1994), and from Hammond *et al.* (1994); the compilations are from Diamond (2002).

do correlate. We have plotted intestine mass (third funnel) and liver mass (an assimilation funnel not illustrated in Fig. 29) against gizzard mass in a range of shorebird species (Fig. 31). As predicted on the basis of the symmorphosis principle, in an allometric comparison between the organ sizes, intestine and liver mass correlate closely with gizzard mass (Fig. 31A, B). Because of the need to crush shells and process shell fragments, the gizzards and intestines of the five hard-shelled prey specialists in the graph (red knot, great knot, surfbird, purple sandpiper, and rock sandpiper), have high workloads relative to other organs in the digestive system (Battley and Piersma 2005). This may explain why several of these mollusc specialists have small livers relative to gizzard and intestine (Fig. 31B).

The intraspecific comparison for red knots (Fig. 31C, D) is even more interesting. In the plots of intestine and liver mass against gizzard mass, the data seem to fall into two groups. As expected, intestine mass (Battley and Piersma 2005), shows a positive correlation with gizzard mass. The slope of this relationship is much steeper during the spring stopover in Iceland than in winter (Fig. 31C). Liver mass does not correlate with gizzard mass at all in winter, but does so during refuelling at the spring stopover site (Fig. 31D). This is an exciting finding, as red knots have been shown to select spring stopover sites on the basis of high food-quality, i.e. places with prey that have a lot of meat and little shell material (van Gils *et al.* 2005c, 2006a). In winter, gizzard size increases are explained to a large extent by the decrease in the ratios of meat to shell mass (van Gils *et al.* 2003a, 2006b).

We suggest that two mechanisms influence the degree of symmorphy within the food-processing chain of red knots and other shorebirds: (1) the need for shell processing (scaled to body mass and size-related energy requirements, see Box 2), determines the size of both gizzard and intestine, but not that of the liver, (2) the quantity of food processed determines the size of the intestinal tract (to facilitate absorption and/or to adjust to proper levels of

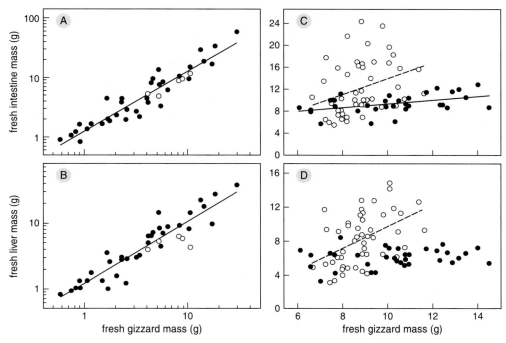

Figure 31. Examination of the symmorphosis design principle predictions for the digestive system of shorebirds, based on comparisons between the size of the gizzard as the first element in the digestive chain, and intestine and liver as subsequent elements, both in an interspecific comparison between shorebird species (A, B) eating either hard-shelled (open circles) or other prey (filled circles) and in an intraspecific comparison for red knots (C, D) that either find themselves in a refuelling situation during northward migration in Iceland (open circles) or in Western Europe in winter (filled circles). Linear regression lines are shown where slopes are significantly different from 0. In A and B, the five hard-shelled prey specialists, from the lightest to the heaviest, are rock sandpiper *Calidris ptilocnemis*, purple sandpiper *Calidris maritima*, red knot, surfbird *Aphriza virgata*, and great knot *Calidris tenuirostris*. Although technically we have worked with stomach masses, in shorebirds the proventriculus (glandular stomach) is so small (Piersma *et al.* 1993b) that we are justified in talking about gizzards (muscular stomachs) here.

Based on data presented in Piersma *et al.* (1999b), Battley and Piersma (2005), and T. Piersma (unpubl. data).

immune defence; Piersma *et al.* 1999b, Baker *et al.* 2004). Greater assimilation rates of nutrients by the liver (e.g. during stopovers) induce increases of liver capacity. Do we really see symmorphosis here? We believe we do, but as usual it is symmorphosis with a qualification.

Synopsis

Symmorphosis, the principle that evolved body designs avoid excess capacity, e.g. in cascades of serial physiological processes, such as

the respiratory chain, is now widely accepted as a useful, heuristic design principle. As we will see later in the book, like other criteria used in optimality-driven evaluations of organismal performance, symmorphosis is better seen as a useful null hypothesis, than as a hypothesis with very precise and rigid criteria for rejection. The discussion of safety factors has demonstrated how an evaluation of cases where simple, economy-based expectations are not upheld, develops our biological insight.

Box 2 Introduction to allometry: assessing relative changes in body parts and whole organisms and their performance levels

Arithmetic offers neat tricks, and one of the neatest and most widely applied set of tricks in comparative biology is the use of logarithms and exponents ('power relationships') in so-called 'allometric scaling' exercises (Dial *et al.* 2008). With the use of logarithms, it is generally possible to unbend curvilinear relationships, to transform them into linear ones of the kind suitable for standard statistical examinations (regression, covariance analyses, etc.). When an organ or a performance measure, such as oxygen-consumption rate, does not change in constant proportion to its body mass, we say that it scales *allometrically* (allo = different, metric = measure). *Isometric* scaling (iso = same) represents the special case of geometrically similar objects that show constant proportional changes with size: at constant shape, surface area will simply scale to length with an exponent of 2, and volume will scale to length with an exponent of 3 (the fitted regression line will go through the origin). As the ground rules of proper arithmetic conduct for the use of logarithms and exponents are presented in several recent ecophysiological textbooks (see, for example, Willmer *et al.* 2000, p. 43, and Karasov and Martínez del Rio 2007, p. 14), there is no need to repeat them here. These books, and all-time or recently published 'classics' like Peters (1983), Calder (1984), Schmidt-Nielsen (1984) and Bonner (2006), can be consulted for many biological applications.

Ever since Kleiber's analysis of the relationship between metabolic rate, MR, and body mass, M (published first in 1932, but Kleiber 1961 is more widely available), ecophysiologists and, at a later stage, ecologists, have been intrigued by the power function (or allometric relationship): $MR = aM^b$ (where a, the intercept, is the scaling constant and b, the slope, is the scaling exponent).

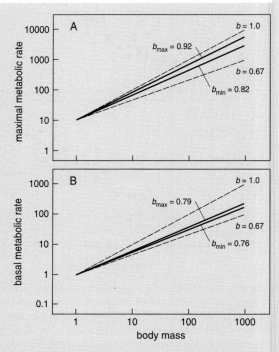

Figure 32. Estimates of b, the scaling exponent, for the allometric relationships of maximum metabolic rate (A) and basal metabolic rates (B) on body mass in mammals using the 'allometric cascade model' (an algebraic model based on Wagner (1993) that integrates estimates of all possible rate-limiting steps determining metabolic rate) implemented by Darveau *et al.* (2002) and Hochachka *et al.* (2003). The range of achieved scaling exponents for maximum metabolic rate and basal metabolic rate shown here represent the outcomes of sensitivity analyses for reasonable range of estimates in up to 10 potential metabolic rate-limiting steps. For reference, slopes of 1.0 and 0.67 are shown. Note that for simulation purposes, both metabolic rate and body mass values are given in arbitrary units.
Modified from Darveau *et al.* (2002).

Much paper has been printed with single-cause explanations for *b*, which, depending on one's sample of data points, is often close to 0.75, but not always. Chapter 3 has made abundantly clear that metabolic or oxygen-consumption rates may reflect many different potential rate-limiting steps. It is this realization that caused Charles Darveau, his supervisor Peter Hochachka and their associates, to try and operationalize b not to reflect a single rate-limiting process, but to reflect the whole cascade of key steps in the complex pathways of energy demand and energy supply (Darveau *et al.* 2002, 2003, Hochachka *et al.* 2003, Suarez *et al.* 2004; the immediate criticisms by Banavar *et al.* 2003, West *et al.* 2003 are also instructive to read and to get a feel for this enormous literature).

The team of Hochachka and Darveau comes to the conclusion that the generally lower scaling exponent b for basal metabolic rate compared with maximum metabolic rate (Fig. 32), can be explained by different rate-limiting steps being most important in the two exercise conditions. For basal metabolic rate, oxygen-delivery steps of the kind discussed in the symmorphosis literature and in this chapter (relatively low exponents), contribute almost nothing to the overall scaling exponent b. In contrast, for maximum metabolic rates, b is controlled entirely by the oxygen delivery steps (relatively high exponents).

Part II

Adding environment

Part II

CHAPTER 4

Metabolic ceilings: the ecology of physiological restraint

Captain Robert Falcon Scott, sled dogs, and limits to hard work

If only Robert Falcon Scott[1] had followed up Ernest Shackleton's suggestion to use sled dogs to reach the South Pole in the cold Antarctic summer of 1911–12, he might have beaten his Norwegian adversary Roald Amundsen to the pole. Amundsen did use dogs to pull sledges, sacrificing some to feed the others as they went. More importantly, if Scott had listened to Shackleton, and not insisted either on bringing back 60 kg of what turned out to be effectively useless 'geological specimens' from the pole on the then almost empty sledges, he and his companions might well have made it back to civilization rather than dying 300 km short of their base, in late March 1912 (Huntford 1985, Solomon 2001, Fiennes 2003, Noakes 2006). Of the five-man party, Evans died in February and Oates in mid-March. By the time Scott, Wilson, and Bowers lost their lives, they had been man-hauling sledges for some 159 consecutive days, of which the last 60 were in exceptionally bitter cold, covering a total distance of 2500 km and expending an estimated 1 million kilocalories (or 4.2 million kJ) each (Fig. 33). The South African sports physiologist Timothy Noakes, author of the runner's bible *Lore of Running* (Noakes 1992), comes to the conclusion that this was 'one of the greatest human performances of sustained physical endurance of all time' (Noakes 2006, 2007).

Scott's team, and other teams travelling across the Antarctic continent at the time (Noakes 2006), certainly spent far more energy in total than endurance runners and cyclists during famous lower latitude sporting events (Fig. 33). However, it was the long duration that made the polar exertions so physiologically and psychologically demanding. Enduring several months of immense physical challenge is beyond what most of us can imagine. Still, from a physiological point of view, it is perhaps fairer to make comparisons of performance levels over intervals of one day, such as the detailed reconstructions of the daily energy balances of Scott, other explorers and endurance athletes, reconstructed by Mike Stroud (1998). An English medical doctor turned exercise physiologist and adventurer (Stroud 2004), Stroud used himself and fellow expeditioner, Sir Ranulph Fiennes, as guinea pigs during sev-

---

[1] It only takes a few small steps to get from Antarctic explorer Robert Falcon Scott to our own world. This is because his son, Sir Peter Scott (1909–89), an avid ornithologist, sportsman, painter and founder of the Wildfowl and Wetlands Trust, as well as the Worldwide Fund for Nature (and designer of its panda logo), was the mentor of Hugh Boyd, now an *eminence grise* in waterbird ecology. Hugh, with a career within the Canadian Wildlife Service, has been a great source of inspiration and friendship to us.

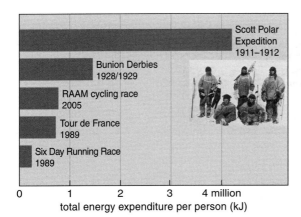

Figure 33. Reconstructed total energy expenditure levels averaged per individual participants for five events that Noakes (2006) considers to represent the limits of human endurance. From bottom to top: (1) the six day pedestrian running races in Madison Square Garden, New York; (2) the 21-stage Tour de France cycling race; (3) the non-stop cycling Race Across America (RAAM); (4) the 'Bunion Derbies' foot race over 4960 km from Los Angeles to New York in 1928 and in the reverse direction in 1929 (Berry 1990); and (5) the heroic performance of Captain Robert Falcon Scott's team who man-hauled their sleds across the Antarctic for 159 days in 1911–12. The inset shows Scott's team at the South Pole on 18 January 1912. Standing from the left are Henry ('Birdie') Bowers, Captain Scott, and biologist Dr Edward Wilson. Sitting from the left are Lawrence Oates and Edgar Evans. None survived the return journey.

Compiled from Noakes (2006).

eral Arctic and Antarctic sledge-hauling trips, and went on to publish the results of an impressively wide range of physiological measurements, often painfully obtained (Stroud *et al.* 1993, 1997).

Stroud's summary (Table 4) makes clear that polar explorers who expended 25–30 MJ per day always incurred substantial energy deficits. They ate only 60–70% of their daily energy requirements and, not surprisingly, lost about a kilo every four days. For this reason, Stroud (1987b)

believes that starvation actually killed the men in Captain Scott's party. In fact, Stroud and Fiennes almost killed themselves during their attempt to ski across the whole of Antarctica in 1992–93 (Stroud 2004). Before their eventual rescue, they had lost over 20 kg each. Although humans can endure several days of heavy physical exercise with limited food rations (Guezennec *et al.* 1994, Hoyt *et al.* 2006), by day 95 of the expedition, Stroud and Fiennes showed considerable reductions in muscle power, reductions that were linked to marked decreases in cytoplasmic and mitochondrial enzyme activities (Stroud *et al.* 1997). Still, Stroud (1998) claims that 'the restriction on food intake was imposed by the self-contained nature of the expedition rather than by the ability to eat or absorb more food'. Two other studies on endurance exercise levels that approach or exceed those of the arctic explorers, dealing respectively with Tour de France cyclists (Westerterp *et al.* 1986) and Swedish cross-country skiers (Sjodin *et al.* 1994), show rather smaller deficits, or even complete energy balance. The data point for the Swedish skiers (Table 4) suggests that an adult human male who expends 30 MJ per day must be close to his 'metabolic ceiling', i.e. the maximum energy output that can possibly be sustained by an organism that eats enough to stay in energy balance.

The energy expenditure of the sled dogs, which in 1911–12 successfully pulled Roald Amundsen from the Antarctic rim to the South Pole and back over a period of 97 days, was reconstructed to be approximately 21 MJ per day (Orr 1966, Pugh 1972, Campbell and Donaldson 1981). Although this is a very high level, Alaskan sled dogs racing for 70 hours in bitter cold conditions expended twice that much (47 MJ/day, Hinchcliff *et al.* 1997).[2] As

---

[2] This expenditure comes with considerable costs, including the death of dogs during the races. The level of performance asked from these dogs, even when trained, leads to gastric ulcers and increased intestinal permeability (Davis *et al.* 2005, 2006), as well as a degree of muscle damage (Hinchcliff *et al.* 2004).

Table 4. A summary of human studies (male subjects only) on energy intake and expenditure per individual during prolonged intense exercise, and the extent to which energy intake during these periods could cover expenditure. Expanded from Stroud (1998).

Event	Duration (days)	Energy intake (MJ/day)	Energy expenditure[1] (MJ/day)	Deficit (MJ/day)	Reference
Captain Scott's polar party	>150	ca.18	ca. 25	7	Stroud (1987b)
South Pole expedition	70	21	25	4	Stroud (1987a)
North Pole expedition	48	19	30	11	Stroud et al. (1993)
Trans-Antarctic expedition	first 50	20	32	12	Stroud et al. (1997)
	last 45	22	21	-1	
Tour de France	20	>25	34	<9	Westerterp et al. (1986)
Military mountain training	11	13	21	8	Hoyt et al. (1991)
Military arctic training	10	>11	18	<7	Jones et al. (1993)
Military jungle training	7	17	20	3	Forbes-Ewan et al. (1989)
Cross-country skiing	7	31	30	-1	Sjodin et al. (1994)
Sahara-multi-marathon	7	15	27	12	Stroud (1998)

[1] Most expenditure values are based on isotope-dilution studies; see Speakman (1997a) for the complete introduction. However, the value for the 70-day South Pole expedition is based on an energy balance evaluation. All values for Scott's party were estimated. In view of the uncertainties around each of the estimates, values presented are rounded to the nearest MJ. Human subjects tend to under-report their own food intake; where such effects were most obvious, reported energy intake estimates are minima.

pointed out by Noakes (2007), with a total expenditure of energy that approached the level of Scott and his comrades (Fig. 33), Amundsen's sled dogs might well have achieved 'the greatest animal "sporting" performance' of all time. But sled dogs weigh only about half as much as humans, and are very different animals anyway. How should we compare the athletic achievements of British polar explorers and Norwegian sled dogs?

The need for a yardstick

While hard at work on the text of this book, we have been deskbound all day. However tired we feel at night, our body machinery has been running at low levels. Moving around in the house with some thinking and writing does not take much more effort than sleeping; in energetic terms, it takes about 70% more effort (Black et al. 1996). The Arctic explorers, cyclists, and skiers that we have just talked about, expended about three times as much as book-writers would, or four to five times their rate of energy expenditure during sleep. This level happens to be close to the multiple expended by animals highly motivated to work; birds taking care of nestlings, for example, or small mammals nursing pups (Drent and Daan 1980, Hammond and Diamond 1997). Ever since Krogh (1916) started measuring metabolic rates of animals—that is, the metabolic rate of homeotherms under standardized conditions: inactive, in the thermoneutral temperature zone, in the absence of digestion, growth, moult, ovulation or gestation—animal physiologists have used this strictly defined level of energy expenditure, named 'basal metabolic rate' or BMR (see Box 3), as the yardstick for comparisons between different activity levels within and between species (King 1974).

Realizing that maximal efforts shown by hard-working animals may well reflect a compromise between physiological capacity (maximal physical output) and the 'willingness' to incur elevated mortality risks due to short and long-term consequences of increased performance, Drent and Daan (1980) introduced the concept of 'maximal sustained working levels', the aerobic capacity for work of animals as it is limited by evolutionarily shaped physiological constraints. They expressed it in multiples of BMR (Fig. 34). Animals that would choose[3] to ignore this 'physiological warning level' would lose body condition and face precipitous increases in mortality risk. A similar evolutionary concept was earlier formulated by Royama (1966). He called it the 'optimal working capacity', defined as the energetic performance level of parents beyond which they would suffer from increased risks due to physical fatigue, infection, and predation. Rudi Drent's and Serge Daan's 1980 paper 'The prudent parent: energetic adjustments in avian breeding' not only set the stage for evolutionary explanations of animal energetic performance levels (Peterson *et al.* 1990, Weiner 1992, Hammond and Diamond 1997), but also formulated a widely used organismal investment model that makes the distinction between 'capital' spenders that rely on stored resources, and 'income' spenders that rely on concurrent intake of nutrients or energy (Tammaru and Haukioja 1996, Jonsson 1997, Bonnet *et al.* 1998, Meijer and Drent 1999, Boyd 2000). Despite its modest outlet—the Dutch ornithological journal *Ardea*—'The prudent parent' has become one of the most highly cited papers in ecology.

Paraphrasing Drent and Daan twice, one could argue that 'maximum sustained working levels' are only achieved by animals that are 'income' spenders, as they maintain energy balance over the period of peak performance.

The expenditure levels of Captain Scott, Mike Stroud, and Sir Ranulph Fiennes clearly do not count during the times they were crossing parts of the Antarctic land-ice on foot (they relied quite heavily on their endogenous nutrient reserves), but those of the Swedish cross-country skiers studied by Sjodin *et al.* (1994) do, and supposedly that of the lumberjack in Fig. 34 does too. We now need data on human BMR to bring the values in Table 4 into perspective. Here we hit a snag, as BMR varies widely as a function of size, sex, and age (e.g. Henry 2005, Froehle and Schoeninger 2006, Wouters-Adriaans and Westerterp 2006), and it may even vary within individuals (at least somewhat, Johnstone *et al.* 2006)—to wit the central theme of this book! Celebrating the 75th anniversary of Scott's epic walk, during their first Antarctic mission in 1986–87, Stroud and Fiennes had their BMR measured on the way back, two weeks after they had reached the South Pole. Despite considerable body mass losses, BMR was 60% higher than before they started the 'walk' (Stroud 1987b).

In the absence of BMR measurements on the same subjects that yielded the empirical data on rates of energy expenditure during prolonged intense exercise (Table 4), we must rely on appropriate data published independently. On the basis of Henry's (2005) compilation of 10,552 human BMR values, for an athletic Caucasian man, a BMR of 6–7 MJ/day would represent a fair estimate. On this basis, polar explorers would have lived at a level of ca. 4 times BMR (Table 4), while the well-fed athletes taking part in the Tour de France and cross-country skiing would have exceeded this lumberjack-level (Fig. 34) and probably came

---

[3] The use of the words 'choose' and 'willingness' reflects the typical 'slang' of behavioural ecology. What we mean to express here and later, when expressions such as 'animals that would choose to do this rather than that' are used, is to say that 'animals that are built to do this rather than that, would have the highest fitness'. The latter is a more evolutionarily respectful use of words, but it is much more elaborate.

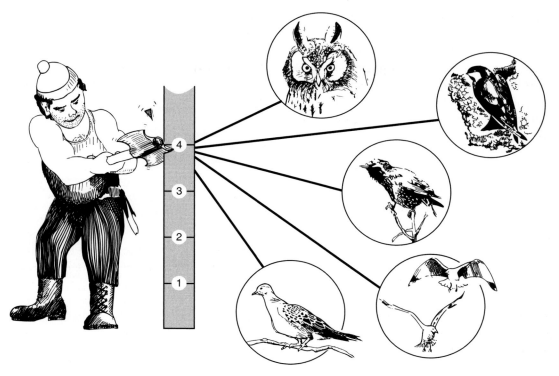

Figure 34. Original version of the cartoon in which Rudi Drent and Serge Daan (1980) indicated that what they called 'maximum sustained working level' of parent birds tending nestlings, usually approached values close to four times basal metabolic rate, a level also achieved by humans doing heavy labour, such as lumberjacks (Brody 1945). Their bird data referred to long-eared owls *Asio otis* (Wijnandts 1984), house martins *Delichon urbica* (Bryant and Westerterp 1980), starlings *Sturnus vulgaris*, glaucous gulls *Larus glaucescens*, and turtle doves *Streptopelia risoria* (Brisbin 1969).

closer to 5 times BMR. A small proportion of estimates of the ratio between sustained metabolic rates and BMR of birds and mammals are higher than 5 (Bryant and Tatner 1991, Hammond and Diamond 1997, Speakman 2000). A 'sustained metabolic scope' (henceforth called factorial scope) of 5 may, therefore, represent a reasonable first guess at a maximal physiological working level of seriously challenged animals that still maintain energy balance.

It was just a matter of time before someone first recognized the usefulness of allometry (see Box 2) for scaling this level of energy expenditure by making double-logarithmic plots of sustained metabolic rates on body mass. Kirkwood (1983) was the first, but seems to have done this rather hastily; his is an odd and incomplete collection. He assembled 21 miscellaneous estimates for metabolizable energy intake in domestic and non-domestic mammals and birds kept under 'intensely demanding conditions'. He found the overall exponent of 0.72 to be close to published exponents for BMR in mammals and birds (i.e. the lines for BMR and maximum metabolizable energy intake run parallel), the level to be on average 3.6 times BMR, and concluded that 1700 kJ/kg$^{0.72}$ per day 'may prove to be a valuable yardstick with which to compare intakes of various species under natural and experimental conditions'. Applied to our human athletes, his equation would predict a 70-kg human to be able to expend 36 MJ/day, and to eat and digest and assimilate enough food to remain in energy balance. Only cyclists in the Tour de France come close, but incur an intake deficit (Table 4). Nevertheless, Kirkwood's preliminary allometric yardstick has been widely used. In the meantime, his yardstick has been refined for specific groups of animals (Lindström and Kvist 1995).

Peaks and plateaus: what is true endurance?

In 1990, Charles Peterson, Ken Nagy, and Jared Diamond drew the first representation of the empirical finding that the longer an activity is maintained, the lower the possible power output can be, with a decline towards some asymptotic value that is only a few times BMR (Fig. 35; Peterson *et al*. 1990). Relying on anaerobic metabolism, humans and other vertebrates can achieve energy expenditure levels up to a 100 times BMR for a few seconds (Bennett and Ruben 1979). This should just be enough to outrun or outfly a predator, which after all faces the same problem, namely the build-up of toxic levels of lactic acid during anaerobic ATP production. Still, exercise-trained animals may maintain energy expenditure levels of 20–30 times BMR for many minutes (Schmidt-Nielsen 1984), though activity levels maintained for an hour or more rarely, if ever, exceed 15 times BMR (Fig. 35).

Exercise physiologists have developed different experimental procedures to assay metabolic peak performance. For example, they may expose animals to severe thermoregulatory demands until they show a decrease of core temperature (yielding a measure called *summit metabolic rate*; e.g. Swanson 1993, Chappell *et al*. 2003, Swanson and Liknes 2006, Vézina *et al*. 2007) or they may chase them for several minutes on treadmills or in running

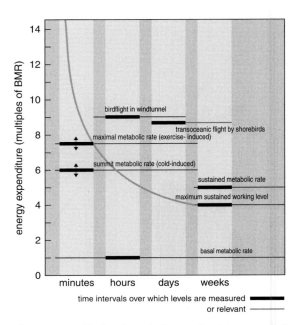

Figure 35. Unlike basal metabolic rate, the power output of exercising animals is a negative function of the duration for which the particular activity level is maintained. Inspired by Peterson *et al*. (1990), the energy expenditure levels in this figure were compiled from Drent and Daan (1980), Dawson and Marsh (1989), Hammond and Diamond (1997), Chappell *et al*.(1999, 2003, 2007), Videler (2005), Vézina *et al*. (2006), and Gill *et al*. (2005, 2009). Note that especially cold-induced summit metabolic rate and exercise-induced maximal metabolic rate can vary a great deal among individuals, species and higher taxonomic levels.

wheels to the point of exhaustion while measuring oxygen consumption (a measure called *maximal metabolic rate* or VO_{2max}; see previous chapter, especially Fig. 25 and Chappell *et al*. 1996, 1999, 2007, Bishop 1999, Bundle *et al*. 1999, Hammond *et al*. 2000). In many such studies, BMR is measured in the same individuals, even though *factorial scopes* (i.e. multiples of BMR) vary a lot. Averaged for different age- or sex-categories or species, summit and maximal metabolic rates usually fall between 5 and 10 times BMR (Fig. 35). Actually, these values seem rather low given the short intervals (minutes rather than hours) over which such peak performance measurements are made.

Another interesting point is that, although both exercise- and cold-induced metabolic rates strongly involve skeletal muscle work (in running and shivering, respectively), animals tend to have higher factorial scopes for exercise-induced power output than for cold-induced power output (Hinds *et al*. 1993). In birds, exercise-induced scopes are higher than cold-induced scopes (Dawson and Marsh 1989, Hinds *et al*. 1993), but in some small mammals, cold-induced factorial scope is higher than exercise-induced scope (Chappell *et al*. 2003). These differences may be due to the fact that birds carry a relatively larger musculature for locomotion than mammals, i.e. the flight muscles, with the discrepancy between the sizes of great and small flight muscles negatively affecting shivering capacity (Hinds *et al*. 1993, and see Piersma and Dietz 2007). In addition, mammals, but not birds, can use 'brown fat' to generate heat for thermogenesis (i.e. 'non-shivering thermogenesis'; see Cannon and Nedergaard 2004, Morrison *et al*. 2008). With their two channels of heat generation, mammals achieve high cold-induced factorial scopes.

Experimental measurements of the cost of bird flight, carried out in wind-tunnels over many hours (Pennycuick *et al*. 1997, Kvist *et al*. 2001, Ward *et al*. 2004, Engel *et al*. 2006, Schmidt-Wellenburg *et al*. 2006, 2008) also vary widely, but gravitate towards values around 9 times BMR (Videler 2005). During such long and intense exercise, it was not the thrush nightingale *Luscinia luscinia* called 'Blue' that flew continuously for 16 h that became exhausted, but the human experimentalists that needed to stand guard at the wind-tunnel from 7 a.m. the first day to 1 a.m. the next (Lindström *et al*. 1999, Klaassen *et al*. 2000, Å. Lindström pers. comm.).

Using the latest satellite-transmitter technology, it has now been demonstrated that bar-tailed godwits *Limosa lapponica* cross the entire length of the Pacific from Alaska to New Zealand—a distance of 11,000 km—in single non-stop flights that may last nine days or even a few more (Gill *et al*. 2009). On the basis of reconstructed fat and protein losses during such flights, the godwits would be flying at energy expenditure levels of about 9 times BMR (Gill *et al*. 2005). This level is consistent with modern wind-tunnel data, but still curiously high compared to the trendline (Fig. 35). It is almost double the long-term levels for sustained metabolic rate in hard-working mammals and non-flying birds, yet takes place over similar lengths of time.

A big difference, of course, is that long-distance migrants are not in energy balance. They do not eat during the flight, but deplete the sizeable stores of fat and proteins that are accrued before they take off (Piersma and Gill 1998, Gill *et al*. 2005). Indeed, despite showing evidence of remarkable protein saving (Jenni and Jenni-Eiermann 1998), such migrant shorebirds not only empty their abdominal and subcutaneous fat stores, but, with the exception of lungs and brain, also burn up belly organs and other proteinaceous tissues in flight (Battley *et al*. 2000). This raises the possibility that sustained metabolic rates are constrained by the capacity of animals to digest and assimilate food, something that even active animals do not have to do as long as they rely on stores. Is there evidence that food processing and assimilation, i.e. the lower funnels that feed into Weiner's (1992) 'animal barrel' (Fig. 29), could indeed determine plateau energy expenditure levels?

Regulation of maximal performance: central, peripheral, or external?

Here let us return to the topic of the design of metabolic ceilings left in the previous chapter, and the question of whether the food-processing chain and respiratory chains show any degree of symmorphy. To set the scene, let us first look at an advanced reincarnation of Weiner's barrel model (Fig. 36). This diagram summarizes the combined insights of University of Aberdeen biologists John Speakman and Elżbieta Król (2005) after many years of experimental studies on metabolic ceilings. Figure 36 makes clear that the energy absorbed into the body (bow-tie B) is not the only process that can constrain sustained energy intake and expenditure; four other processes (bow-ties A and C to E) could do this as well.

In principle the greatest constraint could be the availability of energy in the environment, the energy intake rate (bow-tie A). However, we have already made the point that it is unlikely that intake rate generally constrains maximal performance—it certainly did not in the well-fed Swedish skiers or Tour de France cyclists—since maximum instantaneous food intake rates tend to be so much larger than overall, long-term intake rates (Jeschke *et al.* 2002, van Gils and Piersma 2004, Jeschke 2007). Furthermore, in birds, food-supplementation

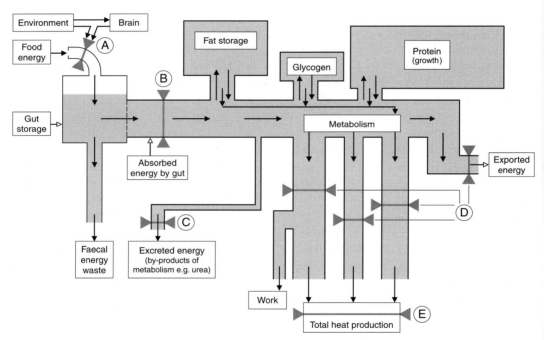

Figure 36. Diagram summarizing energy flows in an animal. Arrows indicate energy flows, and the bow-ties A to E indicate possible rate-constraining processes (see text). Energy is ingested at the mouth and enters the alimentary tract where there is a storage capacity to accommodate excess intake. Some of the ingested energy cannot be absorbed by the gut and is eliminated as faecal waste. The energy that is absorbed into the body may be diverted into different types of storage. Boxes represent the steady-state sizes of stores and arrows the movement of energy into and out of these stores. There is long-term storage in the form of fat deposited into adipose tissue and short-term storage in the form of glycogen in the liver. Energy may also be diverted into growth, which we could call protein storage, and such energy may also be withdrawn. During conversions, some energy is lost as heat. Some energy is lost in materials that are eliminated as by-products of metabolism (such as the nitrogenous end-products of protein metabolism: urea, uric acid, and ammonia). The remaining energy can either be exported from the animal as specific compounds (milk during lactation, for example), or used for metabolism. This includes the energy used to fuel cellular processes (i.e. BMR), thermoregulation, and muscle contraction. Some energy is exported as mechanical work (locomotion, for example).

Adapted from Speakman and Król (2005).

experiments have rarely managed to increase measures of peak performance (Meijer and Drent 1999, Speakman and Król 2005). As was pointed out by John Speakman (pers. comm.), the very existence of storage organs at the entrance to the gut (crop, stomach; see Kersten and Visser 1996) suggests that the rate at which energy can be harvested often exceeds the rate at which it can be processed. In principle, it is also possible that metabolic ceilings are limited by the capacity to excrete metabolic end-products, e.g. the capacity of the kidneys to excrete urea (bow-tie C). In lactating mice fed a high-protein diet, the latter may actually occur (J.R. Speakman pers. comm.).

For a long time there were two main hypotheses to explain the regulation of maximal performance. One, related to the capacity of the alimentary tract to absorb energy (the food-processing chain), is called the *central limitation hypothesis* (bow-tie B). The other, related to the capacity to expend energy at the different sites of utilization (the respiratory chain, with the muscles, testicles, ovaries, and, in mammals, brown adipose tissue and mammary glands as the sites of heat production) is called the *peripheral limitation hypothesis* (bow-tie D) (Ricklefs 1969, Hammond and Diamond 1997, Hammond *et al.* 2000, Bacigalupe and Bozinovic 2002, Speakman and Król 2005). Although the null hypothesis of symmorphosis predicts that neither of them would be limiting on its own (Chapter 3), these two alternatives have attracted a lot of elegant experimental testing, especially in small rodents. The rigour of the tests in this research programme, and the fact that they turned up an exciting new hypothesis, justifies some elaboration.

Rodents breed fast, have fur, but can easily be exposed to temperatures that will elevate their metabolism, and have mammary glands that produce, and teats that deliver, milk. These characteristics make small mammals excellent models for experimental studies of metabolic ceilings (Table 5). Attention has focused on *sustained energy intake* rather than *expenditure* because during reproduction 'considerable amounts of energy are exported as milk' (Speakman and Król 2005). In small mammals, pregnancy increases energy demand by about 60%, while during peak lactation, energy intake quadruples (Johnson *et al.* 2001a). To test whether this represents the metabolic ceiling, rodent mothers have been given more young (Hammond and Diamond 1992, Künkele 2000, Johnson *et al.* 2001a), young for a longer time (Hammond and Diamond 1994), or more hungry young (Laurien-Kehnen and Trillmich 2003). In none of these studies did food intake increase. Pup masses typically went down, which makes sense given that the same amount of food had to be shared by more pups. To test whether this apparent limit was centrally determined, diet quality was reduced (Speakman *et al.* 2001, Hacklander *et al.* 2002, Valencak and Ruf 2009). Mice, and also European hares, responded by eating and assimilating more food, suggesting the ability of the digestive tract to upregulate capacity.

In order to increase selectively the demands of the peripheral systems, Perrigo (1987) forced lactating mice to run for their food, but found that, despite free access to food, neither of the two mouse species tested showed elevations of food intake. Furthermore, when Kristin Schubert and colleagues (2009) applied a similar paradigm in laboratory mice, they found that minimum metabolic rate and daily energy expenditure (DEE) were both actually reduced (see below, 'The evolution of laziness'). When peripheral demands were elevated, by making lactating rats and mice simultaneously pregnant, food intake did not increase either (Biggerstaff and Mann 1992, Koiter *et al.* 1999, Johnson *et al.* 2001c). However, in all studies where rodents were exposed to cold (Table 5), food intake did increase, especially during

Table 5. What can lift the metabolic ceiling? Nine experimental ways to examine the various physiological mechanisms that could limit sustained energy intake in rodents. Expanded from reviews in Speakman and Król (2005), Król *et al.* (2007) and Speakman (2008).

No.	Manipulation	'Taxon' examined	Reference
1	Increase litter size	Swiss Webster mouse	Hammond and Diamond (1992)
		Guinea pig *Cavia porcellus*	Künkele (2000)
		MF1 laboratory mouse	Johnson *et al.* (2001a)
2	Prolong lactation	Swiss Webster mouse	Hammond and Diamond (1994)
3	Increase demand of precocial pups by witholding food early in lactation	Guinea pig	Laurien-Kehnen and Trillmich (2003)
4	Decrease quality of food so that they have to eat and digest a greater volume	MF1 laboratory mouse	Speakman *et al.* (2001)
		European hare *Lepus europaeus*	Hacklander *et al.* (2002)
		European hare	Valencak and Ruf (2009)
5	Force lactating animals to run for food	Deer mouse *Peromyscus maniculatus*	Perrigo (1987)
		Wild-type house mouse *Mus domesticus*	Perrigo (1987)
		Outbred house mouse ICR (CD-1)	Schubert *et al.* (2009)
6	Make them simultaneously pregnant and lactating	Rockland-Swiss mouse	Biggerstaff and Mann (1992)
		Brown rat *Rattus norvegicus*	Koiter *et al.* (1999)
		MF1 laboratory mouse	Johnson *et al.* (2001c)
7	Increase temperature gradient between animal and environment	Swiss Webster mouse	Hammond *et al.* (1994)
		Mouse *Apodemus flavicollis*	Koteja (1995)
		Deer mouse	Koteja (1996)
		Cotton rat *Sigmodon hispidus*	Rogowitz (1998)
		Deer mouse	Hammond and Kristan (2000)
		MF1 laboratory mouse	Johnson and Speakman (2001)
		House mouse	Koteja *et al.* (2001)
8	Surgically remove some mammary glands	Swiss Webster mouse	Hammond *et al.* (1996b)
9	Shave fur of mothers to increase heat loss	MF1 laboratory mouse	Król *et al.* (2007)

lactation! All this suggested that performance limits were not set centrally. That mice could increase food-intake rates when muscles and brown adipose tissue were made to work harder in the cold, suggested that it was actually the capacity of the mammary glands that set the upper limit.

Kimberly Hammond and colleagues (1996b) tested this by surgically removing half of the mammary glands. If the capacity of these tissues were flexible, it would expand, but it did not. This seemed consistent with the peripheral limitation hypothesis until Speakman's group, using the doubly-labelled water technique in novel ways (Speakman 1997a), began to measure milk production (Johnson and Speakman 2001, Król and Speakman 2003a,b, Król *et al.* 2003). As noted earlier, the mice ate more, the colder it got. However, with increased heat loss, milk production also went up and so did the mass of pups at weaning! Król and

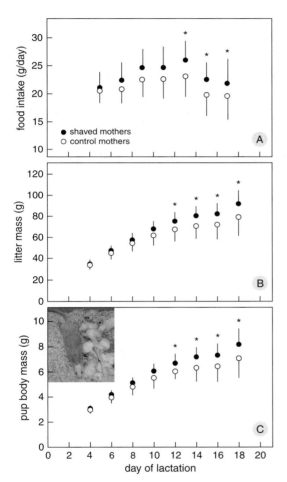

Figure 37. Food intake of 20 shaved and 20 unshaved control MF1 laboratory mouse mothers during (A) lactation, (B) the mass of their litters, and (C) the mass of their individual pups. From days 6 to 12 the mothers had dorsal fur shaved off to increase heat loss capacity. Values are means ± 1 SD. Stars indicate days with significant differences (P < 0.05) between the shaved and control groups.

Compiled from Król *et al.* (2007).

Speakman (2003a, 2003b) concluded that performance in lactating mice must be limited by the capacity of the female mouse to dissipate heat. They put this to a test by shaving some mice during lactation (Król *et al.* 2007). As shown in Fig. 37, this experiment confirmed that shaved mothers ate more and had heavier litters with heavier individual pups.

Król and Speakman were quick to recognize that the problem of heat dissipation is probably rather general, and is likely to constrain metabolic performance in many ecological contexts. Apart from the well-known link between high temperature, solar radiation, and humidity, and the reduction in milk production in dairy cattle (Thompson 1973, King *et al.* 2006, Nassuna-Musoke *et al.* 2007), sheep *Ovis aries* (Abdalla *et al.* 1993), and pigs *Sus scrofa* (Black *et al.* 1993, Renaudeau and Noblet 2001), heat-dissipation limitations might help explain latitudinal and altitudinal trends in clutch sizes of birds (larger where it is colder; Bohning-Gaese *et al.* 2000, Cooper *et al.* 2005, but see McNamara *et al.* 2008). It could also provide a reason why humans, one of the few mammals that can sweat profusely, are able to outrun pretty much any running game in the midday heat (Heinrich 2001). And it could help explain why long-distance migrant birds tend to fly at night (Klaassen 1996), and why fuelling rates of migrant shorebirds are so low in the tropics (Battley *et al.* 2003, Piersma *et al.* 2005).

Temperature and the allometric scaling constant

As most of us know from experience, bodies heat up during exercise. In running horses, (near-lethal) body temperatures as high as 45 °C have been measured (Hodgson *et al.* 1993). Just as the capacity to lose heat can limit sustained performance in lactating mice, rising body temperatures can put a limit to the duration or the level of intense exercise (Weishaupt et al. 1996, Fuller et al. 1998, González-Alonso et al. 1999, Arngrimsson et al. 2004). Noting that the allometric scaling constant for maximal metabolic rate in mammals (ca. 0.80) is consistently higher than that for BMR (0.70–0.75; Taylor et al. 1981, Bishop 1999, Savage et al. 2004, Weibel et al. 2004), Gillooly and Allen (2007) proposed that the steeper size-dependence of maximal metabolic rate than of BMR may be explained by larger mammals achieving the highest muscle temperatures at maximal activity. The larger the animal, the more unfavourable its surface–volume ratio when losing, rather than retaining, heat. In general, chemical reaction rates increase exponentially with temperature, as does metabolic rate (Gillooly et al. 2001).

Gillooly and Allen (2007) suggest that the short-term maximal metabolic rate (technically called VO_{2max}) is a function of body temperature. This led them to the somewhat counter-intuitive prediction that mammals should be able to achieve higher maximal metabolic rates at higher ambient temperatures, because heat loss is more restricted, leading body temperatures to rise more. Craig White and colleagues (2008) tested this hypothesis on the basis of published maximum aerobic performance measurements in rodents, and rejected it. They found no relationship between VO_{2max} and testing temperature, which varied from −16° to +30 °C. Interestingly, they did find that animals that were acclimated at lower temperatures achieved higher maximal metabolic rates. Cold necessitates shivering, and thus a more intense use of muscle tissue. This, as we shall see in the next chapter, usually leads to hypertrophy. In fact, within species, such chronic muscle-mediated thermogenesis can lead to greater increases in aerobic capacity, whether measured as maximal or summit metabolic rate, than endurance training (Schaeffer et al. 2001, 2003, 2005, Vézina et al. 2006).

Protecting long-term fitness assets: factorial scopes and optimal working capacity revisited

When one builds bigger bodies to ingest, absorb, excrete, and do work, the expanding machinery requires ever more energy for maintenance. Such increases in maintenance requirements will be reflected by increasing BMR levels. A summary of the early studies on mice whose energy expenditure had been elevated in a variety of ways (see Table 5), shows BMR to go up with sustained metabolic rate (Fig. 38A). This effect comes with an increase in body mass, due to hypertrophy of the liver, digestive tract, and mammary glands (Hammond et al. 1994, Speakman and Johnson 2000, Johnson et al. 2001b). As a consequence of having put in place increasingly powerful machinery, factorial scope (the ratio between sustained energy expenditure and BMR) climbs to an asymptote of about 7 times BMR (Fig. 38B; Hammond and Diamond 1997).

We have come full circle. During our quest to establish whether maximal aerobic performance is limited centrally, by digestion and assimilation systems, or peripherally, by the power that can be generated by the receiving organs, experimental evidence has not been uniformly consistent with either alternative. In a sense, this celebrates the notion of symmorphosis, the principle of logical, economic animal design (Chapter 3): symmorphic design would have negated evidence in favour of either hypothesis for a single site limiting aerobic performance. That rates of heat loss best explain the limits to aerobic performance in lactating laboratory mice, which only push up their metabolic ceilings when it is cold enough or when insulation is poor enough, does not detract from the fact that mouse mothers can only do this by building a bigger body, a bigger metabolic machine. Thus, they still face a maximum factorial scope of 7.

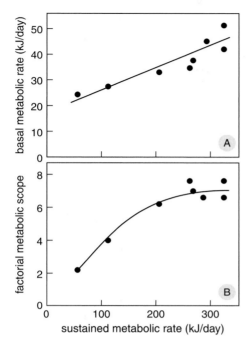

Figure 38. How the laboratory mouse equivalent of basal metabolic rate (BMR) goes up with experimentally increased levels of energy expenditure due to hypertrophy of various organ groups (A), and how simultaneously the factorial scope of sustained energy expenditure on BMR reaches plateau levels of ca. 7 times BMR (B).

Adapted from Hammond and Diamond (1997).

So, if animals are pushed very hard, under some conditions (e.g. when enabling for greater heat loss), they can raise the ceiling from working at 5 times BMR (Fig. 37) to working at 7 times BMR (Fig. 38). As we have seen, endurance athletes in energy balance can push their performance levels to 5 times BMR, but not further. We have also seen that free-living birds, except in the case of marathon migrants, generally do not work harder than 4 times BMR (Fig. 34). The considerable gap between the maximum sustained working level of 4 times BMR that hard-working parent birds are prepared to give (Drent and Daan 1980), and the physiological maxima of 5–7 times BMR that can be achieved under exceptional conditions, makes

evolutionary sense if working hard comes at a survival cost (Valencak *et al.* 2009). If very hard work precipitously increases the likelihood of death (e.g. because of free-radical derived oxidative DNA and tissue damage; Harman 1956, Alonso-Alvarez *et al.* 2004, Wiersma *et al.* 2004, Cohen *et al.* 2008), without leading to compensatory increases in reproductive output (Williams 1966a, 1966b, Lessells 1991, Daan and Tinbergen 1997), evolutionary trade-offs would select for animals that are not prepared to work harder than what we can call the 'optimal working capacity' (Fig. 39; Royama 1966). On the basis of field data on provisioning birds,

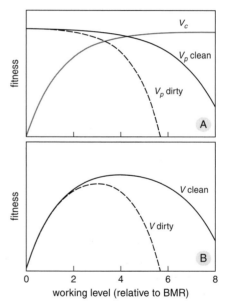

Figure 39. The marginal value of work: Royama's (1966) concept of an optimal working level during reproduction couched in terms of a simple optimality plot of (A) the two main fitness components (the residual reproductive value of the parent Vp and the reproductive value of the clutch Vc) or (B) their sum as a function of the parent's relative metabolic rate. Two sorts of environment are distinguished: a 'clean' one in which the chance of catching a disease is small, and a 'dirty' one with a higher chance of becoming infected.

Inspired by fig. 13 in Drent and Daan (1980) and fig. 1 in Daan *et al.* (1990b).

optimal working capacity is seldom higher than 4 times BMR (Drent and Daan 1980) and most often lower (Bryant and Tatner 1991, Tinbergen and Dietz 1994). Speakman (1997b) found the same in small mammals, stating that these animals 'routinely live their lives at well below their physiological capacities, in the same way that we drive our cars at well below their mechanical capacities'.

The evolution of laziness

As we have seen, there seems abundant evidence that animals, even when confronted with challenges such as experimentally increased brood sizes that might gain them more offspring, are simply not prepared to work harder than 4 times BMR (Drent and Daan 1980) or, indeed, harder than necessary (Masman et al. 1988). Nevertheless, only a few experiments have demonstrated that 'increased daily work precipitates natural death'. An important example is the study by Serge Daan and co-workers (1996) with that very title, in which they analysed time till death of 63 free-living kestrels Falco tinnunculus that had raised broods with manipulated sizes. They were able to demonstrate that kestrels with larger broods incurred increased mortality the following winter compared with those with normal or reduced brood sizes.

Experiments with insects confirmed these trends. When honey bee workers Apis mellifera were forced to fly with extra loads and work really hard, they showed decreased life spans (Wolf and Schmid-Hempel 1989). When house flies Musca domestica were prevented from flying by keeping them in small flasks, so that their energy expenditure decreased, they had increased life spans (Yan and Sohal 2000). Interestingly, extra work load experiments on laboratory rats and various kinds of mice have yielded equivocal results (Holloszy and Smith 1986, Navarro et al. 2004, Speakman et al. 2004b,

Selman et al. 2008, Schubert 2009, Vaanholt et al. 2009). Basically, elevating DEE did not not shorten life span.

Whether or not enhanced energy expenditure has life-shortening effects, there is still evidence that mammals, as well as birds, prefer to work at optimal levels, where metabolic investment in reproduction is concerned. Lactating mouse mothers challenged to work for food support make the 'decision' not to work above a certain level—or indeed, to increase body mass as required for maintaining metabolic machinery—even though this results in a linear decrease in milk energy output and thereby the biomass of weaned litters (Schubert et al. 2009). Perrigo (1987) earlier observed that two wild-type mouse species differed in their willingness to work in support of offspring. He attributed this to their different life-history strategies. While Mus musculus are opportunistic breeders capable of maintaining fertility whenever food is available, Peromyscus maniculatus are strictly seasonal, and presumably face a higher opportunity cost of litter production. This species appeared more determined to save a smaller number of pups when Mus mothers had already abandoned their litters.

Earlier we encountered the élite cyclists who perform at very high levels during the three weeks of the Tour de France (Westerterp et al. 1986, Table 4). Interestingly, in the Vuelta a España, a similarly gruelling cycling race but a week shorter, total energy expenditure was similar (Lucia et al. 2003). This suggests that when expending energy, humans, and probably other animals, account for time. Animals are only prepared to deliver a certain level of effort over certain time periods, a phenomenon called 'pacing' in the world of sport (Foster et al. 2005). In the words of Tim Noakes (pers. comm.): 'Train hard when it is important, and rest when it is important. And don't confuse the two.' We suggest that the time spans over which the accounting of metabolic performance takes place in different animals could be a rich area of experimental research.

Free-living animals are resistant to experimental manipulations to increase their instantaneous work levels, for what we can now see are sound reasons. Unless some mechanistic shortcuts are found to the actual expenditure decision processes (e.g. Speakman and Król 2005), it will be impossible to verify experimentally the optimality model of Fig. 39 along the whole range of work levels. Nevertheless, it seems likely that within taxa the 'death on working level curves' would be determined by ecological conditions, e.g. the general environmental or social stress levels that have become known as 'allostatic loads' (McEwen and Wingfield 2003, Wingfield 2005, Rubenstein 2007), or the levels of pathogens in the environment (Piersma 1997).

Such variations suggest that there is scope for studies in which relationships between energy expenditure levels and survival are studied, either in a strict comparative context (comparing relationships in groups with different environments) or in experimental contexts (by additionally manipulating aspects of the environments in which the expenditure–survival relationships are studied). During such endeavours it remains important to pay attention to constraints other than work level, especially time limitations, that prevent animals from keeping up with the hard work and from eating for very long (e.g. Kvist and Lindström 2000, Tinbergen and Verhulst 2000). When interpretable aspects of the ecological context can be linked to maximal performance, or rather to the precise levels and relationships between experimentally manipulated work levels and both reproduction and survival (Fig. 39, and see Bouwhuis et al. 2009 for a possible field setting), we may achieve deeper levels of understanding of the selection factors driving the height of metabolic ceilings.

We believe that such insights need to come from adequate incorporation of *ecology* into what at first sight may seem a purely physiological problem. The impressive portfolio of studies on highly inbred strains, or genetically modified lab mice, will thus need expansion to a variety of free-living animals, with due attention to ecological context for an appreciation of the evolutionary causes of metabolic ceilings.

Evolutionary wisdom of physiological constraints

Unlike bar-headed geese *Anser indicus* that have evolved particular muscle phenotypes to migrate over the Himalayas at heights of 9000 m (Scott *et al.* 2009), humans clearly are a low-altitude species. When Reinhold Messner and his team were getting close to the summit at 8848 m of Mount Everest on 8 May 1978, experiencing an equivalent of 7% oxygen rather than the 21% at sea level, they found it increasingly hard to stay on their feet (Messner 1979): 'every 10–15 steps we collapsed into the snow to rest, then crawled on'. Their leg-muscle fatigue could not have been due to lactic acid poisoning because, paradoxically, observations on human exercise at low partial oxygen pressures show that, whereas lactic acid levels should go up with increasing hypoxia at high altitudes, they actually go down (Bigland-Ritchie and Vollestadt 1988). Also, paradoxically, rather than going up to compensate for the decreasing partial oxygen pressures at great heights, heart rates actually go down with altitude (Sutton *et al.* 1988). This change is related to blood being shunted away from the muscles (hence the fatigue experienced by Messner and his comrades; Amann *et al.* 2006) and towards the brain—the 'selfish' organ that controls oxygen delivery to the system (Noakes *et al.* 2001, 2004). Noting that, during important games, athletes can deliver an end spurt at the point where their muscles must show greatest fatigue, Noakes (2007) concludes that it is also the brain that regulates exercise performance by arranging an anticipatory response pattern: only the brain has the knowledge that the finish is close. In more general terms, he writes, 'the limits of our endurance lie deeply in the human brain, determined by our heredity and other personal factors yet to be uncovered' (Noakes 2006).

To this we would add that the limits to endurance lie in any animal's brain. Brains have been shaped in the process of organic evolution to deliver sensible, evolutionarily informed, behaviours. The evolutionary algorithms underpinning such sensible behaviour prevent animals from 'overdoing it' and losing life and limb whilst doing so. This was the 'wisdom' that slowed Messner down before he reached Mount Everest's summit. And it is this wisdom that makes kestrels 'lazy' and unwilling to spend much more than 500 kJ/day, or 4 times BMR (Fig. 40; Deerenberg et al. 1995).

Beyond Rubner's legacy: why birds can burn their candle at both ends

In 1908, the German animal physiologist Max Rubner published a comparison between intake rates of six domestic animals (guinea pig, cat *Felis domesticus*, dog, cow, horse, and human) and their life span (Rubner 1908). Despite a 50,000-fold difference in body mass between the smallest and the largest species, he was struck by the finding that there was only a five-fold variation in what we will call here the mass-specific lifetime energy expenditure (the multiplication of mass-specific rate of energy expenditure and maximum life span). Humans had by far the largest mass-specific lifetime energy expenditure within this group of species; taking them out of the comparison, the range of variation was reduced to a factor of 1.5.

About a decade later, the American biologist Raymond Pearl (1922, 1928) used Rubner's conjecture as the basis of his 'rate-of-living' theory: the idea that animals have a certain allotment of energy to spend during a lifetime, or a fixed number of heartbeats (e.g. Calder 1984). In Eric Charnov's words we could call this a 'life-history invariant' (Charnov 1993). According to the rate-of-living theory, the elevated energy expenditure levels of Captain Scott and his comrades during their Antarctic exploits would have shortened their lives, even if they had survived the attempt to reach the South Pole first. From our point of view,

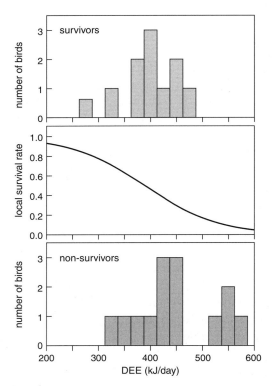

Figure 40. Local survival rates of kestrels *Falco tinnunculus* in the northern Netherlands as a function of daily energy expenditure (DEE) during the raising of chicks. The upper and lower panels show the frequency distributions of experimental birds that were either (top) or not (bottom) recaptured in the study area during the next breeding season. The middle panel shows the logistic regression curve of survival on DEE ($X^2 = 5.66$, df =1, P = 0.02).

From Deerenberg *et al.* (1995).

the rate-of-living theory is attractive because it could provide a link between metabolic performance and typical life-history variables, such as annual survival and life span. But how invariant is lifetime energy expenditure, or rather, at what taxonomic level is it invariant, and what does this tell us about the evolutionary levelling of metabolic ceilings?

The most recent, and the most comprehensive, assessment of Rubner's idea that animals have a certain amount of energy to spend was made by John Speakman and associates (Speakman et al. 2002, Speakman 2005, Furness and Speakman 2008). As summarized in Fig. 41, among species there is about a 20-fold (in mammals, note natural log scales) to 40-fold (birds) variation in lifetime energy expenditure per gram of body mass. It does not seem to matter whether lifetime energy expenditure is (under-) estimated on the basis of BMR, or properly estimated using measurements of daily energy expenditure. There are clear declines of mass-corrected lifetime energy expenditures with body size (Fig. 41) and, therefore, lifetime expenditures can hardly be called invariant. Lifetime expenditures certainly appear taxon-specific, a conclusion that shows most clearly in a comparison between lifetime energy expenditures of mammals and birds (Fig. 41 right panels). Indeed, as related to us by our text editor Marie-Anne Martin, 'my friends in North Wales are always sad when they keep losing their pet rats and ferrets, while my friend in Gloucestershire has just incorporated her 54-year old African grey parrot into her will'. This is the more surprising since birds generally have higher rates of energy expenditure than mammals (e.g. Koteja 1991, Ricklefs et al. 1996). The average bird has 3–4 times more kJ to spend on BMR (Fig. 41), and also on DEE (Speakman 2005), than the average mammal.

Before returning to why an average bird spends so much more energy than an average mammal, we would like to discuss what may actually be pretty good evidence for a certain invariance in lifetime energy expenditure. The case is reminiscent of the levelling-off of athletic performance measures in humans and other domestic animals (Fig. 28). In the endeavour to increase economic productivity of farm animals by 'improved' keeping conditions, better food, and intense genetic selection, it is now widely acknowledged that clever breeding and husbandry schemes cannot prevent the most highly productive animals from being most at risk from behavioural, physiological, and immunological problems (Rauw et al. 1998, Lucy 2001). For example, in the USA, the average milk yields of dairy cows quadrupled between 1950 and 2007, but despite earlier maturation, the overall reproductive life spans of these cows nearly halved (Knaus 2009). Livestock breeders seem to be faced with Charnov's life-history invariance, although they are not quite prepared to accept this reality (e.g. Liu et al. 2008).

That birds tend to have so much more time, and so much more energy, to spend in a lifetime than mammals, is correlated with the finding that rates of bodily decline (i.e. rates of ageing) are lower in birds than in mammals (Holmes and Austad 1995a, 1995b, Ricklefs 1998, 2000, Ricklefs and Scheuerlein 2001, Costantini 2008, Holmes and Martin 2009). Getting a bit ahead of ourselves, this may be because birds, at least the flighted ones (and also bats, the only group of flying mammals), have experienced selection pressures that have allowed the evolution of biochemical mechanisms that reduce the wear and tear on tissues (Barja et al. 1994, Perez-Campo et al. 1998). The mechanisms, which relate to mitochondrial functioning and the ways that electron leakage and free-radical production are balanced under conditions of high and low energy demands, are just beginning to be revealed (Buttemer et al. 2008, Costantini et al. 2008, Dowling and Simmons 2009, Monaghan et al. 2009). We leave

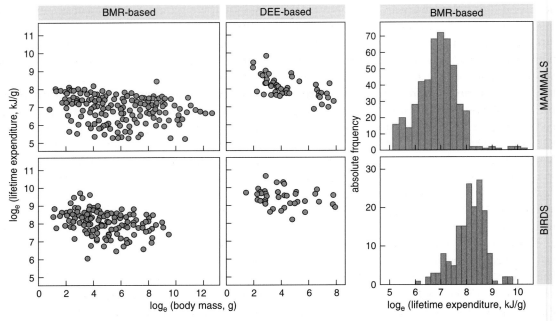

Figure 41. Contrasts between mammals (top panels) and birds (bottom panels) in mass specific lifetime energy expenditure (kJ/g) as a function of body mass (left two sets of panels) or as a frequency distribution (right set of panels). Values were calculated by multiplying life span of a species either by measured BMR or measured daily energy expenditure (DEE).

Compiled from Speakman (2005), where the statistical details can also be found.

Note that mass-specific lifetime energy expenditure declines significantly with body mass (see also Furness and Speakman 2008).

readers to consult Nick Lane (2005), in his splendid *Power, sex, suicide: Mitochondria and the meaning of life*, for a summary of the field and the development of this complicated, but hugely exciting, story. What is most important here is that the capacity to fly seems to go with lower extrinsic (ecological) mortality, such as the likelihood to be depredated (Austad 1997), and that lower extrinsic mortality, in combination with the high metabolic demands of flight, may well be driving the appearance of the mitochondrial mechanisms that reduce intrinsic wear and tear (Lane 2005).

At this point, we have returned to ecology, the driver of what may at first have seemed purely mechanistic, physiological processes that one can study in isolation from the messy outside world (Partridge and Gems 2007,

Monaghan *et al.* 2008). Our reading of the literature suggests that any kind of hard work, perhaps above taxon- (or rather, ecology-) dependent thresholds, does come with wear and tear. A precipitous increase in the likelihood of organ or performance failure, and mortality associated with increases in energy expenditure (Ricklefs 2008), would explain why animals are reluctant habitually to spend as much as they are physiologically capable of (Valencak *et al.* 2009). Indeed, it would explain why special kinds of motivation are needed to lift our own modest human metabolic ceilings.

Hard-working shorebirds

In the 1970s and 1980s, biologists interested in animal energetics were still pre-occupied with the measurement, comparison, and interpretation of BMR in as wide a variety of animals as possible (e.g. Aschoff and Pohl 1970, McNab 1986, 1988, 2002). As a University of Groningen undergraduate, one of us (TP) joined the fray, and, together with Marcel Kersten, started measuring BMR in several species of shorebird held in captivity. It soon became clear that the relatively high values of BMR in shorebirds that we measured, and that had been found by others, were no artefact of method, but a phenomenon typical of the group (Kersten and Piersma 1987), at least of those living in north temperate cold climates (Kersten *et al.* 1998, Kvist and Lindström 2001). Not only shorebirds, but also songbirds, were found to have higher BMRs than other birds (Aschoff and Pohl 1970). Cage-existence metabolism (a 'standardized' measure based on energy intake within the confinements of aviaries; Kendeigh 1970) was consistently high as well. At the time, high BMR values were interpreted as reflecting the high maintenance requirements of the considerable body machinery needed to support high levels of performance in real life (Kersten and Piersma 1987; and see Box 3).

As we have seen, doubly-labelled water ($^2H_2{}^{18}O$) dilution rates in the body fluids of (re-) sampled animals, including athletes, provide robust integral measurements of energy expenditure of animals that go about their own business in the intervening interval (Speakman 1997a). The critical requirement is recapture. During the non-breeding season, when most shorebirds find safety in numbers as they roam around in huge flocks (Piersma *et al.* 1993c, van de Kam *et al.* 2004), recapture within intervals of a few days is impossible (but see Castro *et al.* 1992 for a single example). Thus, assessments of energy expenditure levels during the non-breeding season have to be based on indirect methods (Wiersma and Piersma 1994, Cartar and Morrison 2005). During incubation, however, when shorebirds have made an

investment (the clutch) at a small place (the nest), they may be tricked into entering traps twice. A series of expeditions to the most northerly lands on earth (Alerstam and Jönsson 1999), enabled us to collect enough data on energy expenditure of shorebirds incubating a clutch to test the idea that, during that phase of their annual cycle, they show relatively high levels of energy expenditure (Piersma *et al.* 2003) in comparison with bird species at lower latitudes (Fig. 42). The small shorebirds, for which the costs of thermoregulation will be relatively highest, actually showed energy expenditures that would have been predicted by Kirkwood's (1983) equation to represent metabolic ceiling levels.

These incubating shorebirds were in energy balance (Tulp *et al.* 2009). However, typical for long-distance migrants that make very long, non-stop flights (Newton 2008), shorebirds go through many phases when they are *not* in

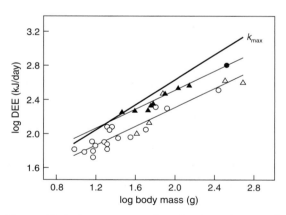

Figure 42. Daily energy expenditure (DEE) as a function of body mass in tundra-breeding shorebird species (closed triangles) and one long-tailed skua *Stercorarius longicauda*, (the largest species) during the incubation phase. These are compared with DEE during incubation of lower latitude bird species (open symbols; from Tinbergen and Williams 2002). Shorebirds are indicated with triangles, other species with circles. k_{max} is the suggested upper limit to daily metabolizable energy intake for homeotherm animals (Kirkwood 1983).

From Piersma *et al.* (2003).

energy balance on a daily basis, particularly during the fuelling phase and when burning off all the stored energy during the flight phase (Piersma 1987). Anders Kvist and Åke Lindström (2003) capitalized on the availability of hungry individuals of a large variety of shorebirds making a short stop during southward migration in the Baltic. These shorebirds have a great urge to fuel up quickly and move onward, and they just love mealworms *Tenebrio* sp. In this setting, Kvist and Lindström were able to show that shorebirds as a group have unrivalled capacities to process food and refuel fast (Fig. 43). BMR was measured concurrently with the intake measurements, and most species had energy assimilation rates higher than 6 times BMR. This enabled them to increase mass by as much as 10% of fat-free body mass per day,

amongst the highest values measured in refuelling birds (Lindström 1991, 2003).

Migrating shorebirds may have the capacity to process food and store fuel at such fast rates because they are usually in a hurry and because they make such long-distance flights. They are in a hurry because they need to get to the arctic breeding grounds in good time (Gudmundsson *et al.* 1991, Drent *et al.* 2003), or because getting back early to their southward staging areas or wintering grounds is beneficial (Zwarts *et al.* 1992, Alerstam 2003). As we have earlier alluded to (Fig. 35), during long-distance flights of many thousands of kilometres, shorebirds spend the equivalent of 8–10 times BMR (Pennycuick and Battley 2003, Gill *et al.* 2005). In the course of such non-stop flights, shorebirds not only deplete much of their fat store,

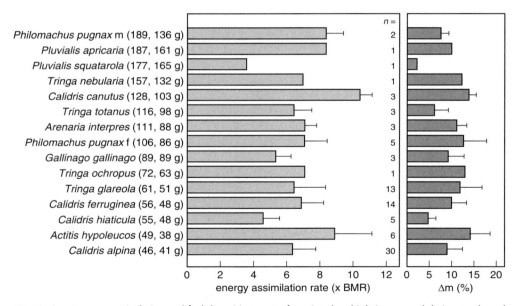

Figure 43. Maximum energy assimilation and fuel deposition rates of captive shorebirds intercepted during southward autumn migration in southern Sweden. Parenthetical values following the species name represent average body mass of the birds on the day (24-h period, see text) when maximum energy assimilation rate was measured and average body mass when BMR was measured. Energy assimilation rates (grey bars to the left) are presented as multiples of BMR to normalize for differences in size and metabolic capacity between animals. The black bars to the right represent the mass increase per 24 h (Δm), the 24-h period with the maximum energy assimilation rate, expressed in % of fat-free body mass. Error bars represent standard deviations.

Modified from Kvist and Lindström (2003).

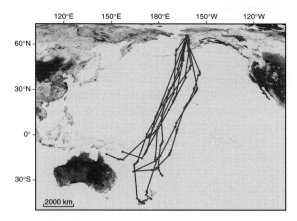

Figure 44. When the sky is the limit: southward flight tracks of nine bar-tailed godwits that maintained estimated rates of energy expenditures of 9 times BMR for up to 9.4 days. Individual godwits were fitted with satellite transmitters and followed during their flights during August–October 2006 and 2007. Small filled circles denote satellite-informed locations collected during 6–8-h intervals, and solid lines show interpolated 24–36-h tracks between reporting periods. The intensity of shading of the land-masses correlates with increasing lack of vegetation cover.

Modified from Gill *et al*. (2009).

but also lose protein, due to atrophy of organs in the body cavity and even the working flight muscles (Piersma and Jukema 1990, Battley *et al*. 2000). Developments in satellite-transmitter technology have recently made it possible to establish that such performance levels are maintained for at least 9 days in bar-tailed godwits (Fig. 44). Maintaining an estimated metabolic rate of 8–10 times BMR for more than 9 days represents a combination of metabolic intensity and duration that is unprecedented in the current literature on animal energetics (Fig. 35).

On the basis of the arguments developed earlier in this chapter, there should be good ecological reasons why these bar-tailed godwits are prepared to work so hard for so long (note that their annual survival is also very high, >96%, P.F. Battley pers. comm.). Gill *et al*. (2009) propose that, by making a non-stop trans-Pacific flight, godwits not only minimize the time and energy allocated to migration, but also minimize the risk of mortality from predators, parasites, and pathogens. Bar-tailed godwits and other shorebirds appear to have been selected to work routinely as hard, and almost as long, as the arctic explorers with whom we started this chapter. They also deplete their bodies to the same extent. However, by making a living along the edges of our planet, so to speak, in places that some parasites, pathogens, and predators have not been able to reach (e.g. Piersma 1997, 2003, Mendes *et al*. 2005, Gill *et al*. 2009), godwits may exploit ecological conditions that do not make them pay the price (of enhanced subsequent mortality) that usually comes with very hard work.

Synopsis

It is clear that animals are not usually willing to perform at levels, or for lengths of time, of which they should be maximally capable. We think that this strategy has been selected by the costs of high performance levels, in the form of subsequently reduced survival or future reproduction. The somewhat lopsided scientific efforts to determine the proximate limits to animal performance, by looking at the metabolic machinery rather than at complete life-histories in naturalistic settings, has made an evolutionary understanding of metabolic ceilings still beyond reach. In relative terms, the long-term metabolic ceiling of ca. 7 times Basal Metabolic Rate in challenged individuals may be real and general, because greater performance requires larger machinery that is ever more expensive to maintain. However, our reading of the literature suggests that absolute limits to performance are mostly relative to ecological context, which sets the severity of the survival punishment for over-exertion. Over generations, the many dimensions of ecology thus determine how far individual animals, including human athletes, are prepared to go.

Box 3 BMR: interesting baseline, epiphenomenon, or both?

'The most remarkable thing about the apteryx (the kiwi) was its small respiratory system, suggesting that in the wild this must be a shy, patient, creepy little bird, with little inclination to exert itself much and therefore little need to breathe heavily'. So wrote Charles Darwin in his early notebook 'D', according to David Quammen (2006). It is not unusual for Charles Darwin's writings to encapsulate many later developments in biology. In the present case, reference to 'little need to breathe heavily' correctly anticipated the finding that the flightless kiwis of New Zealand have the lowest BMR-values reported in birds (McNab 1994, 1996); mention of 'its small respiratory system' reflected the small size of flight muscles and other organs associated with respiration (McNab 1996); and the observation that 'in the wild this must be a shy, patient, creepy little bird' anticipated the general tendency for high BMRs to be associated with high levels of energy expenditure under field conditions (Scholander et al. 1950c, Kersten and Piersma 1987, Daan et al. 1990b, Speakman et al. 2004a).

The rather strictly defined, standardized nature of BMR (the metabolism of inactive homeotherms that are not digesting, growing, moulting, ovulating or gestating under thermoneutral conditions) and the potential to correlate BMR with life-history and biogeographic variables in ever more sophisticated ways, have made BMR immensely popular with physiologists and ecologists alike; according to the ISI Web of Knowledge, well over 1000 published papers have BMR as their topic. See Lasiewski and Dawson (1967), Weathers (1979), McNab (1986, 1988), Elgar and Harvey (1987), Lovegrove (2000), Tieleman and Williams (2000), White and Seymour (2004, 2005), Anderson and Jetz (2005) and White et al. (2007) for examples.

Ricklefs et al. (1996) discussed whether metabolic ceilings should be seen as a direct function of BMR, or whether these quantities should be treated as energetic consequences of different aspects of an organism's performance. They concluded that the relationship between BMR and energy expenditure in the field 'may be fortuitous rather than direct'. We propose that the relationship is indeed indirect, but certainly not fortuitous. In the most general interpretation BMR reflects the size and maintenance cost of the metabolic machinery (much like the fuel usage of an idling motor), an idea first explicitly formulated by our research associate Marcel Kersten in 1985 (Kersten and Piersma 1987). Depending on the ecological context, the machinery may be necessary for animals to achieve high rates of energy expenditure during strenuous exercise (requiring large pectoral muscles and hearts and viscous blood) or to achieve the storage of large amounts of energy rich fuel each day (requiring large guts and livers). Thus, the graphical model presented by Piersma et al. (1996a; Fig. 45A) to show how BMR and field metabolic rate would be functionally coupled is incomplete. In reflecting the size and capacity of the metabolic machinery, BMR does not determine a general ceiling, but rather, by virtue of the size and capacity of particular organ groups, sets *particular kinds* of ceiling (Fig. 45B). Thus, similar levels of BMR can reflect the maintenance requirements of widely varying machinery, not only between and within species, but even within individuals (Piersma 2002, Książek et al. 2009, Fig. 45B). This led another research associate of ours, François Vézina, to reflect with a sigh that 'BMR is a package deal of so many different things'.

Although incomplete because of this compartmentalization, the rationale of using

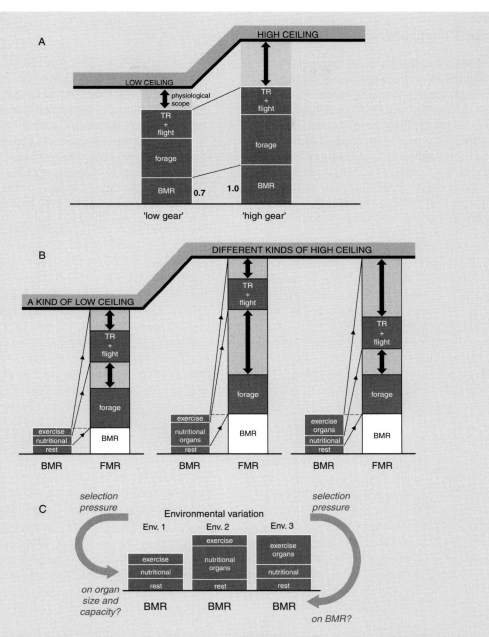

Figure 45. An old (A) and a new (B) version of a diagram to demonstrate the possible functional link between the heights of metabolic ceilings and BMR, and a diagram of the question whether natural selection pressures work on BMR or rather the component parts of the metabolic machinery that require a certain BMR for maintenance (C). In (B) attention is given to the idea that BMR reflects the summed contribution of size-related maintenance requirements of different organ systems (assuming invariable metabolic intensity). The vertical double-arrows in (A) and (B) indicate the 'absolute scope' for enhanced sustained performance given the size of the metabolic machinery.

Panels (A) and (B) are based on Piersma (2002).

continues

Box 3 *Continued*

BMR as a yardstick to evaluate or interspecifically scale maximal metabolic performances, seems nevertheless quite valid as long as the appropriate, context-specific, values for BMR are used (McKechnie *et al.* 2006). However, even though BMR may (1) be repeatable within individuals (Hayes *et al.* 1998, Bech *et al.* 1999, Horak *et al.* 2002, Labocha *et al.* 2004, Rønning *et al.* 2005, Vézina and Williams 2005), (2) have a heritable component (Konarzewski and Diamond 1995, Książek *et al.* 2004, 2009) and (3) be selected for in the wild (Broggi *et al.* 2005), BMR may be too much of an epiphenomenon to be a good trait for natural selection to work on (Fig. 45C). We suggest that natural selection first and foremost targets the size and capacity of different organs and organ groups, and that BMR will follow as a result (cf. Jackson and Diamond 1996, Williams and Tieleman 2000, Sadowska *et al.* 2008).

CHAPTER 5

Phenotypic plasticity: matching phenotypes to environmental demands

Adaptive arm-waving, and more...

Barnacles are peculiar crustaceans. Protected in cone-shaped shells, firmly attached to rock and other hard substrates, they weather the pounding of waves, the attacks by predators, the occasional desiccation that comes with living on the high shore and the high temperatures on sunny days. They feed on small planktonic items, which they fish from the surrounding water by a 'net' made up of what their arthropod ancestry has provided them with: six pairs of strongly modified arms. These arms are called cirri, and the fishing net a cirral net (Fig. 46A, B).

It is easy to imagine that, when the water is calm, even wide and fragile nets will work, whereas under strong water flows such nets would be ripped to the side and become inefficient at best, and damaged at worst. Several species of intertidal barnacles indeed carry long arms on sheltered shores, and short arms on exposed shores (Fig. 46C; Arsenault *et al.* 2001, Marchinko and Palmer 2003). At similar densities of edible plankton, high water flows will compensate for the smallness of the cirral net. Under very exposed conditions, however, the adaptive miniaturization comes to a halt (Li and Denny 2004), as barnacles increasingly begin to limit cirral netting to times of relatively slow water flow (Marchinko 2007, Miller 2007). Barnacles are quite easily dislodged from where they live and can be artificially glued on to other pieces of rock or shell. This enabled Kerry Marchinko (2003) to show that, when displaced to other flow regimes within Barkley Sound on the Pacific coast of Vancouver Island, they will quickly adjust the length and strength of their arms (Fig. 46D), even when mature. Thus, cirral nets appear fully reversibly plastic throughout life.

Their crustacean affiliation also shows when it comes to sex and reproduction. Unlike many marine invertebrates that engage in so-called 'free spawning'—jettisoning sperm and eggs in the water, where they find each other by chance (Luttikhuizen *et al.* 2004)—barnacles get intimate. Indeed, as documented by Charles Darwin (1854), at lengths up to eight times body length, barnacles possess the longest penises in the animal world (Neufeld and Palmer 2008). This goes together with the fact that most are hermaphrodites—combining maleness with femaleness—but apparently they do not do it on the cheap and self-fertilize. Glued to the substrate, barnacles cannot cuddle up; they have to use a long penis to find, reach, and then impregnate fastened but fertile neighbours (Klepal 1990, Murata *et al.* 2001). Now, what was true for the feeding arms forming the cirral net, would also be true for a penis extending out of the father's cone: at high water-flow rates, a long penis would be hard to manoeuvre properly. Building on the previously mentioned barnacle work at Barkley Sound, Chris Neufeld and Richard Palmer (2008) indeed found that penises of *Balanus*

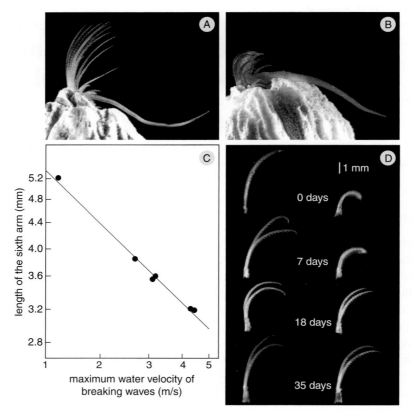

Figure 46. Barnacles *Balanus glandula* extend their arms to form cirral nets with which to withdraw small planktonic food items from the overlying water. Their cirral nets are cast wide in the calm waters of protected shores (A), but the arms are short in the waves of exposed shores (B), these examples representing opposing ends of a gradient (C). On the photos, in addition to folding out their cirral nets, the barnacles have also extended their considerable penises (see Fig. 47). Experimental displacement to protected shores from another protected (left) or a wave-beaten environment (right), shows the arms (in this case the sixth) to adjust adaptively within a couple of weeks.

Compiled from Neufeld and Palmer (2008) (A, B), Arsenault *et al.* (2001) (C), and Marchinko (2003) (D).

glandula became shorter and thicker at the base on the more exposed shores (Fig. 47A, B). For reasons just outlined, the variations in penis dimension made sense (Fig. 47C, D). Perhaps not surprisingly, penis length and thickness were also perfectly reversibly plastic (Fig. 47E, F). Not only do penis size and shape reflect the flow regime of the waters surrounding its owner, in the Eastern Atlantic, lonely barnacles have larger penises relative to crowded ones (Hoch 2008).

A big penis will come at a cost though, as it needs to be grown and maintained, and it may interfere with foraging (Klepal *et al.* 1972, Hoch 2008). While barnacles do not shed sperm and eggs, rather they shed free-living larvae. At temperate latitudes, where larval food supply may vary during the year (Yoder *et al.* 1993, Dasgupta *et al.* 2009), the reproduction of barnacles, it turns out, is also strongly seasonal. Long before the reversibility of penis size was examined, Margaret Barnes (1992) had concluded that some barnacles get rid of their penises altogether outside the mating season. What is left during the 'off season' is a rather cost-free 'stump'.

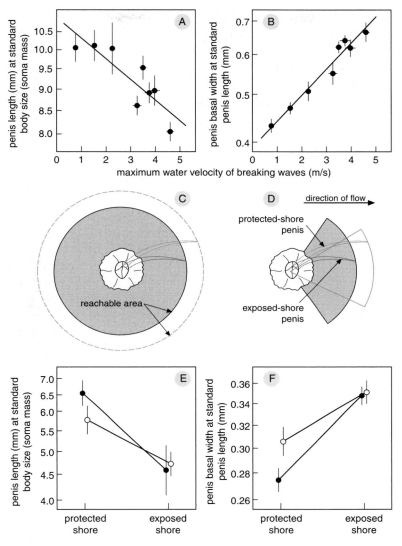

Figure 47. With increasing wave action, the penis of the barnacle *Balanus glandula* becomes shorter (A) and thicker at the base (B). This makes sense because the area within reach of the penis, and therefore the number of potential mates, is a function of its length and strength; in calm water, more area would be reached by a long and slender penis (C) and in strongly flowing water, by a short and strong penis (D). Experimental displacements from either a protected shore (filled dots) or an exposed shore (open circles) shows penis length (E) and basal width (F) have mutually adjusted after 20 weeks. Means with SE.

Compiled from Neufeld and Palmer (2008).

Use it or lose it

Just as reproductively inactive barnacles do away with their penises, astronauts (who speak English) and cosmonauts (who speak Russian) quite unwillingly lose body parts during their time in space (Lackner and DiZio 2000, Payne *et al.* 2007). Remember the sight of weakened astronauts returning from Apollo missions on the TV screen? Without gravity a human environment is less demanding in

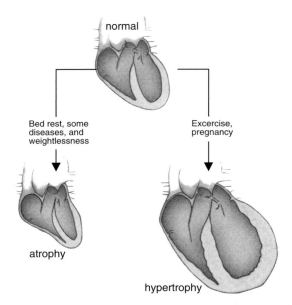

Figure 48. Visualization of the degree of atrophy of human hearts after several weeks of space travel or bed rest, as well as the degree of hypertrophy that comes with exercise and pregnancy.

Modified after Hill and Olson (2008).

several different ways and this shows upon the return to Earth.

One of the first things to be affected is the heart (Fig. 48), which may shrink by as much as a quarter after one week in orbit (Hill and Olson 2008). Heart atrophy is correlated with decreases in blood and stroke volume, blood pressure, and reduced exercise capacity (Convertino 1997, Payne *et al*. 2007). Especially after several months in the International Space Station, and particularly in women, upon return to Earth, astro- and cosmonauts experience dizziness and blacking-out because blood does not reach the brain in sufficient quantities (Meck *et al*. 2001, Waters *et al*. 2002).

Six weeks in bed leads to similar atrophy of the heart as one week in space, suggesting that either weightlessness itself, or the concomitant reduction in exercise, causes heart atrophy (Perhonen *et al*. 2001). For the muscles and bones

of the limbs, the effects of weightlessness are not only easier to predict on the basis of first principles (Fig. 49A), but the predictions are also easier to verify experimentally (Fig. 49B, C). Importantly, some muscle groups are wasted more than others (Vico *et al*. 1998, Payne *et al*. 2007). Because they bear the body's weight, the 'anti-gravity' muscles of thigh and legs would be expected to show atrophy when they are pushed out of work during space flight or during experimental weightlessness on Earth (Fig. 49); and they do (Payne *et al*. 2007). Various types of immobilization experiments successfully mimic the gross effects of weightlessness in space (Fitts *et al*. 2001, Adams *et al*. 2003). Muscles of the lower leg, assayed either by computed tomography or MRI scanning, showed losses, which levelled off at 20% after 40 days (Fig. 50), with little evidence of differences between earth- or space-bound microgravity effects. Despite the best attempts at giving them replacement exercise, after six months aboard the International Space Station, crew members still had lost 13% of their calf muscle volume and 32% of the peak power of their legs (Trappe *et al*. 2009). Not only is muscle volume reduced during disuse, various metabolic changes occur that include a decreased capacity for fat oxidation. This can lead to the build-up of fat in atrophied muscle (Stein and Wade 2005).

In addition to muscle loss, and apart from a deterioration of proper immune functioning during space travel (Sonnenfeld 2005, Crucian *et al*. 2008), arguably the most fearsome effect on bodies is bone loss. Although their hardness and strength, and the relative ease with which they fossilize, give bones a reputation of permanence, bone is actually a living and remarkably flexible tissue. In the late-nineteenth century, the German anatomist Julius Wolff (1892) formulated that bones of healthy persons will adjust to the loads that they are placed under (Fig. 51A). A decrease in load will lead to the loss of bone material at appropriate places, and an increase to thicker

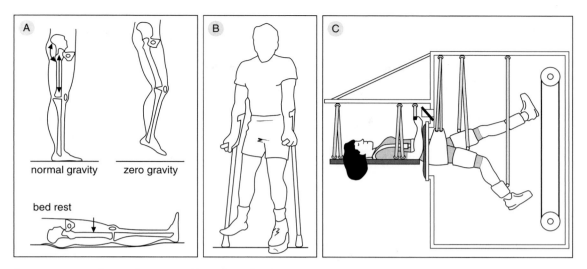

Figure 49. How a change of gravitational forces (size of arrows indicates relative forces) affects the muscles and bones of the leg (A), and some of the contraptions used in human experiments on the effects of use and disuse on muscle and bone size, strength, and functioning (B, C). Diagram (B) shows how loads are taken off the lower limb without constraining the knee or ankle, whereas (C) shows how bed-rested women received compensatory treadmill running in a low-pressure chamber to mimic exercise for astronauts.

Compiled from Lackner and DiZio (2000), Tesch *et al.* (2004), and Dorfman *et al.* (2007).

bone (Fig. 51B). It is no surprise then that, in the gravity-free environments in space, bones will de-mineralize, especially the lower limbs that normally counteract the effects of gravity (Fig.

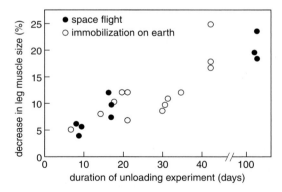

Figure 50. Relative loss of the lower leg muscles (triceps surae, gastrocnemius, soleus, or some combination thereof) as a function of the duration of unloading either during space flight or in immobilization experiments on Earth (this includes limb immobilization by casting, unilateral lower limb suspension as illustrated in Fig. 49 B, and bed rest).

Modified from Adams *et al.* (2003).

51C). With considerable individual variation, during half a year in space, cosmonauts lost up to a quarter of the tibial bone material (Vico *et al.* 2000). Although bone-formation continues under zero gravity conditions—this was established in chicken embryos sent to space (Holick 1998)—relative to bone-resorption, bone-formation rates decline (Holick 2000). What has been of greatest concern is that, unlike muscle loss that levels off with time (Fig. 50), bone loss seems to continue steadily with 1–2% a month of weightlessness. During a 2–3 year mission to Mars, space travellers could lose up to 50% of their bone material, which would make it impossible to return to Earth's gravitational forces. Thanks to Wolff's Law, space agencies will have to become very creative in addressing the issue of bone loss during flights to Mars (Lackner and DiZio 2000, Whedon and Rambaut 2006). So far, boneless creatures, such as jellyfish, are much more likely to be able to return safely to Earth after multi-year space trips than people. Gravity is a Mars bar.

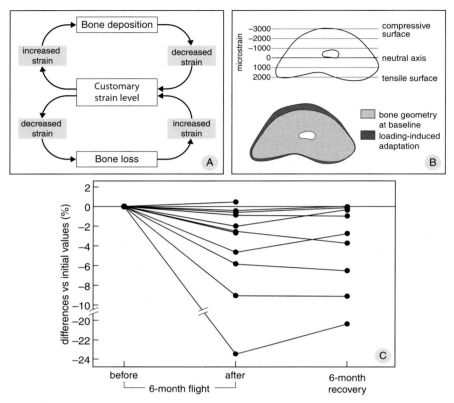

Figure 51. The use–disuse principle formulated as a functional adaptation model for bones (A), an illustration of the principle for an ulna of a rat (B), and a test of disuse–use effects in human cosmonauts (C). In the latter case, the loss of bone mineral density during a 6-month space flight, and the degree of recovery during the subsequent half year on Earth, were established.

Compiled from Ruff *et al.* (2006) (A), Robling *et al.* (2002) (B), and Vico *et al.* (2000) (C).

Bone loss during space travel certainly brings home the point that 'if you don't use it, you lose it', but what about 'if you use it again, do you regain it'? Within six months of their return to Earth, cosmonauts did indeed show partial recovery of their tibial bone mass (Fig. 51C). Obviously most of them needed more time, but even after one year of recovery, bone losses in men who had been experimentally exposed to three months of permanent bed rest were not fully compensated, though their calf muscles had fully recovered much earlier (Rittweger and Felsenberg 2009).

The dynamic gut

It is anybody's guess what happens to the innards of space travellers, but on the basis of what we see in other vertebrates (Starck 2003, Naya *et al.* 2007), in the course of journeys beyond the atmosphere, their digestive tracts may well show change. To get a feel for what is possible, let's look at snakes, animals that go through veritable 'gastrointestinal rebirths' (Pennisi 2005). Pythons, boas, rattlesnakes, and vipers are snakes that employ sit-and-wait foraging strategies (Greene 1997). This makes them go without meals for so long that their stomachs, intestines, and accessory organs shrink and become 'dormant' (Secor *et al.* 1994, Secor and Diamond 1995, Starck and Beese 2001, Ott and Secor 2007). When these snakes do catch the occasional prey, it can be almost as big as themselves. Prey capture, killing by constriction, and swallowing through the extended gape (Fig. 52A), elicits a burst of physiological activity, with drastic upregulation of many metabolic processes (reviewed in Secor 2008). Immediately, the heart starts to grow (Fig. 52C). With a doubling of the rate of heartbeat, and as blood is shunted away from the muscles to the gut (Starck and Winner 2005, Starck 2009), blood flow to the gut increases by an order of magnitude. Within two days, the wet mass of the intestine more than doubles (Fig. 52D): this change is additional to growth, involves the rehydration of previously inactive cells, and the incorporation of lipid droplets (J.M. Starck pers. comm.). This is followed by 100% increases of liver and pancreas mass, and a 70% increase of kidney mass (Fig. 52C). At the same time, stomach acidity increases (a drop in pH from values of 7–8 to 1–2), ensuring that the skeleton of the ingested prey disappears within about six days (Fig. 52B). Many amazing changes, at several different levels of physiological action, take place in the intestine, including the instantaneous lengthening of microvilli (Fig. 52E).

All in all, the stomach operates for less than a week, but the intestines carry on for another week to finish the job. Interestingly, part of the winding down of the gut includes the generation and banking of new cells lining the intestine, thus preparing the pythons for the next meal (Starck and Beese 2001). And what digestively goes for pythons also goes for another reptile, the caiman *Caiman latirostris* from South America (Starck *et al.* 2007). The precise mechanism of the restructuring of the gut and associated organs may differ between mammals, birds, and the sauropsid reptiles to which

snakes and crocodiles belong (Starck 2003), but in all these groups, organ size and capacity are remarkably responsive to changes in demand. We will elaborate examples from birds later in this chapter.

'Classical' phenotypic plasticity: developmental reaction norms

Let's just briefly think about space travel again. Although chicken embryos still show bone formation when taken to space (Holick 1998), on the basis of what we now know it would be impossible for a human child to grow up under zero gravity conditions and still survive the functional demands of a return to Earth. 'Developmental plasticity' is the name for this category of phenotypic plasticity, the one in which environmental conditions during ontogeny determine the size, shape, construction, and behaviour of the eventual mature phenotype. There are innumerable examples of developmental phenotypic plasticity, and this exhilarating variety of ways in which different aspects of environments shape different organismal traits is illustrated in Fig. 53. Picking up the barnacle theme from earlier in this chapter, some of the intertidal acorn barnacles *Chthamalus anisopoma* develop a bent shell when grown in the presence of predatory snails; this form is indeed more resistant to predation than the normal morph (Lively 1986, 1999, Lively *et al.* 2000, Jarrett 2008).

A classic case of developmental phenotypic plasticity is the variation in the degree to which water fleas *Daphnia* show hoods, helmets, spines, and longer tails in response to predators, or to the chemical identifiers (the 'kairomones') that they leave in the water (e.g. Tollrian and Dodson 1999, Caramujo and Boavida 2000, Petrusek *et al.* 2009, Riessen and Trevett-Smith 2009). This variation was first described by the enlightened German zoologist Richard Woltereck (1909) in clones of water

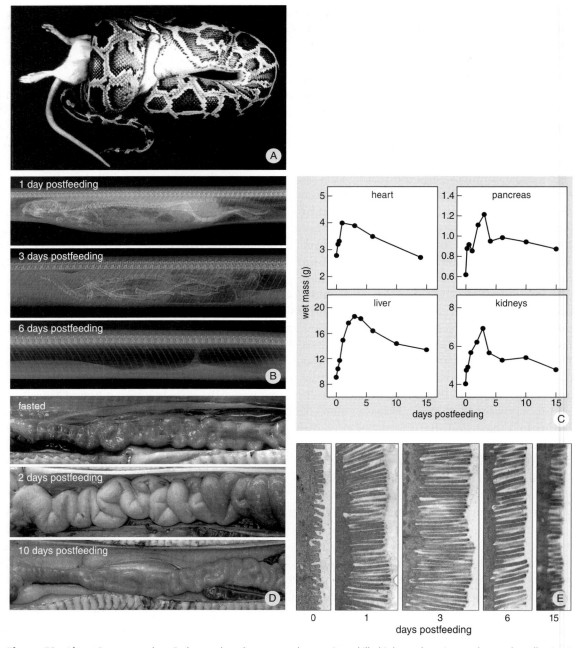

Figure 52. After a Burmese python *Python molurus* has captured a prey item, killed it by asphyxation, and started swallowing it (A), much of the snake's internal physiology gears into action. (B) Shows how the bones of a rat are digested in the acidic stomach within 6 days post-feeding. In the first few days after ingestion, the internal organs show growth spurts and then a slow return to the old levels. Within 2–3 days, the intestine has grown to full capacity (D), including a drastic lengthening of the microvilli (E, the bars in this image represent 1 μm).

Compiled from Lignot *et al.* (2005) and Secor (2008).

fleas (see Harwood 1996, Sarkar 2004). Finding that different pure strains of *Daphnia* show similar variation in hood size, variation that converged under particular environmental conditions, Woltereck began to doubt the distinction between genotype and phenotype, a contrast that was just becoming part of the mental armoury of biologists at the time. Woltereck was straightened out by the Danish plant physiologist and budding geneticist Wilhelm Johannsen (1911). Johannsen suggested, first in two books in Danish and then in an address to the American Society of Naturalists in 1910 (with which he reached the world), that what genotypes do is not to instruct for a single phenotype, but to instruct for the ways that phenotypes respond to environmental variation (Roll-Hansen 1979). This relationship between trait values and specific environmental conditions has since been called what Woltereck named it: 'Reaktionsnorm' or 'reaction norm' or 'norm of reaction' (Pigliucci 2001a, 2001b). The norm of reaction of barnacles to predatory snails, and of water fleas to predatory insects or fish, is now also known as an example of the category of 'induced defences' (Tollrian and Harvell 1999).

Figure 53 shows five more reaction norms reflecting several other types of developmental plasticity. Spiders weaving a coarser grained and bigger web when they can catch large prey (in this case flighted termites), could be said to show 'induced offence' (cf. Padilla 2001, Kopp and Tollrian 2003, Miner *et al.* 2005, Kishida *et al.* 2009). Caribbean sea urchins *Diadema antillarum* have a hard calcite

Figure 53. Examples of phenotypic plasticity in invertebrate animals and in plants, along with some of the common denominators of the particular class of plasticity.

Compiled from (from top to bottom): Lively (1999), Lampert *et al.* (1994), Sandoval (1994), Levitan (1989), Kaandorp (1999), Schlichting and Pigliucci (1998, p. 56), and Nijhout (1999).

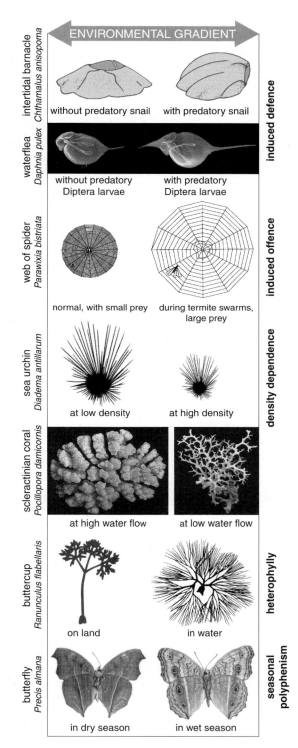

Table 6. A summary of content and scope of some recent books dedicated to the phenomena of phenotypic plasticity.

Reference	Book title	No. of pages	Content and scope
Rollo (1995)	*Phenotypes, their epigenetics, ecology and evolution*	463	An early, but very wide-ranging, synthesis by a molecular biologist of the genetic background and possible hierarchical organization of phenotypic construction, or organismal design, with due recognition of the role of ecological factors.
Schlichting and Pigliucci (1998)	*Phenotypic variation: a reaction norm perspective*	387	*The* textbook on phenotypic evolution from combined genetic and developmental perspectives. Well-illustrated. Norms of reaction are the core idea to explore the basics and evolutionary consequences of phenotypic responses to environmental variation.
Tollrian and Harvell (eds) (1999)	*The ecology and evolution of inducible defenses*	383	Explores and reviews a wide range of cases of developmental plasticity that fall under the banner of 'induced defence mechanisms': from the use of armour, via unpalatability, to immunity and behaviour as environmentally induced mechanisms for bodily defence.
Pigliucci (2001a)	*Phenotypic plasticity: beyond nature and nurture*	328	A solid attempt to develop mainly genetic interpretations of phenotypic plasticity as the core concepts in evolutionary biology.
West-Eberhard (2003)	*Developmental plasticity and evolution*	794	Almost encyclopaedic review of the importance of developmental processes for evolutionary innovation. Develops the notion that in evolution phenotypes change first, and that genes follow.
DeWitt and Scheiner (eds) (2004)	*Phenotypic plasticity: functional and conceptual approaches.*	247	With a loose focus on developmental processes and their genetic interpretation (reaction norms), the chapters explore a wide range of issues to do with phenotypic plasticity.
Turner (2007)	*The tinkerer's accomplice: how design emerges from life itself*	282	Formulates the role of self-organization, in interaction with replicators on one side and the environment on the other, in phenotypic expressions. It develops the idea that organisms show design, not because their genes instruct them to, but because agents of homeostasis build them that way.
Blumberg (2009)	*Freaks of nature: what anomalies tell us about development and evolution*	326	This is mostly a book about development, but one in which the importance of interactions between developing bodies and their environments is strongly emphasized.

| Gilbert and Epel (2009) | *Ecological developmental biology: integrating epigenetics, medicine, and evolution* | 480 | Fantastically comprehensive and well-illustrated textbook on the ways that environments affect development of animals, and why this knowledge is important in medicine and evolutionary biology. |

skeleton but are able to shift body size in response to variations in local food conditions. Levitan (1989) experimentally altered competitor densities, and showed that body size increased and decreased to levels predicted from field-based food abundance–body size relationships. By adjusting structural body size, the urchins reduced maintenance costs and thereby optimized reproduction and survival according to local food availability (Levitan 1989). Although body size is altered, the mouth structures (Aristotle's lantern) of the urchins, and thus their capacity to eat, remained unchanged (Levitan 1991).

Depending on the type of water flow, scleractinian corals build either denser or more spaced out structures (Kaandorp 1999, Todd 2008, who still owe us a term for this category of phenotypic plasticity). Buttercups *Ranunculus* have 'proper' wide leaves when growing on land and almost filamentous leaves when growing in water, representing a category of plant plasticity called heterophylly (Cook and Johnson 1968). The last example in Fig. 53 is quite typical of the insect world. Depending on season, a tropical butterfly *Precis almana* may either show angular wing shapes and a dull brown colour, so that it resembles a dead leaf (in the dry season), or show rounded and colourful wings with eyespots (in the wet season) (Nijhout 1999). This category of phenotypic plasticity is known as (seasonal) polyphenism. A final category of

phenotypic plasticity is the size-, age-, condition-, and/or context-dependent sex change that occurs in some plants, and in animals as diverse as annelids, echinoderms, crustaceans, molluscs, and, most famously, in fish (Polikansky 1982, Munday *et al.* 2005, 2006).

All the 'textbook' examples amalgamated in Fig. 53 represent relatively well-studied cases of phenotypic plasticity. Some, such as the water fleas of Woltereck and subsequent authors, have been important in the development of quantitative genetic-related theories of phenotypic plasticity (Pigliucci 2001a). What are missing from this list of examples are the extensive, but usually reversible, internal and external changes that occur in vertebrate bodies, changes that have played major roles in the previous and in the present chapter, but are almost absent from the burgeoning literature on phenotypic plasticity (see Table 6). With a notable exception (Turner 2007), the phenotypic plasticity literature has mostly passed by the staggeringly large body of phenomena related to internal (physiological), external, and seasonally cyclic (moult) and behavioural plasticity, especially in the higher vertebrates. To make sure that readers will have encountered the full range of plasticity phenomena by the end of this chapter, let's now look at the reaction norms of male and female rock ptarmigan *Lagopus mutus*, showing phenotypes that vary with time of year, snow cover, and, in the case of males, the availability of mates.

Seasonal phenotype changes in ptarmigan, deer, and butterflies

Rock ptarmigan are a species of grouse that lives on the arctic tundra. For most of the year the tundra is covered by snow, and both sexes are camouflaged by white plumage (Montgomerie *et al.* 2001). In the brief arctic spring, when the snow melts and is replaced by summer tundra, the females moult their white plumage in favour of green and brown feathers (Fig. 54). However, the males do not immediately moult their white feathers and are exceptionally conspicuous for some time, being attractive not only to prospecting females, but also easily discovered targets for hunting gyrfalcons *Falco rusticolus*. When opportunities for mating disappear, the male ptarmigan camouflage themselves by soiling their white plumage and so are dirtiest at the start of incubation (Montgomerie *et al.* 2001). Only when a clutch is lost to predators and their mates become fertile again, do the males clean up. They eventually moult into a cryptic summer plumage once the season with mating opportunities really has ended. Thus, the visual aspect of the plumage in females is determined by its role for camouflage only. In males, plumage not only serves the dual role of camouflage in winter (when white) and in late summer (when comprising a cryptic patterning of browns), but also

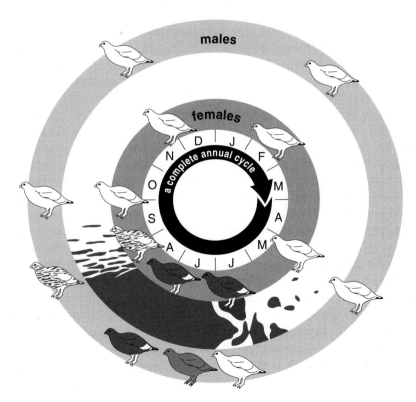

Figure 54. Annual cycle of the external appearance of male and female rock ptarmigan in the Canadian Arctic in relation to changes in the camouflage afforded by the tundra habitat (indicated by the texture of the middle circle). This illustrates the power with which such intra-individual changes can be interpreted in a functional context. The changes in plumage represent a change in feathers, except for the change from a pure white to a dirty white colour of males in late June–July, which is the result of soiling.
From Piersma and Drent (2003) based partly on data from Montgomerie *et al.* (2001).

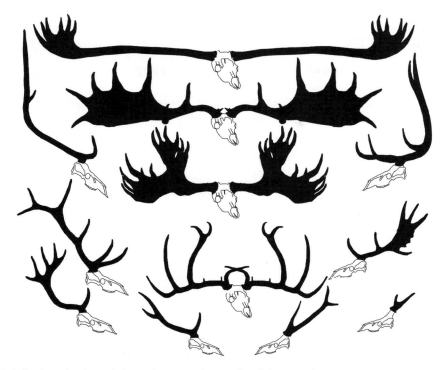

Figure 55. Variation in antler size and shape of extant and extinct 'true' deer Cervidae.

Compiled from Emlen (2008).

the sexually selected, ornamental role (white) in the mating season. The soiling of the white plumage when mating opportunities are no longer present (but when the moult into a cryptic plumage has not yet begun) provides a behavioural 'quick fix' to a critical phenotypic allocation problem.

Showing off is especially important in males that are in a position to monopolize more than one female during their fertile periods, and this is what deer do, at least as long as they live in semi-open habitats. The true deer (Cervidae) are unique among animals in having males that show temporary bony outgrowths on their heads, the antlers (Fig. 55).[1] Antlers are ornamental weapons that should impress both sexual competitors and prospective mates (Geist 1998,

Emlen 2008). What makes deer antlers so different from functionally similar horns and tusks in other ungulates is that they are as seasonal as the plumages of ptarmigan. Antlers first grow at puberty, and successive sets of antlers developed by individuals tend to increase in size and complexity (Goss 1983, Bubenik and Bubenik 1990, Lincoln 1992). Whereas the small forest deer of the tropics may regenerate antlers at any time of the year, at temperate latitudes deer show synchronized seasonal cycles of growth, calcification, cleaning, casting, and regeneration (Lincoln 1992, Goss 1995). Regeneration of antlers occurs outside the reproductive season, in a period of low testosterone, through a process that is different from wound healing in that it is a stem cell based process (Kierdorf *et al.* 2007). Growing

[1] In reindeer and caribou *Rangifer tarandus*, females also have antlers, as they have to compete with the males and amongst each other for scarce fodder under arctic winter snow (Lincoln 1992).

antlers are covered by a 'velvet skin', which is shed when the underlying bone calcifies and dies, yielding the insensitive antlers that can be used as weapons during the rut. Hard antlers remain attached to the skull as long as the stags have raised testosterone levels (3–9 months, depending on species), after which the connection is severed and the antlers are cast (Lincoln 1992). Overall, the phenotype of individual male deer is as seasonally variable as it is in male ptarmigan.

Butterflies only live a couple of months. However, just like ptarmigan and most other organisms, butterflies experience seasonal environments and, in response, many butterfly species show seasonal polyphenism: within single genotypes, but in successive generations and depending on the time of year, they may show strikingly different phenotypes (Kingsolver 1995a, 1995b, Nijhout 1999). Well-studied examples are the *Bicyclus* butterflies from East Africa. Like other tropical butterflies, they are exposed to an alternation of dry and wet seasons. From June to October it is dry and mostly relatively cold, but from November through April it is wet and initially quite hot (Fig. 56). In the dry season, the butterflies are large and cryptically coloured. During the wet season, they show abundant eye-spots and a somewhat smaller body size (Windig *et al.* 1994, Brakefield *et al.* 2007). The alternative morphologies can be generated from split broods reared at either high or low temperatures (Kooi and Brakefield 1999). The butterflies of the dry season emerge around the end of the rains and have to survive almost half a year before they can lay eggs on fresh green grass at the start of the next wet season. To survive that long, they rely on their cryptic colouration and a quiescent lifestyle, including low metabolic rates (and a long life span under laboratory conditions). The ornamented butterflies of the wet season come in two generations. The first, which hatch from eggs laid by their cryptic parents, immediately mate and lay again, thus

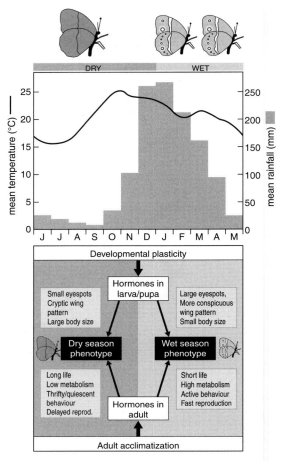

Figure 56. Dry and wet seasonality in Malawi, East Africa, and the phenotypic responses by *Bicyclus anynama* butterflies that have one generation of inconspicuous large morphs in the dry season, and two generations of more conspicuous and smaller morphs in the wet season. Larvae of the last wet-season cohort develop with declining temperatures and produce the cohort of dry-season morphs without eye-spots. The diagram illustrates that this seasonal polyphenism is steered by both developmental plasticity (leading to the conditional metamorphosis induced in the larvae and expressed at pupation) and adult acclimatization (maternal, e.g. hormonal, contributions to the eggs laid in whatever season).

Compiled from Brakefield and Reitsma (1991), and Brakefield *et al.* (2007).

producing a slightly longer lived cohort, that then produces the cryptic cohort that endures the dry conditions, and so on. The wet season butterflies show the distinct eye-spots; these may thwart attacking birds (Kodandaramaiah *et al.* 2009) and are important during mate choice (Constanzo and Monteiro 2007, Oliver *et al.* 2009). The colourful wet season butterflies are also much more active, have a higher metabolism and a shorter life-span (even when not exposed to predation in the laboratory). A wonderful combination of field and laboratory research enabled Paul Brakefield and co-workers (2007) to establish that the seasonal plasticity shown by *Bicyclus* butterflies represents aspects that are typical of both 'classical' developmental plasticity (hormonal effects in larva and pupa) and physiological acclimatization (induced in the reproductive adult stage and maternally contributed to the eggs) (Fig. 56).

The extents of phenotypic change shown by ptarmigan and butterflies are rather similar, but their life spans (and life-cycle characteristics) are not. Individual ptarmigan may experience many seasons, individual butterflies a single season at best. What is called seasonal polyphenism in insects is called life-cycle staging in birds and mammals. And whereas the genetic-developmental architectures behind the phenotypic expressions are likely to be very different in birds and butterflies, in both groups seasonally varying ecological shaping factors to do with feeding, the avoidance of predation, and mating, can functionally explain their strikingly variable phenotypes.

Environmental variability and predictability, and the kinds of phenotypic adjustments that make sense

Different activities related to reproduction and survival (breeding, moult, migration, hibernation, etc.) are usually separated in time within individuals, tend to occur at predictable times of the year in seasonal environments, and are accompanied by changes in the reproductively active phenotype (Murton and Westwood 1977). To perform optimally under a wide range of environmental conditions (variations that are often cyclic), a long-lived individual must track or anticipate the external changes by regulating gene expression to adjust its morphology, physiology, and behaviour (Willmer *et al.* 2000). The cyclically varying phenotypic expressions of these adjustments within an individual rock ptarmigan, for example, are thus called 'life-cycle stages' (Jacobs and Wingfield 2000, Piersma 2002, Ricklefs and Wikelski 2002). The seasonal template for such sequences might be provided by the natural photoperiodic rhythm and/or by an endogenous circannual pacemaker (Gwinner 1986). In addition, temperature, rainfall, food, or densities of conspecifics might give supplementary information, which individuals could use to 'fine-tune' the timing of their phenotypic transformations (Wingfield and Kenagy 1991). The polyphenisms of short-lived organisms, as in many insects, reflect a similar strategy (Shapiro 1976, Danks 1999).

The accuracy with which future environmental conditions can be predicted would determine the kind of phenotypic plasticity that one might expect to evolve (Levins 1968, Moran 1992, Padilla and Adolph 1996). In unpredictable environments, where there is insufficient environmental information, or if the wrong environmental information is used, an organism might end up with a phenotype that does not quite match its current environment (Hoffman 1978). Indeed, such mismatches are considered to be an important cost of developmental plasticity (see Table 7). However, this potential cost would disappear for organisms that are capable of fast and reversible phenotypic change. We will get back to this issue after we have addressed 'trade-offs'.

Carrying on from our introductory classification in Chapter 1 (Table 1), Fig. 57 represents the best attempt at present to place the different

Table 7. Inventory of costs of, and limits to, phenotypic plasticity. Developed from DeWitt *et al.* (1998), Pigliucci (2001a), Lessells (2007), Valladares *et al.* (2007), and Vézina *et al.* (2010).

Costs of plasticity	Limits to plasticity
Maintenance: energetic costs of sensory and regulatory mechanisms, and of any additional structures.	*Phylogenetic 'constraint'*: however adaptive a kind of plasticity would be, in any organism some forms of plasticity are impossible because of historic genetic, developmental or sensory constraints.
Production: excess costs of producing structures plastically rather than as fixed genetic responses.	*Damage-related constraints*: in plants, phenotypic plasticity may be limited by herbivory because grazing damage prevents achievement of the optimal phenotype.
Information acquisition: investments (risk, time, energy) for sampling the environment, including lost opportunities (e.g. mating, foraging).	*Phenotypic compromise*: resource constraints may bias functional adjustment to one trait at the cost of another.
Developmental instability: phenotypic imprecision may be inherent for environmentally contingent development, which can result in reduced fitness under stabilizing selection.	*Information reliability*: the environmental cues may be unreliable or changing too rapidly.
Genetic: due to deleterious effects of plasticity genes through linkage, pleiotropy, or epistasis with other genes.	*Lag time*: the response may start too late compared with the time schedule of the environmental change, leading to maladaptive change.
	Developmental range: plastic genotypes may not be able to express as wide a range of adaptive phenotypes as a polytypic population of specialists would, e.g. because of limitations in the workings of neuroendocrine control system.
	Epiphenotype problem: the plastic responses could have evolved recently and still function like an 'add-on' to the basic developmental machinery rather than as an integrated unit; this may compromise their performance.

classes of phenotypic plasticity within a two-dimensional space, bounded by axes that represent (1) the degree of environmental variability (relative to the organism concerned) and (2) the degree to which this variability is predictable (Drent 2004). Highly predictable changing environments would select for *polyphenism* in short-lived organisms and *life-cycle staging* in long-lived organisms. The lower the predictability of environmental variation, the better it is for organisms to respond opportunistically, rather than seasonally scheduled. *Developmental plasticity* would then describe the kind of variable responses: organisms encountering unpredictably variable environments in the course of their life would benefit from plasticity being reversible, i.e. showing *phenotypic flexibility*. If environmental variation cannot be predicted, organisms might go into *bet hedging* (generating differently adaptive phenotypes at random) in shorter-lived organisms. In theory at least, especially longer-lived animals could also cope with extravagant changeability of the environment by not adjusting the phenotype at all, i.e. show *robustness*.

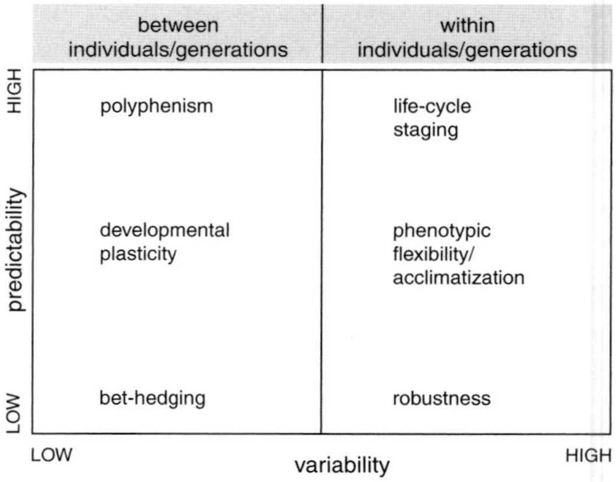

Figure 57. Different classes of phenotypic plasticity (or the lack of it in the case of robustness) arranged according to an axis of environmental variability over time and an axis of environmental predictability over time. The categories on the left show plasticity between individuals (and thus, between generations), and on the right within individuals.
Inspired by Meyers and Bull (2002) and adapted from Drent (2004).

Degrees of flexibility

In the previous chapter we encountered sled dogs that, along with human polar explorers, were the true champions of endurance exercise. Arctic sled dogs hard at work would be expected to show many physiological and internal morphological changes compared to sled dogs on holiday. In summer, the absence of snow and ice puts sled dogs out of work. Nadine Gerth and co-workers (2009) used the contrast between summer and winter for a comparison of muscle size and ultrastructure in Greenlandic sled dogs. During the summer holiday, dogs had thinner muscle, the slimming representing changes in mitochondrial numbers, lipid droplet sizes, and numbers of contractile myofibrillar elements. Gerth *et al.* (2009) were surprised that capillary networks in the muscles did not change with season, and did not explain it. Could there be inherent differences between tissue types in the degree to which they can adjust to changing functional demands (Lindstedt and Jones 1987)? And what measure would best reflect a capacity for change? Cell turnover perhaps?

A fruitful approach to look at cell turnover in different tissues, especially the longer-lived ones, was the serendipitous use of peaks in ^{14}C in the atmosphere generated by nuclear bomb testing in the 1960s. This ^{14}C is incorporated into body tissues. Spalding *et al.* (2005a) used DNA from stored tissues to estimate their age at the moment of sampling (often after death). They verified the method with the enamel of teeth that, once formed, are not rebuilt (Spalding *et al.* 2005b). Whereas the lining of the gut and the outer skin show fast DNA turnover, indicating fast cell-replacement (Fig. 58), the DNA in fat cells, muscle cells, and bone cells is not replaced in many years. The oldest tissues in any person are the heart (cardiac muscle cells) and various components of the brain.

The high turnover of gut epithelium tallies well with the high flexibility of the intestine

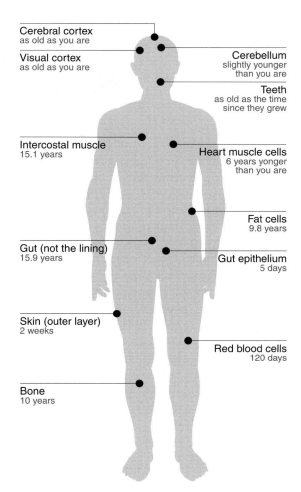

Figure 58. Most of your cells are younger than you are (at least the DNA in the cells is), but there is a lot of variation between cell types.
Modified from Vince (2006), partly based on Spalding *et al.* (2005a, 2005b 2008) and Bergmann *et al.* (2009).

that we discussed earlier. However, we have also shown that hearts show fast changes in size in response to changes in workload, so how does that tally with the cardiac muscle cells living so long (Bergmann *et al.* 2009)? The discrepancy is due to cell components other than the nucleus being built and broken-down, rather than whole cells being renewed and removed. Myofibrillar volume may change, and the densities and total volume of mitochondria may quickly change too (Hoppeler and Flück 2002). Note that the brain is mainly composed of very old cells (Fig. 58). Brains thus beautifully combine permanence (constancy of cell numbers) with change (new neural connections made all the time) (Edelman 2004).

A study on Andean toads *Bufo spinulosus* suggests that there may be generalizable, tissue-specific, capacities to respond to changing conditions. Daniel Naya and co-workers (2009) compared the sizes of various organs as a function of season (comparing summer, when the animals are most active, with winter) and their feeding state (summer animals were exposed to a two-week fast or fed). The degree of hypertrophy of the active over the inactive stage in the two situations was clearly correlated (Fig. 59): heart, lungs, and the large intestine never showed significant change, but stomach, small intestine, and liver showed great change in both contexts. The highly flexible small intestine fits our expectations, as does the inflexible lung (see Chapter 3). Nevertheless, we seem far from clearly understanding why some tissues, or components of

these tissues, show greater and faster responses to changes in demand than others.

Direct costs and benefits, their trade-offs, and other layers of constraint

Do you still remember that a big penis might hinder the waving of cirral nets to catch planktonic food items by barnacles (Hoch 2008)? Then here is a somewhat similar example, where the trade-off between benefits and costs of a sizeable reproductive organ has been mapped in much more detail. Enter arachnids, or spiders. In most species, males are smaller than females, but in some species, males are dwarfs (Vollrath 1998). This might be because such small males can use ballooning on silky threads to find distant females, or because they would be no threat or competitor for the females, which would ease the access to what, in spiders, is the discerning and ruling sex. In cobweb spiders of the genus *Tidarren*, the males are so tiny, about 1% of the female's mass, that it raises 'a delicate problem' (Vollrath 1998): if the mating organs, a paired structure with the name pedipalps (which they fill with the sperm to be pumped into the female), have shrunk in proportion to the body, they would no longer fit snugly—key to lock—into the female's receptive organ, the epigyne. *Tidarren* spiders have solved this problem by the tiny males carrying relatively massive copulatory organs, to the extent that this makes the spiders pretty clumsy and reduces their chances in the scramble to get to a fertile female before others (Ramos *et al.* 2004). The tiny *Tidarren* males have a radical solution, as they actively break off the carrying arm of one pedipalp before filling it with sperm and before they approach a female. The loss of one pedipalp speeds them up considerably, and also makes them more enduring athletes (Fig. 60). This partial emasculation has clear dividends and may not even be as costly as it would seem at first sight.

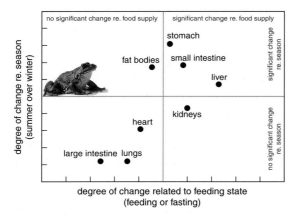

Figure 59. Degree of change in the dry mass of various organs of mature Andean toads *Bufo spinulosus* as a function of either feeding and fasting (X-axis) or season (summer compared with winter, Y-axis). The degrees of change reflect the first two principal components in a factor analysis.

Based on Naya *et al.* (2009).

Female *Tidarren* tend to consume males they have mated with and mostly give them the chance to insert only a single pedipalp. Still, these spiders may trade insemination capacity for speed and endurance, providing an example of a trade-off that leads to the severing of a body part that gets in the way at some stage.

Trade-offs are about alternative investments, the allocation problem of where, and in which parts of the phenotype, to invest scarce resources (e.g. Stearns 1989, 1992, Gervasi and Foufopoulos 2008). Many dung beetle males develop large horns on the head or on the thorax, with which they compete for females. In a manipulative study where larvae of *Onthophagus nigriventris* were prevented from

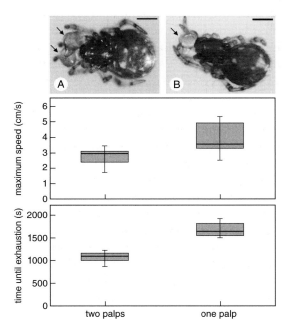

Figure 60. Minute males of the cobweb spider *Tidarren sisyphoides* before (A) and after (B) removal of one half of the paired ejaculatory organs, the pedipalps. The pedipalps are indicated by the two (A) or single (B) arrows; the scale bar represents 1 mm. Box plots of maximum speed before and after pedipalp removal (n = 16) and time till exhaustion in two groups of spiders still having two pedipalps (n = 15), or a single (n = 17) one, are shown below the photos.

Compiled from Ramos *et al.* (2004).

developing such horns, Leigh Simmons and Douglas Emlen (2006) could show that the growth of these horns actually occurred at the expense of an organ that also helps to secure progeny—large testes. All the forms of phenotypic plasticity embody such trade-offs. Theory has it that for phenotypic plasticity to evolve, no single phenotype can be optimal in all environments experienced by the organism (Via and Lande 1985, Moran 1992). Thus, there *must* be trade-offs (and reliable cues to inform the organism about the state of its variable environment). Investments in one direction necessarily, due to time, energy, and nutritional limitations, prevent investments in another direction (for a model and a test with *Daphnia*, see Hammill *et al.* 2008). Only fantasy creatures called 'Darwinian demons' would be able to respond continuously and at many levels to changes in trade-offs by adjusting phenotype. Dung beetles that are able to predict that they can monopolize the female once they have secured it, would invest in horns; those that predict they would not, would play the sperm-competition game and invest in testes. The Darwinian demonic dung beetles would be able continuously to adjust the relative sizes of their horns and testes in response to fluctuating ecological and social environments. Although dung beetles show considerable developmental versatility (Emlen *et al.* 2007, Rowland and Emlen 2009), Darwinian demonic dung beetles do not exist. There are always limits to plasticity.

This is all we might have said on trade-offs and phenotypic plasticity, except that in the pertinent literature there exists another layer of concern: the trade-off between being phenotypically plastic, or not being plastic at all (Newman 1992, Via *et al.* 1995, DeWitt *et al.* 1998, Pigliucci 2001a); this is related to the idea that plasticity itself is a trait with a genetic basis (Relyea 2005). Perhaps there are true costs to plasticity itself (Table 7). Such costs could include the energetic costs of the maintenance of sensory and regulatory structures enabling plasticity, or the excess costs of producing a plastic rather than a fixed (robust) structure. Adaptive plasticity makes it necessary to acquire information about the environment, and such acquisition will cost time. Plastic structures may (or may not) be more unstable than developmentally fixed structures, and if there were to be such things as plasticity genes, in some combinations with other genes they *could* be deleterious. In a review, Pigliucci (2001a) could not come up with much evidence in favour of any such costs. But perhaps all these theoretically possible costs of plasticity are not as far-fetched as they might at first sight appear. In the first specific study on the costs of plasticity (Relyea 2002a), evidence for such costs appears 'pervasive'. Comparing sibships of larval frogs that possessed different degrees of phenotypic plasticity, depending on the trait examined and the environment experienced, increased plasticity could either have positive, negative, or no effects on fitness components, such as degrees of trait development, mass gains, and survivorship. Thus, now that the appropriate studies are being done (Lind and Johansson 2009), there is increasing evidence that plasticity per se may come at a cost.

The mere fact that Darwinian demons do not exist, actually demonstrates that there *are* limits to plasticity, and we provide a list of possibilities in Table 7. Let us quickly run through them. Phylogenetic constraints are obvious, damage through herbivory may prevent plants from achieving locally optimum phenotypes (Valladares *et al.* 2007), organisms may have to compromise between different adjustments in the light of resource limitations (Vézina *et al.* 2010), environmental information is bound to be unreliable at times, and lag-times between environmental changes and phenotypic responses must be an issue.

However, the last two limits-categories are somewhat controversial. As Massimo Pigliucci (2001a) points out, for an intra-individual phenomenon it is odd to juxtapose individual constraint with population variation, as DeWitt *et al.* (1998) did in their formulation of a 'developmental-range' limit. We left it in Table 7 because it is quite possible that control mechanisms (e.g. of a neuroendocrine nature) limit the way in which environmental information is fed into the functionally adjusting phenotype (Lessells 2007).

The 'epiphenotype problem', the issue that plastic responses may be recent add-ons that are not yet genetically and developmentally hard-wired and may therefore compromise performance, is interesting on two accounts. First, it is quite possible that many forms of plasticity represent self-organized rather than developmentally tightly orchestrated structures (Turner 2007). We shall get back to that in a minute. Second, 'a plastic response may start as an epiphenotype and subsequently be integrated by natural selection attempting to reduce the costs and limits of the new response' (Pigliucci 2001a, p. 176). In our final chapter we will return to plasticity as a creative force in evolution (West-Eberhard 2003). In a celebrated class of developmental plasticity that so far has escaped notice, all these elements of costs, benefits, and limits nicely combine. Let's look now at frogs and newts, and especially their tadpoles!

Phenotypes of fear

That amphibians live up to their name, that they never fully made the transition from living in water to living on land, actually gives them a lot of ecological opportunity. By their enormous life-history versatility, they can capitalize on the resources offered by ponds, pools, and puddles, and even the underlying soil (Newman 1992). Toads can survive years in dry desert soil, frogs can survive deep-frozen ground, and when the rains come, or the spring arrives, the sleeping beauties wake up, display, and quickly mate and lay eggs, so that the larvae can use the temporary, ephemeral, pool- and puddle-habitats to mature and metamorphose in a life-history stage that can endure the subsequent lean times (Wilbur and Collins 1973, Wilbur 1990, Relyea 2007). Amphibians in Paradise, salamanders that have found permanent and predator-poor, deep-water bodies, forego that robust land-living stage altogether. Most famous among these reproductively active larvae is the Mexican axolotl *Ambystoma mexicana* (Gould 1977, Ryan and Semlitsch 1998).

What is of most concern here though, is not the metamorphosis into the adult stage, but life as a larva. Look at a tadpole: it is a cute little package of developing protein. Tadpoles make high-quality, easily digested food for anybody that is interested. Apart from hiding (and thereby refraining from foraging and thus not growing), their only defence is an ability to flee fast, to confuse, or, in the case of many toads, possess a distasteful skin. Under threat it helps to have a relatively big and muscular tail (Fig. 61), as it may enable the tadpole to swim fast (Dayton *et al.* 2005). A large tail fin could also act as a lure to attract the attention of predators away from vulnerable parts of the body (Doherty *et al.* 1998, Johnson *et al.* 2008). In many species of frog, tadpoles from ponds with predators are smaller, with relatively shorter bodies and deeper tail fins than tadpoles from predator-free environments (Relyea 2001a). Indeed, one of the most common tadpole predators, dragonfly larvae, preferentially kill tadpoles with relatively shallow and short tail fins, and narrow tail muscles (Van Buskirk and Relyea 1998). Tadpoles lend themselves well to experimentation, as artificial ponds, either stocked or not stocked with enclosed or free-swimming tadpole-predators of various kinds, and with different levels of food or food-competitors, are easily set up. The

work by Rick Relyea and his team has turned the American wood frog *Rana sylvatica* (Fig. 61) into the best-understood example of what it means to have a very plastic larval stage.

Relyea (2003a) showed that when tadpoles were exposed to the smells and actions of dif-

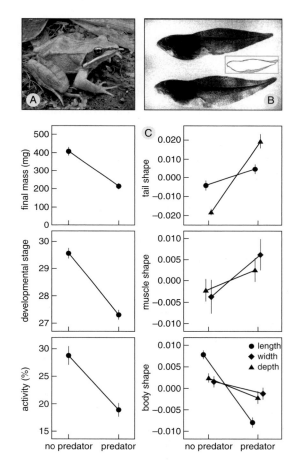

Figure 61. An adult wood frog *Rana sylvatica* (A), typical responses of wood frog tadpoles to either the presence (B, bottom) or the absence (B, top) of predatory dragonfly larvae, with a drawing by Rick Relyea, where the difference in tail depth between the two morphs is shaded (B, inset). The compilation in (C) shows the multiple phenotypic effects of caged dragonfly predators on growth, development, activity, and relative morphology of wood frog tadpoles. The morphological traits were size-adjusted by taking mean residual values (± SE) from regressions on mass.

Compiled from Relyea (2001b) (B) and Relyea (2002a) (C).

ferent predators (there was a choice of four) or combinations thereof, they discriminated and produced predator-specific larval phenotypes. When predators were combined, tadpoles generally developed phenotypes that were typical of the more risky type, suggesting that they perceived the risk of combined predators as the risk of the most dangerous one. Tadpoles not only have to survive, they also have to grow. Wood frogs that face high levels of competition from other tadpoles increase their foraging activity and develop larger bodies. These larger bodies contain relatively larger scraping mouthparts and longer guts, which improve the animal's ability to find, consume, and digest food, but this ability comes at the cost of a large tail, which makes tadpoles more susceptible to predators (Relyea and Auld 2004, 2005).

The extraordinary ability of tadpoles to sense changes in their environment and respond in precise and often appropriate ways, is illustrated, for example, by wood frog tadpoles in experimental ponds that appeared to respond differentially to variations in the *per capita* food levels and to changes in conspecific density alone (Relyea 2002b). When European common frog tadpoles *Rana temporaria* were exposed to non-lethal caged predators (dragonfly larvae) at low tadpole densities they showed morphological as well as behavioural defences. At high densities and with severe competition for food they remained active and only adjusted their morphology (Teplitsky and Laurila 2007). Similarly, when tadpoles were exposed to fish (that pursue their prey), they built deeper and longer tails and bigger tail muscles. In the presence of dragonfly larvae (sit-and-wait predators), they built only deeper tails (Teplitsky *et al.* 2005a, Wilson *et al.* 2005), suggesting that tails as lures were the selective force behind oversized tail fins. Indeed, in a Neotropical tree frog *Dendrosophus ebraccatus*, the presence of dragonfly larvae induced

highly visible, large, and colourful tail fins, whereas the presence of fish induced achromatic tails (Touchon and Warkentin 2008). Tadpoles of the tree frog *Hyla versicolor* altered their defences when 10 different prey (themselves being one of the types) were either crushed by hand or consumed by dragonfly larvae. Across all prey types, crushing induced only a subset of the defences that were induced by consumption, suggesting once more that tadpoles can assess quite precisely the degree and type of threat and respond accordingly (Schoeppner and Relyea 2005).

Nevertheless, even the plasticity of tadpoles finds a limit. When Schoeppner and Relyea (2008) exposed wood frog tadpoles, and when Teplitsky *et al.* (2005b) exposed tadpoles of a European frog *Rana dalmatina*, to a gradient of densities of a particular predator, all measured traits exhibited graded responses that levelled off with increased predation risk. This suggests that there is either an organizational limit to plasticity or a 'functional limit' that reflects the high costs of the defensive phenotype. Part of these costs might be incurred later in life. Wood frog tadpoles reared with caged dragonfly larvae, rather than without, showed relatively large limbs and narrower bodies as fully grown frogs (Relyea 2001b); the survival value of these later adjustments was not established. However, four months after metamorphosis, common frogs that developed from predator-exposed, rather than predator-free, tadpoles swam more slowly with less endurance: the opposite of what happened in the larval stage (Stamper *et al.* 2009). This suggests that the larval adjustments may come with a long-term fitness cost.

We round up with the amazing story of two competing amphibian larvae studied in Japan—a frog *Rana pirica* and a salamander, or newt, called *Hynobius retardatus*. Tadpoles of many a salamander or toad have tendencies to become cannibalistic, but they only do so at high conspecific densities (Pfennig 1992, Michimae 2006), and, preferably, if these conspecifics are not closely related (Pfennig and Frankino 1997). They then develop a broad head and gape large enough to swallow tasty nieces and nephews. They also turn predatory when there are many frog tadpoles around, as the mechanical vibrations of flapping tadpole-tails are sufficient for this transformation to take place (Michimae *et al.* 2005). In a wonderful example of a phenotypic arms race, in response to salamanders becoming broad-headed and dangerous, frog tadpoles developed thicker bodies and a comb on their backs to prevent ingestion by the salamander larvae; to which salamanders responded by further enlarging gape size (Kishida *et al.* 2009)! Interestingly, frog tadpoles do not become bulky and high-bodied solely in the presence of salamander tadpoles, they only adjust when the salamander tadpoles become dangerous, i.e. express the predacious phenotype (Kishida *et al.* 2006). A final subtlety is that when the threat of either predatory larval salamanders or dragonflies is taken away, the specific morphological defences are lost (Kishida and Nishimura 2006). This reversal indicates once more that induced defences can come with serious costs (Hoverman and Relyea 2007), and thus that flexibility, or reversibility, pays if it is still an option developmentally (Relyea 2003b).

Plasticity: the tinkerer's accomplishment?

With a few exceptions (e.g. Newman and Müller 2000, Blumberg 2009), the literature on phenotypic plasticity is permeated with the idea that adaptive phenotypic responses, be they reversible or irreversible, cyclic or not, are instructed from the genome. Of course, they will be, at some level at least. In his eye-opening book *The tinkerer's accomplice*, J. Scott Turner (2007) builds on early lines of reasoning, including the famous 'spandrels'-paper by Stephen Jay Gould and Richard Lewontin (1979), and assembles many convincing arguments that good adaptive design is usually the result of the self-organization that logically follows from growing structures interacting with their immediate physical environments. In his own words: 'organisms are designed not so much because natural selection of particular genes has made them that way, but because agents of homeostasis build them that way'. Natural selection, the tinkerer, has an accomplice.

For an interesting exposition of this notion we go back to antlers. Unlike skeletal bones that are adaptively moulded because of the particular strains that they experience (Fig. 51), antlers grow in the 'assiduous avoidance of strain' by virtue of a specialized, blood-rich, and highly innervated type of skin called velvet. Protected and nurtured by the velvet, cartilage grows until specialized bone cells called osteoblasts, moving in from the blood, eventually mineralize it. As long as the cartilage stays abreast of the osteoblasts, the antlers grow, but, as soon as the mineralizing osteoblasts have taken the upper hand, growth stops. The velvet then dies and, with its naked ossified antlers, the stag is ready for the rut. The ratio between ossification and cartilage growth determines the size and shape of the antlers. In small deer the ratio is high, so antlers stay small and pointed (Fig. 62A). In large deer the ratio between ossification and cartilage growth is low, so antlers can grow as 'expansive broad plates' (Turner 2007). After the rut, the antlers are shed. The new set usually resembles the old set except that, if a deer is well fed, it is a bit larger. That antlers grow larger with age is probably self-organized. Reproductive and stress hormones combine with direct diet cues to affect the relative rates of ossification and cartilage growth, and thereby antler size and shape.

Now comes a surprise. When, during some stage in a deer's life, antlers incur damage at one side because a bit of velvet got damaged during antler growth, this damage will be vis-

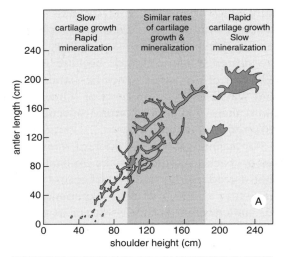

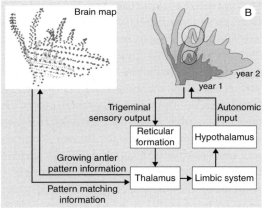

Figure 62. (A) Antler length is a function of the shoulder height of the deer stag that carries it. The outcome is determined by a 'race' between how rapidly cartilage grows and how rapidly it mineralizes, a ratio that is body-size dependent. In small deer, mineralization relative to cartilage growth rates is high, which leads to small, single-point antlers. In the largest deer, the ratio is small and this produces the broad plate-like antlers carried by moose *Alces alces* and the extinct Irish elk *Megaloceros giganteus*. (B) A diagram outlining the ways in which antler shape may be memorized in the brain and how this memory feeds back into the shape of the antler during the subsequent growth cycle.

Put together from figures in Turner (2007).

ible again the next year, even if the velvet is fine then! The injured antlers are cast, but the memory of their shape apparently is not. Turner (2007) argues that the rich infiltration of the velvet with both sensory fibres (routed through the trigeminal nerve and the thalamus) and so-called autonomic fibres, which control a whole range of physiological functions, suggests that 'antlers are in the middle of a feedback loop in which the trigeminal sensory nerves convey information to the brain about the growing antler's shape, while pattern of growth is mediated by information streaming out on autonomic nerve fibres, perhaps through controlling patterns and rates of blood supply to the various parts of the growing antler' (Fig. 62B). This interaction would not only explain why growth deformities are copied in successive years, but also why antlers show such exquisite bilateral asymmetry.

There are no genes that directly instruct for antler size and shape. Antler form naturally results from variations in the relative rates of cartilage growth and ossification (which may partly be under genetic instruction), and the build-up of shape memories in the brain. Now think of the amphibian phenotypes of fear. Could the tail shape of tadpoles 'simply' result from differences in the kinds of exercise induced by the presence of particular tadpole predators?

Phenotypic flexibility in birds

In the most celebrated case of evolution in the world of birds, the radiation of Darwin's finches on the Galápagos Islands (Lack 1947, Grant 1986, Weiner 1994, Grant and Grant 2008), the trait of most interest is beak size. Depending on rainfall, plant growth, and the production of seeds of different sizes, in different series of years either big-beaked or small-beaked finches have survival advantages. As beak size has a genetic underpinning, selection events show up in the beak-size trajectories of the relatively isolated island populations. Immersed, as we are here, in examples of phenotypic plasticity, one cannot help but wonder why small-beaked Darwin's finches, living during times when bigger beaks would be advantageous, do not themselves build bigger bills?

Although oystercatchers *Haematopus ostralegus* can and do modify bill size and shape (Hulscher 1985, van de Pol *et al.* 2009), Darwin's finches apparently cannot. In their very advanced synthesis of 34 years of work on natural selection in action in Darwin's finches, Peter and Rosemary Grant (2008) do not even use the word plasticity! Perhaps the material with which, and the way that, bills are built only allows for limited plasticity (Gosler 1987). This contrasts with most other parts of bird bodies that do show plasticity, mostly in the categories of life-cycle staging (Murton and Westwood 1977, Wingfield 2008) and phenotypic flexibility (this book). In fact, one of the few examples of developmental plasticity in birds is the one for oystercatcher bills where, to an extent at least, youngsters learn from their parents to feed on a particular diet (Sutherland 1987, Sutherland *et al.* 1996). Depending on the prey types that they select, their bill is worn and strengthened in particular ways (yes, self-organization!), even though there remains flexibility later in life.

Rather than structural body size, what predominantly varies in birds are body mass (an indicator of nutrient stores) and the size of most of the organs (Table 8), with the exception of the lungs (Piersma *et al.* 1999b). This is especially the case in those birds exposed to seasonally widely different conditions, either because they migrate from arctic or temperate to tropical climates, or precisely because they stay put at highly seasonal northern or southern latitudes. There is also considerable intra-individual variation in correlated metabolic attributes (Vézina *et al.* 2006, 2007) and aspects of the immune system (Buehler *et al.* 2008a).

Female red knots even show seasonal variation in the mass of their skeleton: they store calcium before going to the trouble of laying a clutch of four eggs (Piersma *et al.* 1996b). A phenotypic aspect not studied in shorebirds, but carefully documented in songbirds, is the seasonal waxing and waning of song nuclei in the brain (Nottebohm 1981, Smith *et al.* 1997, Tramontin *et al.* 2000; note that even modern humans leading domesticated lives show seasonal change in parts of the brain, Hofman and Swaab 1992). Within a week after shifting Gambel's white-crowned sparrows *Zonotrichia leucophrys gambelii* from short to long days, the number of neurons in a brain area called the High Vocal Centre increased by 50,000, or 70%, as did the volume of this nucleus (Tramontin *et al.* 2000). Perhaps surprisingly, the morphological changes were faster than the development of the song control circuitry and the degree of stereotypy of the songs themselves. These took some extra weeks to mature fully.

In view of the focus of much of the rest of this book on foraging, distributional decisions, and time-allocation problems, we round up by examining in greater detail the machinery for digestion and assimilation in birds, and their plasticity. The digestive system of birds can be divided into several main components (Table 9). In any one taxon, each of these digestive organs will have morphological characteristics that functionally reflect the evolutionary history and contemporary ecology of the taxon (Stevens and Hume 1995).

A large gut may provide the ability to use foods of low quality that are difficult to handle mechanically or chemically. The size of the digestive tract should, therefore, reflect the benefits of having organs of that size, in balance with the costs of maintaining that system. Before assessing degrees of variation in organ size, it is worth considering some general factors that may set upper bounds to the size of digestive systems. For example, it is obvious

Table 8. Many phenotypic traits of shorebirds (red knot, great knot, bar-tailed godwit, and others) show considerable intra-individual phenotypic change (C; relative to minimum) in the course of the seasons (S) and/or in relation to ambient environmental conditions (E).

Trait	C	S/E	References
Body mass	2.5 ×	S, E	Piersma et al. (2005)
Skeleton	1.4 ×	S	Piersma et al. (1996b)
Fat stores	150 ×	S, E	Piersma et al. (1999b)
Flight muscles	3 ×	S, E	Dietz et al. (2007)
Blood haematocrit	1.5 ×	S, E	Landys-Ciannelli et al. (2002)
Heart	1.4 ×	S, E	Piersma et al. (1999b)
Liver	2.5 ×	S?, E	Piersma et al. (1999b)
Gizzard	7 ×	E	Piersma et al. (1993b)
Intestine	4 ×	E	Battley et al. (2005)
Skin	1.7 ×	S	Battley et al. (2000)
Testicle	30 ×	S	T. Piersma unpubl. data
Feather mass	1.3 ×	S	Piersma et al. (1995a)
Basal metabolic rate	2.5 ×	S, E	Piersma et al. (1995a)
Peak metabolic rate	1.6 ×	S?, E	Vézina et al. (2006)
Basal corticosterone level	20 ×	S	Piersma et al. (2000)
Daily peak melatonin titer	10 ×	S	Buehler et al. (2009c)
Total leukocyte count	4 ×	S, E	Buehler et al. (2008a)
Microbial killing efficiency	10 ×	S, E	Buehler et al. (2008a)
Preen wax composition	Complete qualitative shift	S	Reneerkens et al. (2002, 2007)

Table 9. Subdivision of the major components of the digestive system in birds. Based on Klasing (1998), Battley and Piersma (2005), and J.M. Starck (pers. comm.).

Organ	Alternative name/components	Primary functions
Oesophagus + crop		Food storage, movement of food toward proventriculus
Proventriculus	Glandular stomach	Gastric secretion
Gizzard	Muscular stomach	Crushing or grinding food, mixing food with gastric secretions
Small intestine	Intestinum tenue, composed of duodenum and the combination of jejunum and ileum (the 'jejunoileum')	Enzymatic digestion, absorption of digestive end products
Ceca		Microbial fermentation, water and nitrogen absorption, immunosurveillance
Rectum	Colon, large intestine	Electrolyte, water and nutrient absorption
Cloaca		Storage and excretion of urine and faeces
Liver		Metabolism of absorbed nutrients, production of bile acids and bile salts
Pancreas		Secretion of digestive enzymes

that there is a physical limit to the volume available for organs within the abdominal cavity. Fat deposits will further reduce the abdominal space for the digestive tract. For example, migratory shorebirds deposit a thick layer of fat around the gizzard, covering the intestines and rectum, and pressing up against the tail end of the skeleton. In great knots *Calidris tenuirostris* about to depart on northward migration from Australia, fat in the abdominal cavity made up 38% of the total abdominal tissue mass (Battley and Piersma 2005). At such times the profile of the abdomen changes, with fat birds showing a 'bulge' behind the legs (Owen 1981, Wiersma and Piersma 1995). Conflicts for space between fuel stores and organs may limit the maximum size of the digestive organs.

Internal organ mass could directly affect flight costs and performance of birds. Because the costs of flight increase with body mass (Kvist *et al.* 2001), and as manoeuvrability is impaired at higher masses (Dietz *et al.* 2007), minimization of digestive organ mass will be an important consideration. In a comparative study of raptors, Hilton *et al.* (1999) showed that 'pursuing' species (that require fast and manoeuvrable flight) had shorter guts than 'searching' species, presumably reflecting the pursuers' need to minimize mass. The shorter retention time of these species would reduce the period of time they had food in their guts, also keeping body mass low, though at a cost of lowered digestive efficiency. Sedinger (1997) showed that grouse have caeca 4–5 times longer than waterfowl feeding on the same diet, and suggested that the small caeca in waterfowl reflect a balance between the costs of flight for waterfowl and the benefits of the caeca for nutrient balance. And even in animals at rest, mass-specific maintenance costs of the intestine and liver are greater than those of muscle and adipose tissue (Krebs 1950, Blaxter 1989, Scott and Evans 1992). Reductions in metabolically active tissues, therefore, dramatically

reduce basal energy requirements (Battley *et al.* 2001, Konarzewski and Diamond 1995).

Under the simplifying assumption that size (or mass) indicates functional capacity, Fig. 63 summarizes the magnitude of changes in three major gastrointestinal components, the gizzard, gut or small intestine, and liver. Data are separated into four taxonomic groups: galliformes (primarily grouse and quail), songbirds, shorebirds, and waterfowl. Even though only few of the studies in the review explicitly focused on documenting such changes, organ mass variations of 20–80% were common across a wide range of species; length changes of the intestine typically were lower, generally 10–20%. Earlier we discussed whether tissue-related differences in flexibility could be due to differences in cell turnover, based on carbon-isotope signatures of historically spiked DNA (Spalding *et al.* 2005a). A method that informs about the whole cell content, rather than DNA only, may be more appropriate, and measurements of retention times of carbon-isotopes supplied in food fulfil this condition. In a study on zebra finches *Taeniopygia guttata*, Bauchinger and McWilliams (2009) found retention times varied between 8 days for small intestine, 10–13 days for stomach, kidney, liver, and pancreas, 17–21 days for heart, brain, blood, and flight muscle, and 26–28 days for leg muscle and skin. Thus, they found that tissues with a tendency to show large changes (notably the small intestine and liver) had fast turnover rates, whereas less dynamic tissues, such as the flight and leg muscles, had slower turnover rates. Tissue-specific turnover rate may thus partially determine the magnitude of organ flexibility.

Clearly, if we keep an open mind and our eyes open, pretty much all parts of bodies appear to vary with time, age, demand, and environment. To ascertain the functionality of these changes

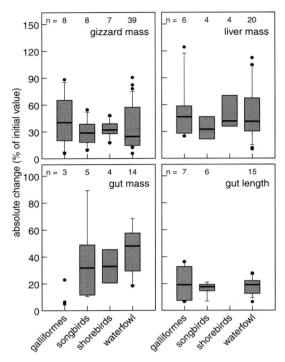

Figure 63. Absolute change (either increase or decrease) in digestive organs of birds. Data represent maximum percentage changes recorded in studies for age- or sex-classes of birds, due to factors including diet type or quality, migration, breeding, and food intake. Measurements from wild and captive birds are included. The gut category represents data for both the entire gut (including caeca and rectum) and the small intestine only. Sample sizes for each taxonomic group are given above the box or points. Boxes represent the 25–75 percentiles (divided by the median), whiskers represent the 10 and 90 percentiles, and outliers are shown as dots. For groups with only a few values, individual data points are plotted.

From Battley and Piersma (2005).

we need to do much more than 'just' record the morphological changes. In the next few chapters we considerably expand our view beyond bodily structures to examine active free-living organisms, especially birds, as an integrated whole.

Synopsis

'A single genotype can produce many different phenotypes, depending on the contingencies encountered during development. That is, the phenotype is the outcome of a complex series of developmental processes that are influenced by environmental factors as well as by genes. Different environments can have an effect on the outcome of development that is as profound as that produced by different genes'. To this succinct description of phenotypic plasticity by Nijhout (1999), we would like to add the possibility that much of the adaptedness shown by phenotypes is actually driven by self-organizational processes (Turner 2007). Reproductively successful phenotypes will satisfy ecological demands by optimizing the balance between various cost and benefit functions of specific kinds of phenotypic (trait) variation. In brief, most traits of all organisms are responsive to particular environment features over shorter or longer time scales. Consequently taking ecological context into account will enrich biology on all fronts.

Part III

Adding behaviour

CHAPTER 6

Optimal behaviour: currencies and constraints

When the going gets tough, the tough get going

Rapidly rising concentrations of atmospheric carbon dioxide help to heat up our globe. Not only that, but the increase in carbon dioxide also means that the seas and oceans around us are becoming more acid. Compared to pre-industrial times, the pH of seawater has dropped by 0.1 units and is predicted to be 0.4 units lower by the end of this century (Orr *et al*. 2005). The corresponding increases in dissolved hydrogen ions (25% and 150%) will have a heavy impact on sea life, especially on those creatures that build a calcareous exoskeleton or protective shell around themselves (Vermeij 1987). Think of the python that we discussed earlier, which lowers the pH of its stomach fluids in order to dissolve the bones of the ingested prey. Calcareous material simply dissolves under acid conditions, and the current acidification of our world seas is apparently enough to create problems for corals, molluscs, and some plankton (e.g. Riebesell *et al*. 2000). Not perhaps as serious as for a rat in a python's stomach, but sufficient to turn their daily lives upside down.

Enter periwinkles *Littorina littorea*, small gastropods living on rocky coasts. Just as in most other molluscs, their calcareous shell acts as a protective shield against predators. Although hard and seemingly unchangeable, their shell is actually quite plastic. Even when exposed to only the taste of shell-crushing crabs when they are reared in 'crab effluent', these snails adaptively increase the thickness of their shells (Trussell and Smith 2000). However, as Bibby *et al*. (2007) discovered, this only occurs when the water is not acidified. Periwinkles tasting crab-juice under normal pH conditions increased their shell thickness by 0.05 mm after 15 days, whereas those exposed in acidified crab effluent were unable to thicken their shells (Fig. 64A). This would bring unresponsive periwinkles into real danger if they were exposed to real crabs, rather than merely to crab effluent, but in fact they have a simple way out of this potential problem. Snails may have a reputation of being slow and living virtually sedentary lives, but they do have *behaviour* as an option. The acidified snails compensate for their lack of morphological defence by increasing their avoidance behaviour, i.e. by crawling out of dangerous water (Fig. 64B).

This is exactly the point that we shall make repeatedly in this chapter: animals behave, and their behaviour is the most flexible and fastest way to respond to environmental change. Think about the ptarmigan encountered in the previous chapter (Montgomerie *et al*. 2001), that soiled their snow-white feathers, thereby camouflaging themselves: it takes weeks to moult into an inconspicuous tundra-like plumage, but only minutes to take a dirty mud bath. There are numerous examples where behaviour outruns morphological change as a way to cope with environmental change.

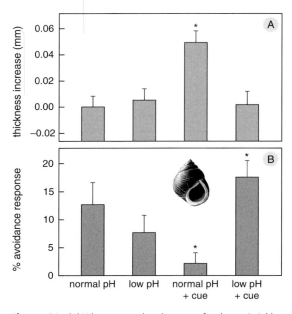

Figure 64. (A) When exposed to the taste of crabs, periwinkles increase the thickness of their shell, but are only able to do so under normal pH conditions. In acidified seawater, the snails cannot build thicker shells, but (B) use a behavioural way to avoid predation by crawling out of the water. Asterisks indicate treatments that are significantly different from other treatments.
Modified from Bibby *et al*. (2007).

Loading leatherjackets

In this chapter we will discuss behavioural flexibility, and we will approach it from a functional angle. It was ethology-godfather Niko Tinbergen who taught us that one (out of four) ways of looking at behaviour is to consider *function*. We can ask how much an observed behavioural choice contributes to an animal's fitness, and also how much the alternative, unchosen options would have contributed (Tinbergen 1963). For example, if a female kestrel lays a clutch of five eggs, we would like to know how much this decision[1] contributes to her lifetime reproductive success. Would she have done better if she had produced four eggs (because she would not have had to work so hard, thus yielding an extra breeding season), or would half a dozen eggs have maximized her contribution to the population's gene pool? This way of thinking about behaviour gave birth to a new field of science, behavioural ecology, which flourished from the late 1960s onwards, and is still very much alive and kicking (see textbooks by Krebs and Davies 1978, 1984, 1991, 1997 and more recently by Danchin *et al*. 2008).

Attempting to determine the contribution of a five-egg clutch, or any alternative clutch size, to a kestrel's fitness is a daunting task. That this is possible is shown by the elegant field experiments of Serge Daan and co-workers (Daan *et al*. 1990a). They manipulated brood size by moving eggs around between different nest boxes. Some pairs had to raise fewer young than their original clutch size, others had to raise enlarged broods, while the brood size in a third group was left untouched. The fate of the adults and their young was followed over the subsequent years, enabling the estimation of survival and reproductive rates. After many years of data collection it became evident that the fitness-value of the unmanipulated brood sizes was highest, because of a reduced parental fitness in enlarged broods and a reduced fitness of the clutch in reduced broods.

However elegant and stimulating such field-based estimates of the fitness-consequences of life-history decisions may be, they do not allow estimates of the fitness-consequences of the more short-term decisions that animals may make many times per day. Think of a parent

---

[1] Behavioural ecologists use the word 'decision' to indicate that an animal has behaved in one of several possible ways (McFarland 1977). They do not suggest that the animal has consciously calculated which behaviour would be optimal in any particular case. Possible mechanisms that animals may use to reach such 'decisions' are considered later in this chapter.

starling *Sturnus vulgaris* that has just delivered a 'beakful' of five, rather than four or six, leatherjackets (*Tipula* larvae) to her offspring waiting in the nest. Parent starlings line up multiple prey items in their beak before returning from the feeding site to the nest (so-called 'central-place foraging'; Orians and Pearson 1979), and they thus need to decide on exactly how many prey items to upload into their beaks (Fig. 65A). While collecting the food at the foraging site (mostly leatherjackets; Tinbergen 1981), they face diminishing returns, because each extra prey item in the beak means a reduced efficiency for finding and handling the next one (Fig. 65B; this has been coined 'the loading effect' by Kramer and Nowell 1980). Would the starling that flew off with five leatherjackets have done better (in fitness terms) if she had continued foraging and collected another prey item before taking off, or should she already have stopped after she had found the fourth leatherjacket? It may not matter at all, and certainly not for a single round-trip, but what if she always loaded five prey items rather than four or six, every breeding season, her whole life long? With about 6000–8000 prey deliveries per brood (Tinbergen 1981, Kacelnik 1984), an occasional second brood within the same season, and 2–3 breeding seasons per lifetime, this adds up to roughly 20,000 such prey-delivery decisions per lifetime. The fitness-consequences of each of these short-term decisions are small and this makes it difficult to quantify their payoffs. For this reason, behavioural ecologists work with measurable surrogates for fitness, so-called 'currencies'.

Just as we all use currencies in our daily lives to express the value of our goods in Euros, pounds, dollars, or yen, behavioural ecologists use currencies to express the fitness-value of behavioural options. As there is likely to be a premium on collecting as much energy per unit time as possible, energy intake rate is often used as a currency to explain foraging behaviour (Stephens and Krebs 1986, Sih and Christensen 2001). Going back to the starling delivering food to its hungry young, a candidate currency to maximize is the *daily amount of food delivered to its offspring*. While doing so, the starling faces a trade-off between, on the one hand, minimizing the daily time flying back and forth between the foraging site and the brood (by collecting as many leatherjackets as it is possible to stuff into the beak) and, on other hand, minimizing the daily time spent at the foraging site (by returning each time to the brood with just a single leatherjacket). Because the variable to maximize is known and quantifiable (the currency), the optimal number of prey to deliver can now be calculated (Fig. 65B). Interestingly, the optimal number depends on the average time it takes to travel between the nest and the feeding patch: the shorter the time it takes to travel (on average), the fewer prey should be brought back upon return (Fig. 65C).

Alex Kacelnik (1984) took advantage of this quantitative prediction in order to find out whether the suggested candidate currency best described the number of prey uploaded as a function of the distance flown, or whether alternative currencies performed better. On the Dutch isle of Schiermonnikoog, he trained free-ranging parent starlings to come to a wooden tray on which they could collect mealworms for their young. By dropping single mealworms through a long plastic pipe, at increasing intervals, Kacelnik generated the birds' own loading curve and recorded the number of mealworms that the trained starlings would take at each visit. Each day, he would randomly move the tray to a different site, such that, by the end of the experiment, he had manipulated the starlings' travel distance, ranging from 8 to 600 m. The results were astounding (Fig. 65D). The number of mealworms went up as a function of travel time. What is more, the results also fitted closely

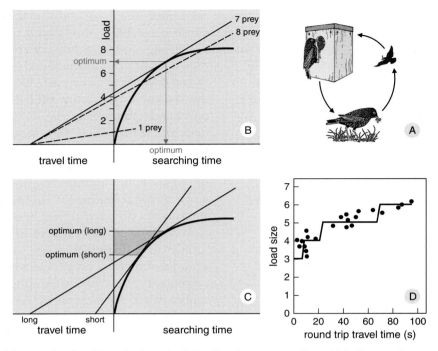

Figure 65. (A) The complete foraging cycle of a central-place foraging parent starling, which alternates between collecting prey items at a feeding site and delivering them to its young in the nest. (B) The loading or gain curve at the feeding site describes the number of prey items loaded in the starling's beak as a function of search time. As the beak-load increases, the efficiency of finding and loading the next item decreases, resulting in a decelerating loading curve. The optimal number of items to load at the average site (seven in this example) depends on the average travel time between patches and can be found by the tangential solution shown. (C) The less time it takes to travel (on average), the fewer prey should be brought back upon return. (D) Kacelnik's (1984) data (dots) fitted best with the prediction based on net family gain rate maximization (line).

Compiled from Shettleworth (1998) (A) and Krebs and Davies (1991) (B–D).

with the prediction based on the maximization of the *net amount of energy gained by the entire family as a whole,* thus subtracting the metabolic rates of the parent and the chicks from the gross amount of energy collected. However, we should add that the predictions of the other three currencies were not too far off this mark either!

This example shows the use of decisions, currencies, constraints, and optimality in behavioural ecology. We must take into account that the starling needs to make many more decisions, which are not treated here (e.g. where to feed, how many prey items to con-

sume itself rather than to deliver to the young, etc.; see Tinbergen 1981), but the *decision* studied here is: how many prey items to load before taking off? Then, after defining the *currency,* the unit in which the pay-offs of the taken decision are expressed, it becomes clear what *constraints* the parent starling faces: travel time, search time for each consecutive prey item, 'loading time' required to pick up each newly encountered prey item, and the metabolic rate in both the parent (in the feeding patch, in the air, and in the nest) and its offspring (while begging and while resting). If any of these constraints could be alleviated (shortened in the

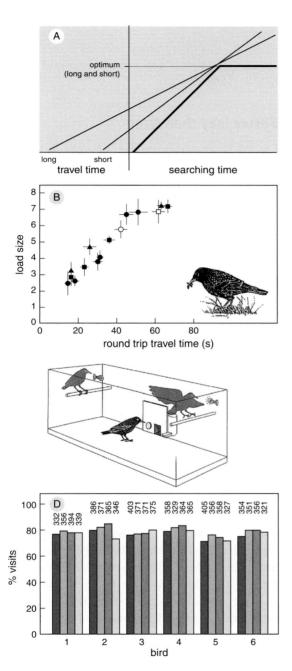

case of time factors, lowered in the case of metabolic demands), the starling would then obtain higher net benefits. Once the constraints are know, then the *trade-offs* can be defined. In the starling's case, the major trade-off is finding the right balance between the daily time spent in the feeding patches and the daily time devoted to travelling. Having identified the trade-offs, the *optimal* solution to the decision of interest can be calculated. This way of studying behaviour gives the behavioural ecologist predictions, and as we shall see next, empirical falsification of clear-cut predictions is an effective way to make progress in science (Popper 1959).

A couple of years after Kacelnik carried out his elegant field experiment in The Netherlands, a remarkable finding would shake up his earlier conclusions. At the Oxford University Farm in Wytham, Kacelnik's PhD student Innes Cuthill repeated the Schiermonnikoog experiment (Cuthill and Kacelnik 1990), but this time offered free-ranging trained starlings a linear, rather than a decelerating, loading curve (Fig. 66A). Thus, once it had arrived at the tray, a starling was again offered one mealworm at a time, but now with a *constant* time interval in between two consecutive mealworms (so-called 'sudden-death patches'; Brunner *et al.* 1996). No more diminishing returns, but a constant supply of food up until a total of eight mealworms had arrived on the starling's dish (which represents the maximum

perch with the light off to the one with the light on; the light turns off at the perch at which it arrives, turning on the light at the other end. This goes on until the travel requirements have been met, after which food is offered at the feeder in the centre of the cage. (D) Percentage of patch visits at which individual starlings consumed all eight mealworms that were offered. Shadings indicate different treatments (travel distance and steepness of the gain curve were varied), numbers indicate sample sizes.

Compiled from Krebs and Davies (1991) (A-B), Shettleworth (1998) (C), and Cuthill and Kacelnik (1990) (D).

Figure 66. (A) When the loading curve is linear, load size should always be maximal, irrespective of travel time. (B) Surprisingly, parent starlings brought home larger loads when travelling for a longer time. Different symbols denote different individuals. (C) Experimental set-up mimicking patch use. In order to travel between 'patches', the starling hops from the

number of mealworms that a starling can possibly carry at any one time). The optimality model now predicts that the starling should always collect all eight prey items before taking off, irrespective of travel distance (Fig. 66A). So what did the starlings actually do?

They ignored the theoretical prediction by bringing home more prey when travelling further (Fig. 66B). Imagine the amazement on the faces of Cuthill and Kacelnik! Yet, unabashed, they continued with another experiment that would eventually bring back light into the new darkness. The linear loading curves were now offered to self-feeding starlings that had no brood to care for, but still needed to travel to move between 'patches' (Fig. 66C). Guess what? Those self-feeding starlings did exactly what they were 'supposed' to do: they gave up virtually every food patch after collecting all eight mealworms, irrespective of the distance travelled (Fig. 66D). The only difference between these latter two experiments was the fact that chick-feeding starlings had to 'upload' and carry their prey items. Cuthill and Kacelnik, therefore, concluded that, with increasing load size, some cost function would increase, whether it was the energy expenditure while in the patch, or the energy expenditure while in flight, or the time it takes to return to the nest. Such costs had been overlooked when Kacelnik (1984) suggested that parent starlings should maximize *net family gain*. Even though this currency did account for parental energy expenditure, it did so in a load-size independent manner.

Such is the power of optimality models. As we have just demonstrated, a clash between theory and facts can lead to new ideas about the true constraints and currencies that underlie the observed behaviour. Optimality models are by definition over-simplistic, but, when falsified, they form a guide for further research. Just as in everyday life, making mistakes is the best way to learn! Having seen an example of

constraints that had to be redefined, the next scientific discovery story tells us how failing optimality models can inform us about alternative *currencies*.

Better lazy than tired

For the most part, foraging theoreticians have included 'time' as the denominator in their favourite currency, meaning that something ought to be maximized *per unit of time*. For example, the number of leatherjackets collected *per minute*, or the net amount of Joules gained *during the total day*. This makes much sense, since foraging takes time, and animals also need to devote some of their daily time to activities such as finding a mate, building a nest, incubating eggs, and so on. It also makes sense when animals are in a hurry—when migrants need to fuel up to reach the breeding grounds in time, for example. In such cases, they devote all the time possible to foraging (Piersma *et al.* 1994b) and maximize the (net) energy income per day.

For this reason, it did not make sense when foragers were often observed feeding at a slower pace than was possible, even foragers that were trying to fuel up as quickly as they could, as you will see shortly. Dungeness crabs *Cancer magister* were eating smaller bivalves than expected from rate-maximizing principles (Juanes and Hartwick 1990), Lapland longspurs *Calcarius lapponicus* were not maximizing the delivery rate to their offspring because they flew slower than they 'should' (McLaughlin and Montgomerie 1990), as did black terns *Chlidonias nigra* (Welham and Ydenberg 1993). Honey bee *Apis mellifera* workers only partially filled up their crops, whereas they should have filled them up completely if they were to maximize the rate of delivering nectar to the hive (Schmid-Hempel *et al.* 1985). This list could be extended further, but in all these examples it turned out that another energetic currency was in fact being maximized, namely, the gross amount of energy gained per unit of energy expended, called 'efficiency', and formally written as:

$$b/c,$$

where b denotes gross energy intake rate and c denotes metabolic rate.

Initially, the reasons for this efficiency maximization were not understood. In fact, efficiency maximization was dismissed as a nonsensical currency by the first foraging theoreticians (Stephens and Krebs 1986). They reasoned that, if 2 J can be gained for 1 J of expenditure, and the alternative would yield 5 J while expending 3 J, then the second option, the least efficient option ($5/3 < 2/1$), should be preferred, simply because it yields a higher net gain ($5-3 > 2-1$). Only if expending energy bears additional costs should it pay off to expend energy parsimoniously. We refer back to Chapter 4 for an in-depth treatment of non-energetic costs due to energy expenditure (such as shortened life span, increased risk of immediate death, hampered immune response). Although such non-energetic costs may explain efficient and prudent behaviour, more recent foraging models have shown that there may nevertheless be purely energetic, rate-maximizing reasons for being an efficient forager. Let us now examine this idea more closely.

More haste less speed

Even when in captivity, red knots fuel up rapidly in spring and deposit the same amount of fat as do their free-ranging conspecifics (Cadée et al. 1996). It was at this time of year that we performed an experiment with four red knots, which were allowed to exploit depletable patches in a large outdoor intertidal aviary (van Gils et al. 2003b). In the same way as the starlings we discussed earlier, the knots faced diminishing returns when emptying their patches (because searching through the patches occurred in a non-systematic way, leading to longer time intervals between consecutive prey encounters). This gave rise to the critical question of this study: at what (diminishing) intake rate would knots give up a patch and switch to a new patch (Fig. 67A)?

Given the time of year, we expected knots to give up their patches so that they maximized their short-term net energy intake rate, i.e. their net intake calculated over total foraging time, taking into account the time and energy cost required to make brief flights between patches. This meant that, on the one hand, patches should not be exploited for too long, as the birds would lose too much time to increasingly long probing intervals, whereas, on the other hand, giving up the patches too soon would cost too much travel time, and especially travel energy, when moving between different patches (Fig. 67B, C). To our initial surprise, the red knots probed much too long in their patches and were more reluctant to move on to a new patch than we had anticipated. Initially, we interpreted this as 'lazy behaviour', since the time devoted to the most costly activity, the short flights between patches, was less than half of what we expected.

Were the knots maximizing their efficiency by obtaining every Joule of energy income as cheaply as possible, irrespective of the time it took? If so, would this increase their life-expectancy at a time of year when they actually need to fuel up as quickly as possible? This could be the case, but it seemed more likely that it would be preferable to speed up the fuelling process, the more so because the birds weighed less than was normal for the last days of April.

The answer was provided by new theoretical work from McNamara and Houston (1997), but also by Ydenberg and co-workers

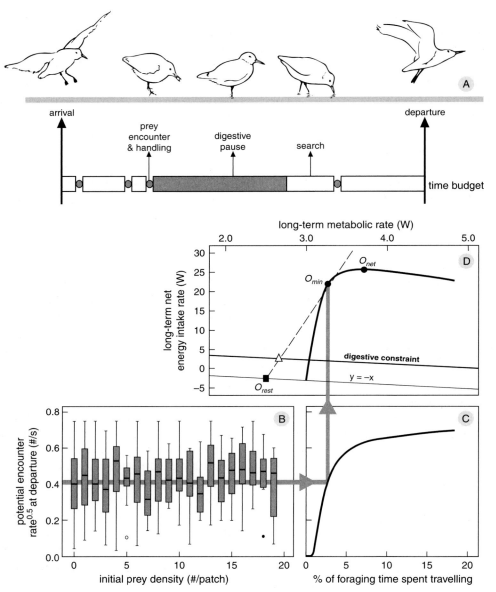

Figure 67. (A) Typical patch-feeding situation for a red knot: between arrival and departure from a patch, the bird alternates between searching for prey (open bars), handling prey (grey dots), and digesting prey (grey bar). (B) The potential prey encounter rate at which patches were given up was independent of their initial prey density. (C) The higher the potential prey encounter rate at which patches are given up, the more time knots will devote to travelling. (D) Long-term net energy intake rate is determined by the giving-up policy, peaking at relatively high quitting intake-rates (O_{net}). However, because red knots face a digestive bottleneck on their gross energy intake rate, their long-term net intake rate cannot exceed the diagonal line indicating this constraint (which has a slope of −1 because net intake (Y-axis) = gross intake (intercept) − cost (X-axis)). Therefore, they should not opt for O_{net}, but rather feed at a slower pace by alternating between resting (O_{rest}) and giving up their patches at lower potential intake rates (O_{min}), in which case their long-term net intake rate is maximized at the triangle. This is exactly what the birds did, as indicated by the grey arrow going from (B), via (C) to (D).

Modified from van Gils *et al.* (2003b).

(Ydenberg *et al.* 1994, Ydenberg and Hurd 1998), and Hedenström and Alerstam (1995). Energy-limited foragers, i.e. animals that face a constraint on the daily amount of energy that can be spent or assimilated, cannot feed during the entire day, as they need to 'respect' their energetic bottleneck (see Chapter 4). Such foragers, and this may sound counter-intuitive, should actually maximize (a modified form of) efficiency *while foraging*, in order to maximize the net daily energy uptake *across the entire day*. The form of efficiency that should be maximized has been coined the 'foraging gain ratio' (Hedenström and Alerstam 1995) or simply the 'modified form of efficiency' (Houston 1995), and takes account of the metabolic rate when not foraging, i.e. the resting metabolic rate r:

$$b/(c - r).$$

This apparent paradox has a relatively simple explanation. If the gross amount of energy that can be assimilated cannot exceed a certain threshold, then it pays to gather this amount of energy as cheaply as possible, taking into account r, the metabolic rate while not feeding (the faster the threshold is reached, the more energy is lost to non-feeding behaviour). The less energy that is expended in total, the more energy remains. With respect to our knots, they were indeed energy-limited: the amount of energy they could assimilate during any one day was constrained by the rate at which they were able to process their shellfish prey. As knots ingest their prey whole, they must internally crush and process the bulky shell-material, which then needs to pass rapidly into the digestive system in order to make room for new food. The crushing of the shell takes place in the muscular stomach, called the gizzard (Piersma *et al.* 1993b), and it is especially the size of this part of the 'gut' that determines the daily amount of shell mass that the birds can process (van Gils *et al.* 2003a), and thus the amount of flesh that can be assimilated daily (about which we shall learn much more in the next chapter).

The intake rate at which our birds left their patches did indeed maximize the foraging gain ratio while they were feeding. To investigate how this strategy would maximize overall net energy intake, we used the graphical approach advocated by McNamara and Houston (1997) as shown in Fig. 67D. The triangle in that graph (the long-term net energy intake rate) cannot exceed the diagonal line indicating the digestive constraint. The more this point is located to the left on the diagonal, the higher the long-term net energy intake rate (since the diagonal is going 'downhill'). This left-most location on the diagonal is achieved by alternating between resting (option O_{rest} in the graph) and maximizing the foraging gain ratio while feeding (option O_{min}, which is found by drawing a line from O_{rest}—dashed in this case—that is tangential to the thick curve defining the continuous range of feeding options).

Of course, the argument for efficiency maximization only holds as long as there is enough time to reach the digestive bottleneck. When there is only a limited amount of foraging time, animals are said to be *time constrained* rather than *energy constrained*. When 'heavily' time constrained, the best thing such animals can do, when aiming to maximize net daily energy intake, is to maximize their net intake rate while feeding. It is as simple as that and means that animals may switch 'tactics', depending on whether they face a *time* constraint or an *energy* constraint, as exemplified below.

Oystercatchers pressed for time

The aviaries in which we offered depletable patches to knots were used to study foraging oystercatchers *Haematopus ostralegus* 17 years earlier by Cees Swennen and others (1989).

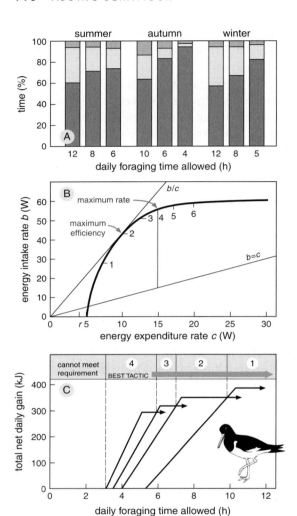

Figure 68. (A) Oystercatchers devoted more of their time to foraging (lower bars) with a predictable shortening of the low-tide period (middle bars denote resting and upper bars denote preening). (B) Harder work (horizontal axis, with *r* indicating resting metabolic rate) leads to diminishing foraging returns (vertical axis). Numbers indicate different rates maximizing different currencies (net energy intake rate is maximized in 4, efficiency is maximized in 2, and the foraging gain ratio is maximized in 1). (C) Under a digestive constraint, total net daily intake is maximized by maximizing the foraging gain ratio (tactic 1 in (B)). However, this strategy requires much foraging time. With a shrinkage of the available foraging time, an oystercatcher does better to switch strategies, with maximization of the net energy intake rate (tactic 4 in (B)) being the best tactic for the shortest available feeding times.

They wondered how animals deal with shrinking foraging time, such as often happens to birds exploiting the intertidal zone during low tide. Strong onshore winds and periods of frost may dramatically shorten a shorebird's foraging time (Zwarts *et al.* 1996). If such time restrictions are predictable and animals can foresee them, can they compensate by working harder? During the 1980s, rate-maximization was still *the* dominant currency in foraging theory, and animals were expected to obey this 'rule' under all circumstances.

Seven individual oystercatchers were allowed to exploit cockles *Cerastoderma edule*, but had to retreat into their roosting cage when the tide came in. As it was possible to manipulate the tidal length in these cages, the birds were offered normal tides with 6 hours of low tide, but also shorter tide intervals of 5, 4, 3, 2.5, and even 2 hours. These tidal regimes were made predictable by offering the same tidal length for a period of about a week (Swennen *et al.* (1989) also performed a series with unpredictable feeding periods). Seemingly violating the rate-maximizing paradigm, by working harder when their available foraging time was shortened, the oystercatchers showed that they did not always work at maximum rates. The birds spent proportionally more time foraging during shorter low tides (Fig. 68A), found their prey items faster, and it also took them less time to handle and ingest them. At the time of his experiment, Swennen may have wondered why his oystercatchers did not always work at their maximum rate; nowadays the surprise may be the other way around

Modified from Swennen *et al.* (1989) (A) and Ydenberg and Hurd (1998) (B–C), making a slight correction in (B) by setting resting metabolic rate *r* at 4 rather than 5 (otherwise the foraging gain ratio would be maximized while resting, a point that was overlooked by Ydenberg and Hurd).

for us. Why do oystercatchers work harder when foraging times are shortened? As we have just seen, energetically bottlenecked foragers should maximize their (modified form of) efficiency in order to maximize their net daily gain. Oystercatchers are a classical example of digestively bottlenecked foragers (Kersten and Visser 1996, Zwarts *et al.* 1996), and should thus feed at a relaxed pace under all circumstances, whether the tide intervals are short or long (a further consideration being that feeding at a high rate may incur additional non-energetic costs, such as bill damage; Rutten *et al.* 2006). Again theoretical work solved this mystery, this time by Ydenberg and Hurd (1998).

The simple explanation is that, with a progressive shortening of their foraging time, oystercatchers did get rid of their digestive constraint, but acquired a new *time* constraint in return. Modelling gross food intake as a function of workload (Fig. 68B), Ydenberg and Hurd (1998) made clear that, in order to continue maximizing net daily energy gain, oystercatchers should switch currencies when faced with a progressive shortening of foraging time. When there are more than 9.5 hours of feeding per day (i.e. almost 5 hours per tidal cycle), oystercatchers should maximize the modified form of efficiency, the foraging gain ratio (tactic 1 in Fig. 68C). Under a tighter regime, this policy becomes time limited and the birds do better by switching to efficiency maximization (tactic 2 in Fig. 68C), allowing themselves once more to obtain the maximum amount of gross energy that they can assimilate during the day. However, when there are fewer than 7 hours of daily feeding (each low tide lasting 3.5 hours at most), birds do better to switch to tactic 3 (Fig. 68C), and they should switch to instantaneous rate maximization when feeding times fall below 6 hours per day (low tides of 3 hours or less; tactic 4 in Fig. 68C).

Informational constraints: getting to know your environment

The case of the oystercatchers exemplifies two major constraints in a single story—energetic ceilings and time constraints—but there is, in fact, a third, more hidden constraint that also underlies the entire study. This is the degree to which the oystercatcher's brain is able to absorb, process, and remember information about the current tidal regime. Swennen and his colleagues trained the birds for a couple of days before they got acquainted with their latest tidal regime. As a subsidiary test, they also performed some trials in which each day a new, completely randomly chosen tidal regime was offered. Not surprisingly, the oystercatchers did not respond to tide length and were unable to meet their daily energy demands when exposed to shortened low tides. Knowledge about your environment can be a matter of life and death!

The starlings that were the earlier stars of the show should also be aware of an important environmental parameter: their long-term *average* intake rate, determined by feeding policy and also by the average distance between their patches and their nests. The rate-maximizing starlings should leave each patch when the instantaneous, or *marginal*, intake rate drops below the long-term average intake rate. For this reason, this model has been coined the 'marginal value theorem' (MVT; Charnov 1976a). Thus, the graphs in Fig. 65B and C, represent the *average* situation in the forager's environment. Only the *average* travel time to patches and the *average* gain curve predict the optimal *average* number of leatherjackets loaded. However elegant the graphical approach of the MVT, as exemplified in Fig. 65B and C, the implication is that this approach does not work when predicting, *for a single environment* (Fig. 69A), the optimal leaving policy for different patches at different distances from the central place (and this has often been overlooked, as recently stressed by Olsson *et al.* 2008). In fact, within a single environment with no variation in patch quality, because all patches should be left at the same marginal intake rate, similar load sizes should be brought back home from all possible distances. When the cost of carrying a load increases with load-size (as should be the case for the parent starlings; Cuthill and Kacelnik 1990), but also when there are additional, load-size independent costs of travelling, *negative* load-distance correlations are expected (Olsson *et al.* 2008).

Most studies on central-place foraging have been carried out *between* different 'environments' (Fig. 69B) and generally find positive load-distance correlations (reviewed in Stephens and Krebs 1986). However, in a recent *within*-environment study, the negative load-distance effects are confirmed for overwintering Bewick's swans *Cygnus columbianus bewickii* (van Gils and Tijsen 2007). Although these birds do not bring food to a central place, they do get together at a central place, each night, when roosting at a lake or a flooded field. Exploiting a mosaic landscape of recently harvested sugarbeet fields, the birds feed for less time in distant fields, thus quitting those fields at higher intake rates (Fig. 70). More examples of negative load-distance effects were detected in situations where predation risk increased further away from a central refuge (Brown and Kotler 2004). We will come back to such anti-predation issues in Chapter 8.

To keep up with changes in your environment, it is critical to be able to perceive these changes. In fact, by *each day* placing the artificial

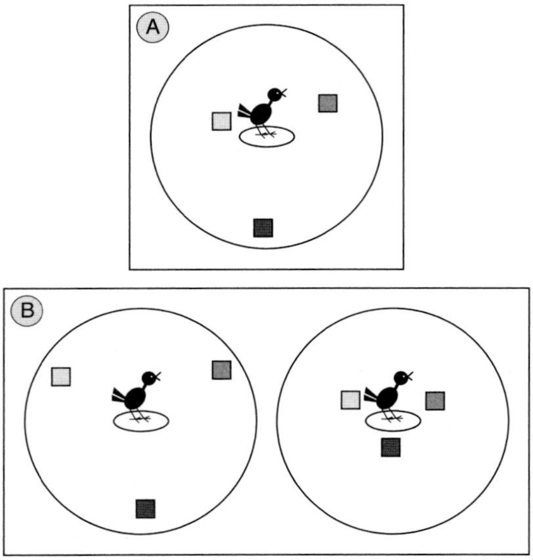

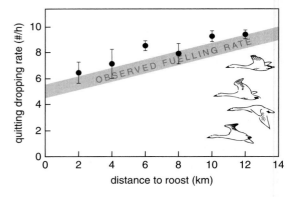

Figure 70. A within-environment effect of central-place foraging in Bewick's swans: the birds gave up their foraging sites (fields of sugar beet remains) at much higher feeding rates (expressed as easier to observe dropping rates) when these were located further from the roost (dots are means, bars are standard errors). Marginal value theorem predictions were met since the marginal net intake rates at quitting (calculated from the quitting dropping rates) precisely matched the long-term net intake rates (calculated from the observed rates of increase in their abdominal profiles), as indicated by the thick grey bar.

Modified from van Gils and Tijsen (2007).

Figure 69. Cartoons of central-place foraging, with the central place in the middle (bird on the circle), surrounded by patches (squares) of various quality (shading). The effects of travel distance on foraging decisions can be quite different when studying them (A) within environments or (B) between environments. Most studies comply with the between-environment comparison.

Modified from Olsson *et al.* (2008).

feeding patch at a different distance from the nest, Kacelnik (1984) changed the starlings' environment from day to day, making this study a between-environment study. As suggested by the positive load-distance effects (Fig. 65D), the starlings did perceive these rapid changes as environmental changes and responded adaptively. How quickly starlings can learn about their environment was examined in a follow-up study by Cuthill *et al.* (1994). In a laboratory set-up, starlings had to get used to new travel distances every two days. After the sudden increase or decrease of travel distance, the birds gradually adjusted their load size in the correct direction (larger loads for longer distances). After six full cycles of travel and patch use, the birds reached the asymptotic behaviour that they maintained until the next distance-change. Quite a steep learning curve! But note that in this indoor laboratory study, one thing was kept simple for the birds: the type of patch offered was always the same. Out in the field, patches differ in their quality and the amount of prey they contain, and in the next section we will discuss how birds get to know their patch.

Informational constraints: getting to know your patch

The function describing the cumulative energy gain when exploiting a patch is usually drawn as a smooth curve (e.g. take a look at Fig. 65B and C). Such a continuous, uninterrupted intake may be applicable for foragers sucking up liquid food, such as nectar-eating hummingbirds (Roberts 1996). However, for many foragers, food comes in discrete lumps at rather stochastic moments in time (e.g. Fig. 67A), yielding stepwise gain curves (Fig. 71A). We have assumed that many foraging decisions are based on instantaneous intake rate, but how can and how should such foragers *measure* their own intake rate if there are bouts

of no intake between prey captures? And, secondly, if they are able to get some measure of instantaneous intake rate, should they really base their foraging decisions on this measure, or are there better alternatives?

Before the marginal value theorem was published (Charnov 1976a, Charnov and Orians 1973), there were ideas about how discrete-prey foragers could estimate their own intake rate and how they could decide when to quit a patch. A simple and attractive idea was that such foragers should give up a patch after a fixed time-interval of not finding a prey item, called the Giving-Up Time (GUT). Niko Tinbergen measured it in carrion crows *Corvus corone* searching for camouflaged eggs (Tinbergen *et al.* 1967), and Croze (1970) came up with a strong and intuitive argument why animals may use GUTs: 'the GUT is taken to be a quality of the crow's persistence—an expression of the amount of effort the predator is willing to allot in pursuing one more of a particular prey'. In a classical experiment, Krebs *et al.* (1974) concluded that black-capped chickadees *Parus atricapillus* adopted a constant GUT when giving up their patches (but others noted that too few prey were taken to be conclusive about this, e.g. Iwasa *et al.* 1981).

However elegant and natural the GUT concept may seem, it did not fit with the equally elegant marginal value theorem (MVT). First, the MVT requires a continuous, uninterrupted intake in which GUTs cannot exist. Second, even if the MVT is slightly modified, such that it *can* make predictions about GUTs, then a constant GUT turns out to be suboptimal. In that case, foragers should be more patient (apply longer GUTs) in richer patches, provided that they can recognize patch quality instantly (McNair 1982). But even this latter rule does not actually work in practice, as it requires patch quality to be recognized instantly. Most foragers, especially those that feed on cryptic prey, need sampling time to

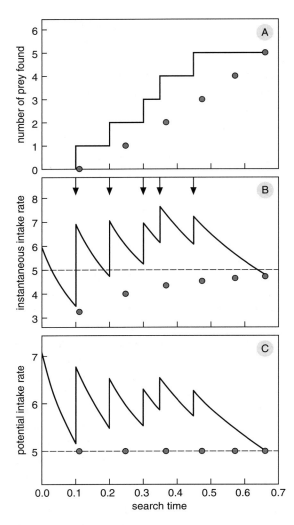

Figure 71. (A) Feeding on discrete prey yields a stepwise gain curve (line), with moments of prey-encounter alternating with bouts of no intake of rather unpredictable length. Optimal moments to leave such a patch (grey dots) depend on the combination of the number of prey items found and the total time it took to find them. (B) Fluctuation of the estimated instantaneous intake rate for a Bayesian forager in a clumped environment experiencing the gain curve plotted in (A). This estimate goes down with search time, but jumps up with each prey encounter (arrows). In order to maximize its long-term intake rate, the forager should not leave its patch at a certain critical instantaneous intake rate (dashed line), but should instead be a little more patient and leave when the estimates have reached the grey dots. (C) By doing so, it leaves at constant estimated potential intake rates (i.e. the intake rate that potentially can be achieved during the rest of the patch visit).

Modified from Dall *et al.* (2005) (B–C).

identify patch quality and can, therefore, not adopt McNair's 'flexible GUT rule'. Such naïve foragers should find some way to integrate the information that their foraging success contains, so that they can estimate patch quality (i.e. prey density) while feeding and determine a good moment to give up the patch. They would do really well if they could combine such 'patch sample information' with an *a priori* expectation about prey density, based on the environment-wide frequency distribution of prey densities, i.e. if they updated their foraging information (Oaten 1977, Iwasa *et al.* 1981, McNamara 1982, Olsson and Holmgren 1998, Green 2006, Olsson and Brown 2006, van Gils 2010). The patch-sample information need not be complicated; it only needs the total number of prey found in a patch and the total search time elapsed (Iwasa *et al.* 1981; Fig. 71A). In most natural environments, where prey densities follow a clumped frequency distribution, the perceived prey density at which such updating foragers give up their patches would not be constant, but rather an increasing function of patch exploitation time (Fig. 71B). This is because, in such environments, every encounter with a new prey makes it more likely that there is more prey, and thus perceived prey density increases with each consecutive prey encounter. Therefore, updating foragers should be more patient during the initial phase of depleting a patch and 'wait for good news' about patch quality in the form of prey encounters. If they do so, they would leave patches at a constant *potential* intake rate (Fig. 71C), where the potential intake rate is the intake expected over the remainder of the patch occupation (Olsson and Holmgren 1998). Everything comes to those who wait.

Do updating animals really exist?

If animals can combine prior information about the environment with new information about a current option (be it a patch, a prey, a mate, or whatever), they can markedly improve their assessment of the value of the current option, and thus improve fitness (Olsson and Brown 2010). This is the essence of 'Bayesian' updating, named after Thomas Bayes (1702–61), an Anglican priest interested in probability theory. Just to give you a simple example of Bayesian updating in practice: imagine a patch in which a forager has found two prey items during the first minute of search. Should it continue searching or should it leave this patch? If the environment is structured such that a patch can only contain two prey items at most, then the forager should definitely move on. By contrast, if patches can contain many more than just two items, then the forager should stay, especially if it took little time to find these two items. Thus, knowing your environment (in a statistical, probabilistic sense) greatly improves your assessment abilities; without knowledge about your environment, it is much harder to make the right choices. A popular example of Bayesian updating in humans is the so-called 'Monty Hall problem' (Selvin 1975), named after an American quizmaster. Imagine you are participating in a TV game in which you have to select one out of three closed doors (Fig. 72). Behind one of these doors stands a car, behind each of the other two doors stands a goat. You will take home either a car or a goat, depending on which door you choose. Once you have selected your door, the friendly quizmaster, who knows what stands behind each door, helps you by opening one of the other two doors for you, with a goat behind it. Knowing this, or, in Bayesian terms, *updating your prior expectation with new sampling information*, should you switch to the other closed door? The answer is yes; by switching, your chance of winning the car increases from 1 in 3 to 2 in 3. The intuitive, but incorrect, answer is that there is no need to switch doors, since your chances would only increase from 1 in 3 to 1 in 2, irrespective of whether you switch or not!

Since the seminal paper on Bayesian foraging by Oaten (1977), relatively few tests of this rapidly expanding family of models exist. Nevertheless, a recent review paper strongly suggests that animals *are* capable of Bayesian updating: in ten out of the eleven studies that were compared, behaviour was consistent with Bayesian-updating predictions (Valone 2006). Most tests were performed on foraging animals, but there is also the case of female peacock wrasses *Symphodus tinaca* searching for males (Luttbeg and Warner 1999). Even estuarine amphipods *Gammarus lawrencianus* use Bayesian updating when selecting their mates (Hunte *et al.* 1985). As echoed by Valone (2006), the strongest support comes from studies in which prior knowledge about the environment is experimentally manipulated, such as in the recently published work on Bayesian bumblebees *Bombus impatiens* (Biernaskie *et al.* 2009). In this study, bees were each offered patches of 12 flowers, of which five flowers contained nectar (a 30% sucrose solution). The bees were fooled by the other seven flowers, as they contained only water, and no sugar. Optimal foraging in such a uniform environment would indicate a greater tendency to leave the 12-flower patch with each successive encounter of a nectar flower (Fig. 73A), because each encounter means that the patch will harbour one nectar flower less in the future, as the bee empties each encountered nectar flower. And that is indeed what they do (Fig. 73B).

Now comes the evidence that the bees were indeed Bayesian bees: when offered a different environment, with highly variable patches (in which the 12 flowers contained either one or nine nectar flowers), the bees switched policies according to the theory. The tendency to stay in the patch should (Fig. 73C) and did (Fig. 73D) peak when encountering a second

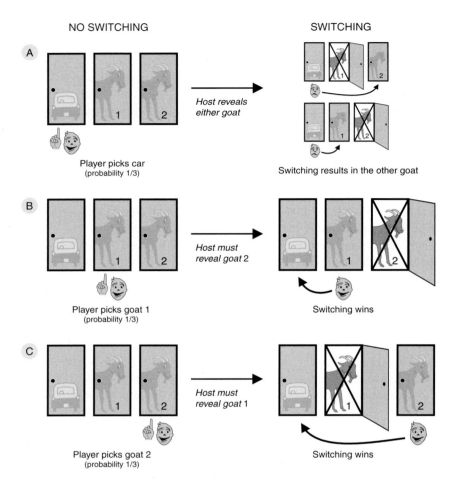

NO SWITCHING   SWITCHING

A   Host reveals
either goat

Player picks car
(probability 1/3)

Switching results in the other goat

B   Host must
reveal goat 2

Player picks goat 1
(probability 1/3)

Switching wins

C   Host must
reveal goat 1

Player picks goat 2
(probability 1/3)

Switching wins

Figure 72. How to win a car that is behind one of three closed doors? Pick a door, and you will have a 1 in 3 chance of finding a car behind your door, provided you do not switch (three cartoons on the left). If you decide to switch after the quizmaster reveals one of the two goats (three cartoons on the right), your chance of winning the car increases to 2 in 3. The crucial point is this: when the quizmaster opens a door to reveal a goat, this does not give you any new information about what is behind the door you have chosen, so the chance of there being a car behind a different door remains 2 in 3; therefore, the chance of a car being behind the remaining door must be 2 in 3.

Modified from http://en.wikipedia.org/wiki/Monty_Hall_problem.

nectar flower in that patch, since from then on the bee can be sure that it is exploiting a rich patch containing nine nectar flowers (note that because of depletion, every subsequent encounter with a nectar flower in the same patch reduces the potential value of the patch). The bees learned about the new environment in a matter of hours; a good example of the effectiveness, speed, and flexibility with which a phenotype can respond to a changing environment through behaviour!

Red knots can also behave like Bayesians, as revealed by the four captive birds exploiting depletable patches in our large outdoor intertidal aviary (van Gils *et al.* 2003b). Though we did not manipulate their prior environmental expectations, we were able to distinguish between the best rule for Bayesians, in which

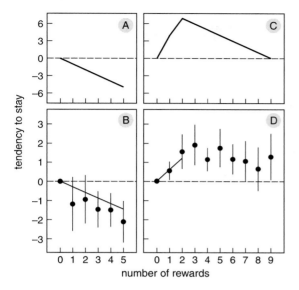

Figure 73. Tendencies to stay in a patch of flowers related to the number of nectar flowers encountered within that patch: (A) as expected for the uniform environment; (B) the observed tendency decreases with each successive encounter; (C) as expected for the highly variable environment; (D) the observed tendency peaks at the second encounter.

Modified from Biernaskie *et al.* (2009).

they 'wait for good news' and allow low estimates of current prey density during the initial phase of patch exploitation (dots in each of three graphs in Fig. 71; the so-called 'potential value assessment rule'; Olsson and Holmgren 1998), and the second best rule for Bayesians, in which patches are left at a constant estimate of current prey density (dashed horizontal line in Fig. 71B; the so-called 'current value assessment rule'; Iwasa *et al.* 1981). The knots turned out to use the best rule, and thus to have the patience to stay long enough in seemingly poor patches that initially rendered nothing because of bad luck. By doing so, they left their patches at a constant potential intake rate (Fig. 67B as conceptualized in Fig. 71C), rather than at a constant instantaneous intake rate.

In a further remarkable study, lesser spotted woodpeckers *Dendrocopos minor* were shown

to forage like Bayesians. Ola Olsson and co-workers (1999) cut off 354 branches on which individual woodpeckers had been foraging for known times. By subsequently X-raying these branches, they were able to count the number of prey items remaining in them (the so-called 'Giving-Up Density' GUD; Fig. 74A). They found more prey items in those branches that were occupied longest (Fig. 74B), a diagnostic for Bayesian foragers adopting the potential assessment rule when exploiting clumped prey distributions (Fig. 71B). In fact, such positive correlations were not only found within individual birds, but also between birds. Realizing this, Olsson *et al.* then made an important step. They reasoned: if animals respond more or less optimally to the quality of their environment (e.g. travel distances, food availability), we can flip the optimality argument around and use the observed behavioural variation *between* woodpeckers to make inferences about the quality of their individual environments. High marginal intake rates at patch departure (equivalent to high GUDs) should, according to MVT, reflect high long-term average intake rates. The idea that optimal foraging theory could be used in a reversed manner had been suggested much earlier by Wilson (1976), but was only picked up after the seminal paper by Brown (1988). And it yielded sensible patterns too: woodpeckers leaving their branches at high GUDs did indeed raise more fledglings than woodpeckers leaving them at low GUDs (Fig. 74C).

The psychology of decision-making

Animals are not statisticians, and it is difficult to conceive that animals can calculate probabilities. Even comparing the least amount of patch-sample information, the number of prey captured and the cumulative search time, with the optimal stopping points may be a bridge too far (Fig. 71A). Nevertheless, we have just

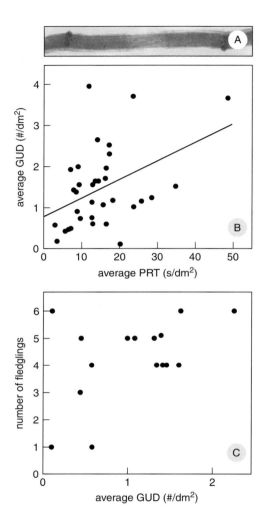

Figure 74. (A) X-ray image of a dead branch containing larvae (small, dark wormlike shapes) that remained after woodpeckers had searched for them. (B) Giving-up density (GUD) of larvae increased as a function of patch residence time PRT, with each dot denoting the averages per individual bird. (C) Breeding success, expressed as the number of fledglings per nest, increased as a function of an individual's average GUD.

Modified from Olsson *et al.* (1999).

reviewed some evidence for Bayesian decision-making in animals, so the obvious next question is how they do this. Have animals evolved simpler 'rules of thumb' that may approximate to the optimal policies 'invented' by theoretical ecologists (McNamara *et al.* 2006)?

When prey densities follow a clumped distribution among patches (i.e. many poor patches and rather few good ones), a Bayesian forager's estimate of the local prey density would decrease with elapsed search time, but would make an upwards jump each time a prey is encountered (Fig. 71B and C and Fig. 73C; or a fall, as in Fig. 73A in a uniform environment). This idea is based on purely mathematical models, but, interestingly, the same idea occurred in an early empirical paper in which the author speculated about the psychology of patch departure in a parasitoid wasp *Nemeritis* (Waage 1979; Fig. 75). Waage did not use the terminology of a forager's estimate of prey density, but rather of a forager's 'tendency to stay in the patch', a motivational variable in the terminology of psychologists (Geen 1995). In the words of Dick Green (1984), think of this tendency 'as being determined by a simple wind-up toy that originally is wound up enough to run for one unit of time; but each time a prey is found the toy is wound up enough to run for two additional units of time'. Evidence for such prey-induced increments/decrements in motivation has accumulated during the last few years, especially in the literature on parasitoid foraging (Wajnberg 2006, Biernaskie *et al.* 2009). In the form of survival analysis (notably Cox's proportional hazards model; Cox 1972), the incremental (or decremental) effect of each prey encounter can be detected and estimated in terms of decreases (or increases) in the chance of leaving the patch at a given time (Pierre and Green 2007). The realistic aspect of the proposed incremental/decremental mechanism is that it does not require counting abilities on the part of the animal: the forager can forget about previously encountered prey, it just needs to 'wind up its toy' upon catching a new prey item (for which it does not need much brain capacity; Holmgren and Olsson 2000). What it does require though, is a good sense of time, since the leaving tendency goes down over time. Animals do have an idea of time (Aschoff 1954), but how accurately can they assess the duration of short time intervals such as occur during patch exploitation?

Again it is starlings, those lab-rats of behavioural ecologists, who can provide some of the answers. Brunner, Kacelnik, and Gibbon (1992) wanted to know if animals were able to assess the duration of time intervals. So they trained captive starlings to reproduce time intervals behaviourally. In the forage-and-travel cages discussed earlier (Fig. 66C), the birds could earn food by pecking at a so-called 'pecking key'. However, food always came at fixed time intervals (Fig. 66A), and it made no sense to peck before this interval had passed. Each patch ended with a sudden depletion, which was made unpredictable by offering a variable number of food items per patch. Brunner *et al.*

measured the pecking intensity after the last prey of the patch had been offered, and showed that the starlings had a good sense of time. Pecking intensity peaked exactly when the food-item-that-never-came should have come (Fig. 76). This result was upheld for different interval durations (Fig. 76), which provided insight into the underlying mechanism.

The most popular mechanistic view of timing in animals is presented by the so-called 'scalar expectancy theory' (SET; Gibbon 1977, Shettleworth 1998, Malapani and Fairhurst 2002). This assumes that a time interval is perceived, encoded, and then compared to a stored representation of similar intervals. However, the memory for these similar intervals is imperfect, as the animal draws a random sample from a *distribution* of values centred on the currently perceived interval. SET predicts that the variance of the memorized distribution increases with the length of the perceived time interval, but in such a way

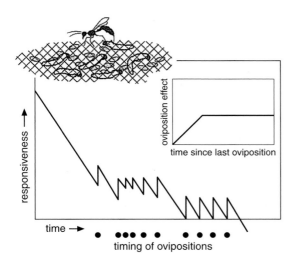

Figure 75. The proximate mechanism of patch-departure after oviposition in the parasitoid wasp *Nemeritis* as envisioned by Waage (1979). The responsiveness or tendency to stay declines over time, but increases with each encounter with a host. How much it increases, depends in a simple manner on the time since the last oviposition (inset). When the responsiveness drops to zero, the wasp leaves the current patch.

Based on Waage (1979).

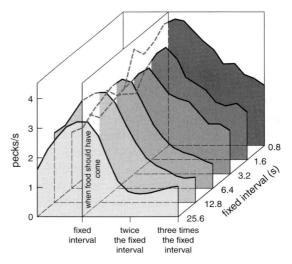

Figure 76. Starlings have a good sense of time. In an indoor experiment where they had to peck for food, average pecking intensity peaked when the food item that never came should have come, and this result was repeated for different fixed intervals.

Based on Brunner *et al.* (1992).

that the coefficients of variation of all stored distributions remain constant. Indeed, in the starling experiment, the coefficient of variation did not change with the duration of the time interval (Fig. 76). Note, however, that competing alternative ideas about timing in animals do exist. For example, support for a mechanism called 'learning to time' (LeT) has been found in the psychologist's lab rats—pigeons (Machado and Keen 1999). Irrespective of the precise underlying mechanism, the message here is clear: animals can keep track of time quite accurately.

Ideal birds sleep together

In this chapter we have so far considered foraging animals as solitary individuals living in uncertain environments in which other individuals play no role. However, the opposite is often the case: individuals of many species feed together, travel together, roost together, and breed together. Although there are obvious costs involved in living gregariously—think of competition for space, food, and mates—there are also many benefits, and improved information gain is probably one of them (Danchin et al. 2004, Bonnie and Earley 2007, Seppänen et al. 2007, Valone 2007). For example, think of the Bayesian bird using its own foraging success ('personal information') to assess patch quality. If this very same animal shares this patch with others and is able to keep track of their success too ('public information'), it may reach much better estimates in a much shorter time (Valone 1993, Valone and Giraldeau 1993, Templeton and Giraldeau 1996, Smith et al. 1999). Scaling things up, using public information may be essential to keeping track of continuously changing environments.

Let us stick to foraging: over time (days, weeks), food stocks change due to prey growth, reproduction, emigration, and mortality. In a large home-range, there is no way a solitary forager can keep up with all these dynamic resource changes just on its own. What may be a good patch today, can be a poor patch tomorrow, and where to find good alternatives in that case? Joining knowledgeable individuals on their daily foraging trips will greatly enhance survival probabilities in temporarily dynamic environments, and a communal roost or a breeding colony may be a good starting point for naïve foragers opting for a 'guided food tour' (Ward and Zahavi 1973). Think about scavenging birds, such as ravens or vultures, feeding on highly ephemeral food 'patches' that may only be around for a day or two. These birds, particularly ravens Corvus corax, have been known actively to recruit conspecifics from a central roost upon encountering a sizeable carcass (Wright et al. 2003). But roosts may also be places where plenty of 'passive information' is at hand: by-products of the performance or activities of others may mirror the quality of the resources having been fed upon. For example, fatness, plumage quality, or roost arrival time may be indicative cues that are passively being advertised (so-called 'inadvertent social information'; Danchin et al. 2004), but may also provide the naïve individual with useful information on the past foraging success of others (Bijleveld et al. 2010). Red knots roost together too: thousands of individuals get together on the shore whenever the incoming tide floods their muddy foraging grounds. The following story suggests that they too may greatly benefit from each other in finding their hidden food.

The null model of animal distribution, the so-called 'ideal free-distribution' model (Fretwell and Lucas 1970), assumes ideal foragers, meaning that they are omniscient with respect to the spatial distribution of their food. They thus know where to go, and pay no price in getting there (the latter referring to the free aspect of the model). Both assumptions are of course over-simplifications, but the model has been very useful as a stepping stone in studies of interference competition (van der Meer and Ens 1997). We wanted to see how far this model would bring us when trying to explain where red knots would feed in our 'backyard' study area, the Wadden Sea (van Gils *et al.* 2006c).

The western part of the Wadden Sea (Fig. 77A) is home to about 40,000 *islandica* knots, which feed during low tide on the intertidal

mudflats in between the mainland and the barrier islands (Kraan *et al.* 2009). At high tide, these birds get together at Richel, a wide, open sandflat of about 1 km² in the mid-north of the area (Fig. 77A). During five consecutive late summers (1996–2000), we caught red knots at this roost, radio-tagged them, and followed their whereabouts throughout our study area. At the same time, we mapped their food resources in great detail by sampling each

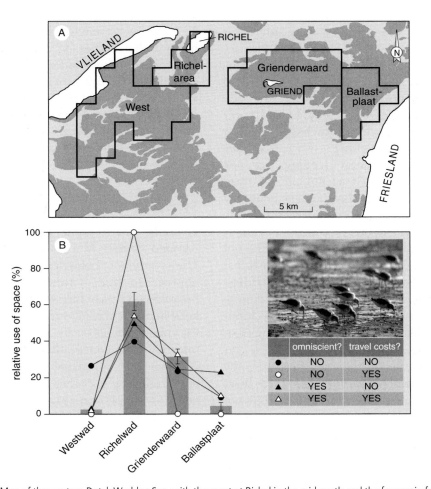

Figure 77. (A) Map of the western Dutch Wadden Sea, with the roost at Richel in the mid-north and the four main feeding grounds distinguished (West, Richel-area, Grienderwaard, and Ballastplaat). (B) The use of these feeding grounds by red knots, predicted by the four different scenarios (symbols connected by lines), and as observed by radio-telemetry (grey bars, including standard errors). Modified from van Gils *et al.* (2006c).

year the macrozoobenthos at nearly 2000 stations laid out in a grid at 250-m intervals. Combining measured tidal heights with fine-scaled elevation data, we could predict the daily exposure time for each station. Food densities and exposure time were the main input variables in a so-called 'stochastic dynamic programming model' in which we simulated red knot movements. We considered four possible scenarios. If knots had no knowledge about their environment but were to travel for free, they would distribute themselves randomly, only respecting the differential availability of mudflats due to differences in elevation (filled dots in Fig. 77B). If, instead, these naïve birds travelled at a cost, they would not get too far from the roost and all hang around near Richel during low tide (open dots in Fig. 77B). If they 'would know better', have complete knowledge of their 'foodscape', and travelling was for free, they would go to the best feeding grounds (Ballastplaat), as much as the tide allowed them (this particular mudflat lies relatively low and is therefore exposed relatively briefly; closed triangles in Fig. 77B). On the other hand, if such knowledgeable birds did not travel for free, they would get to Ballastplaat much less often, as this site is located furthest from the roost, and would instead spend more time feeding nearby (but skipping the food-poor area called 'Westwad'; open triangles in Fig. 77B).

The last prediction matched best with the movement data of our radio-tagged birds (bars in Fig. 77B): knots seem to be ideal, but not free. It was no surprise to us that they were not free, since they have to pay a price for travelling, but we did not expect that the fit with the ideal/omniscience prediction would be so close. With a highly dynamic food supply that is consumed by many, grows fast, and hides in the mud, it is out of the question that only personal sampling information can keep knots up-to-date. They must somehow exchange public information, be it actively or passively, and their central roost is likely to play an important role as an information centre (Bijleveld *et al.* 2010).

With this report from the Wadden Sea, we have set the scene for the next chapter(s), in which 'Wadden Sea knots' will feature again. There will be one important difference, however. So far we have assumed that all knots have similar bodies. In the next chapter, we will focus on these individual red knots, and we will see that different individuals do different things. Much of this behavioural variability can be explained by their flexible bodies, bodies that we 'looked into' before releasing them once again into their wide open windy landscape.

Synopsis

Behaviour is a rapid way for animals to respond flexibly to a changing environment. The study of behaviour has often been approached from a functional angle, which has allowed researchers to make clear-cut, falsifiable predictions about behaviour. Knowing the constraints faced by an animal (including informational constraints), and assuming a currency that approximates fitness, researchers can qualify decisions as *optimal* or *suboptimal*. This chapter provides a stepping stone to the next chapter, where we adopt this successful optimality approach to interpret and predict flexible organ size changes within shorebirds.

CHAPTER 7

Optimal foraging: the dynamic choice between diets, feeding patches, and gut sizes

Eating more by ignoring food

In the previous chapter we demonstrated the intellectual elegance of behavioural optimality reasoning, a concept that enables ecologists to make stepwise scientific progression by means of Popperian falsification of quantitative predictions. In this chapter we will show how our foraging work on red knots was built on this approach, step by step. Initially we thought that the standard optimal-diet model of behavioural ecology would be sufficient to explain the composition of the menu of red knots functionally. However, this proved not to be the case, and only after including the process of digestion into the diet model were we able to explain diet selection. This in turn led to a deeper understanding of the knots' related selection criteria, notably of the habitats they made use of, and the strong differences between different phenotypes in these selection criteria. Not only could we then understand why different phenotypes responded differently to the different food patches on offer, but also how the idea worked in reverse—the food supply moulding the phenotypes. Elaborating these observations will be the aim of this chapter. But let us begin first with a brief review of the standard diet model that we abandoned so soon after embracing it.

Back in 1966, Robert MacArthur and Eric Pianka developed a simple but effective model that showed how a forager, trying to maximize its long-term average food-intake rate, should refrain from consuming every prey item that it encountered. A decade later, the model was refined by others (e.g. Pulliam 1974, Werner and Hall 1974, Charnov 1976b), but the general structure and predictions remained unchanged. The basics are these: a forager can either search for prey, or handle an encountered prey item, but it cannot do both things at the same time. Therefore, handling a prey item carries a time cost, as searching has stopped for a moment. Sometimes it may be more fruitful to neglect an encountered prey item and to continue searching for a better one. This is the case when the 'intake rate while handling the item' is less than the maximum possible intake rate averaged over a long period of foraging. The 'intake rate while handling', usually referred to as 'profitability', is the key parameter in the model and equals the ratio between the energy content of a prey item and the required handling time. The main prediction of this simple diet model is thus clear cut and intuitively appealing: in rich environments, where high long-term average intake rates can be achieved, foragers should be more selective; in poor environments, diets should be broader. Its simple elegance also appealed to the godfather of 'behavioural ecology', Lord John R. Krebs, who tested this model in captive great tits *Parus major*.

Krebs *et al.* (1977) offered their tits segments of mealworms as prey. These came in two forms: long pieces (eight segments long) and short pieces (four segments). By adding a small piece of plastic tape to the short 'worms', the handling times of both types were kept more or less similar (the birds had to keep this prey under one foot to peel off the tape). In this way, the profitability of each type varied by a factor of two—either four or eight segments of energy per prey. The prey items were offered sequentially by putting them on to a conveyer belt one-by-one. The eight-segment long—most profitable—prey passed by at three different rates. The crux of the experiment is that only at the highest encounter rate would the tits' long-term intake rate be above the profitability of the shorter worms. Thus, only at this highest delivery rate should the birds ignore the shorter worms. And that is indeed more or less what they did. At the two lower prey encounter rates the birds were not selective and took small and large worms in equal proportions, while at the highest encounter rate the birds

ate many more large ones than small ones (Fig. 78). However, even at this high intake rate, they did not ignore the smaller worms completely. Krebs and co-workers suggested that the birds were actually sampling the smaller prey every so often as it passed by because they needed to test its profitability. Such sampling was an activity not included in the original model.

This study by Krebs *et al.* is one of the very many examples that show that the classical optimal-diet model is a 'strong' model, a model that has successfully been 'confirmed' many times since its conception in 1966 (reviewed by Schluter 1981, Stephens and Krebs 1986, Perry and Pianka 1997, Sih and Christensen 2001). Perry and Pianka (1997) conclude that the narrowing down of a forager's breadth of diet with an increase in food abundance 'may be optimal foraging theory's most robust theorem to date'. On the other hand, the classical optimal-diet model has also been refuted many times. Sih and Christensen (2001) conclude that empirical studies on *immobile* prey yield results that are usually in line with the predictions of the model, whereas studies on *mobile* prey often refute them. Perhaps this should not have surprised us. An important aspect of prey behaviour, the possibility of escaping when encountered, is not dealt with in the model, and thus the standard model should not have been tested in cases involving mobile prey. We should have realized that the standard model captures the trade-off between wasting time and gaining energy for a specific situation, and cannot be applied directly to all possible foraging scenarios. As it was, with red knots feeding on highly sessile mollusc prey, we initially expected a close match between the standard theory and our observations. In fact we found the opposite: a clash between the theory and the facts. Could the prey's hard-shelled armour be the possible explanation for these discrepancies?

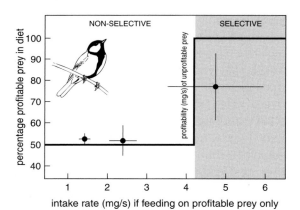

Figure 78. In an experimental situation, profitable and unprofitable prey were offered to great tits. The tits largely obeyed the optimal diet hypothesis by being non-selective under relatively low intake-rates, whilst selecting (mostly) for the profitable type under relatively high intake-rates. The thick stepwise line and the shading give the theoretical prediction, where the shift from being non-selective (non-shaded) to being selective (grey shaded) occurs when the intake rate on the profitable type only would be equivalent to the profitability of the unprofitable type, which is just over 4 mg/s. Dots are means and error bars give 95%-confidence intervals.

Modified from Krebs *et al.* (1977).

A hard nut to crack

Red knots are mollusc-specialists (see Chapter 1), but, remarkably, molluscs are not their favourite meal. If captive knots are offered the choice between 15-mm long hard-shelled cockles *Cerastoderma edule* and 50-mm long shrimps *Crangon crangon*, they will go for the shrimps (van Gils *et al.* 2005a). Only after depleting the tray of shrimps will the birds switch to the cockles. With only MacArthur and Pianka's (1966) diet model in mind, this diet choice cannot be explained, as a shrimp is more energy rich (1–2 kJ of metabolizable energy) than a cockle (less than half a kJ) but takes proportionally more time to handle (more than half a minute rather than the five seconds or less required to handle a cockle). This makes shrimps half as profitable (40–50 Watts) as cockles (80–90 Watts). Knots should like cockles better than shrimps! Why does this not seem to be the case?

The answer lies in the knots' habit of ingesting their prey whole, even shelled prey. Because of this, they need to crush their food in their guts, and they do this in their muscular stomach, the gizzard, before digestion can commence (Piersma *et al.* 1993b). The crushing and the subsequent processing of the shell material take time. In fact, in an experimental setting where knots were offered no choice, but only one prey type at a time in large quantities, we found that this time-cost scaled directly with the amount of indigestible shell material in any prey item: the more shell material a prey contained, the longer the retention time in the digestive system (van Gils *et al.* 2003a). This constraint turned out to be very robust: whichever prey type that we offered, our hungry knots, 'with eyes bigger than their stomach', could not ingest their prey faster than a rate of 2.6 mg of dry shell material per second (Fig. 79). Nevertheless, this ultimate processing rate need not necessarily constrain

the intake rate in a field situation. Here knots need to search for their food rather than to pick out items from a tray. Digestion can take place during foraging, and as long as food is found at a slower pace than the digestive system can process it, there is no digestive constraint (Jeschke *et al.* 2002). However, when prey items are found and handled faster than they can be processed (as in 'luxurious' experimental settings), knots face a digestive bottleneck. In this case there is a premium on picking out those food items that contain the highest amount of metabolizable energy per unit shell material. The cockles in the choice-experiment had about 400 mg of shell mass each; the large shrimps had only 80 mg of indigestible carapace mass. This makes the ratio of metabolizable energy per mg ballast material, the so-called 'digestive quality', 20 times higher for shrimps (1500 J per 80 mg) than for cockles (350 J per 400 mg). This potentially enables

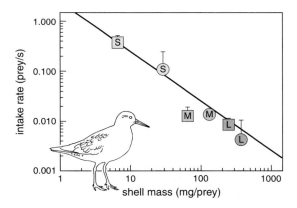

Figure 79. No matter whether a prey type contained a little or a lot of shell mass per item (horizontal axis), the knots' intake rate (vertical axis) would be constrained by a constant maximum rate of processing shell mass of about 2.6 mg/s (fitted regression line with a slope of −1 for log-transformed axes). Six different prey-types were offered: Baltic tellins *Macoma balthica* (squares) and cockles *Cerastoderma edule* (dots), both in three different size classes (S = small; M = medium; L = large). Symbols give means, bars give standard errors.

Modified from van Gils *et al.* (2003a).

knots to metabolize energy from shrimps (2.6 mg per second × 1500 J per 80 mg = 48.8 Watts) at a rate that is 20 times faster than from cockles (2.6 mg per second × 350 J per 400 mg = 2.3 Watts). Are you still following us? We have just given you the explanation for the knots' preference for shrimps over cockles when either come in excess quantities. Our observations finally knocked the classical diet model on the head, at least in situations where knots were eating invertebrates (van Gils *et al.* 2005a).

If knots like shrimps so much better than cockles and other bivalves, why have they become mollusc-specialists and eat cockles and other hard-shelled bivalves most of the time? Why are they not shrimp-specialists or, more generally, crustacean-specialists? Red knots are sensorily specialized to detect hard-shelled prey in soft wet sediment (Piersma *et al.* 1998), but the answer lies in a crucial difference between our laboratory studies and the field situation. In the wild, food does not occur in excess quantities. It is not laid out conspicuously in plastic trays, but scattered and hiding in the sediment. Knots need to search for it. What is more, shrimps are relatively rare, while cockles and other low-quality molluscs are relatively abundant. In addition, shrimp patches contain rather few cockles and typical cockle patches harbour low shrimp densities (Fig. 80). For this reason, patch choice comes before prey choice: if knots want to feed on shrimps, they need to go somewhere other than if they plan to feed on cockles. Due to the relatively low abundance of shrimps and other high-quality crustaceans (Fig. 80C), knots' intake rates when feeding in a typical shrimp-patch are constrained by the low density of the shrimps (grey dot in Fig. 81). In contrast, due to the relatively high abundance of cockles, and due to their heavy shells, knots' intake rates when feeding in a typical mollusc-patch are constrained by the capacity to process shell

material digestively (black dot in Fig. 81). Because knots have such a large gizzard (relative to body mass their gizzards are among the largest of any shorebirds; Battley and Piersma 2005), this capacity is relatively high (2.6 mg of

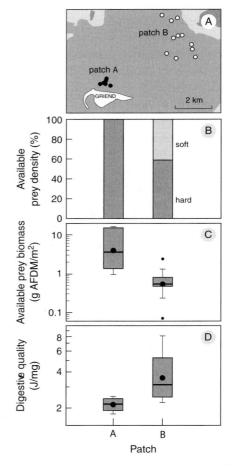

Figure 80. (A) Hard and soft food is often found in different localities, exemplified here for the late-summer situation in 1998, near the isle of Griend, western Dutch Wadden Sea. (B) Virtually all prey in patch A were hard-shelled, while in patch B about half were soft bodied. (C) Patch A contained an order of magnitude more ash-free dry mass (AFDM) of food per square meter than patch B, (D) but of much lower quality, largely due to the absence of soft-bodied crustaceans. Box-and-whisker plots in (C) and (D) give mean (large dot), median (horizontal line within box), inter-quartile range (box), range (bars) and outliers (small dots). From van Gils *et al.* (2005b).

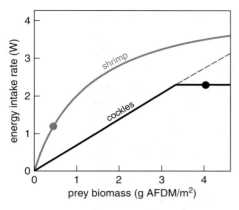

Figure 81. Functional responses of knots to densities of shrimp (grey line) and cockles (black line). Cockles usually occur in much higher densities than shrimps. Therefore, long-term intake rate in a typical cockle patch (black dot) is higher than in a typical shrimp patch (grey dot), even though the intake rate in the cockle patch is constrained by the digestive processing rate (black horizontal line as opposed to the black dotted line, which would be the intake rate of cockles without a digestive constraint; note that in the latter case the asymptote of the 'cockle functional response' would exceed the asymptote of the 'shrimp functional response', as cockles are more profitable than shrimps). Scaling of functional responses is based on experimentally determined parameters.

dry shell material per second). This means that the digestively constrained intake rate when they are feeding on cockles and other molluscs is usually higher than that when they are feeding on high-quality crustaceans, as the latter is constrained by the density of the crustaceans (Fig. 81). This would explain why knots under field conditions usually ignore the soft food that they like so much in the laboratory.

However, there is yet another element in this fascinating story! Each year during late summer, the time when knots return from their tundra breeding grounds to the Wadden Sea, we find flocks of knots that *do* feed in the relatively poor shrimp- and crab-patches. These birds seem to ignore the much more abundant cockles and other molluscs. In order to explain this, we need to call organ flexibility into the equation at this point.

It takes guts to eat shellfish

Every year, during the late-summer arrival period, we catch red knots at Richel, their main high-tide roost in the Dutch Wadden Sea. The usual aim is to colour-ring them in order to estimate year-specific survival rates (see Chapter 9). On some of these occasions, we also glued tiny VHF-radios (1.3–1.8 g) to the back of individual knots. This enabled us to ascertain which individuals visited the shrimp- and crab-patches and which did not.

Initially, we found no patterns in the behaviour. In terms of body size and body mass or sex, the 'odd' crustacean-feeders were indistinguishable from their shellfish-feeding counterparts. However, things became clearer when we began to use ultrasonography. By measur-

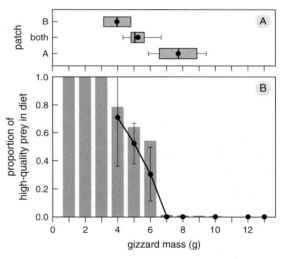

Figure 82. (A) The hard-shelled-prey patch A (Fig. 80A) was visited by knots having a large gizzard, while the soft-bodied-prey patch B was visited by knots having a small gizzard. Birds that alternated between both patches had intermediate gizzard sizes (for explanation of box-and-whisker-plots see Fig. 80). (B) As predicted by the Digestive Rate Model (grey bars, see p. 138), and as empirically confirmed by dropping analyses (dots and error bars, representing means and standard errors), the proportion of soft-bodied prey in the diet declined as a function of the knots' gizzard mass.

From van Gils *et al.* (2005b).

ing gizzard width and height ultrasonographically, we were able accurately to assess gizzard mass in live knots (Dietz *et al.* 1999). This turned out to be a methodological breakthrough. These measurements revealed that the birds with the smallest gizzards flew to the crustacean-patches, birds with the largest gizzards visited the cockle banks, whereas birds with intermediate gizzard masses alternated between both (Fig. 82; van Gils *et al.* 2005b)! We had discovered phenotype-related habitat selection. However, these observations did not yet explain *why* crustacean-patches were visited (in fact, we simply reasoned that feeding on hard-shelled food is always more beneficial). Moreover, it left us with a new issue, a true 'chicken-and-egg problem'. Did the soft-food feeders eat soft food because their gizzards were so small, or were their gizzards so small because they ate soft food?

To answer this question, we needed to go back to the laboratory. We already suspected that gizzards become smaller when birds eat food that is easy to process. For example, our captive knots are kept on a diet of artificial, protein-rich trout-food pellets, a form of soft food that does not require crushing. Dissecting some of those captive birds revealed gizzards of around 3 g, while the average knot in the wild has a gizzard that is three times as large (Fig. 83; Piersma *et al.* 1993b). Nevertheless, the proof of the pudding is in the eating! Thus, we performed a controlled experiment to test our 'gut feeling' about food affecting gizzard size. By alternately offering captive knots soft and hard food, and monitoring their gizzards ultrasonographically, we found that the gizzards varied in mass synchronously with the dietary regime: over a period of about a week's time they became smaller when birds fed on soft food and became larger when birds ate hard-shelled food (Fig. 84; Dekinga *et al.* 2001). Did some knots develop a small gizzard simply because they happened to feed on soft

food? This may be the case, but it did not seem to make sense from an optimality point of view, since intake rates are much higher when feeding on abundant hard-shelled food than on scarce crustaceans (Fig. 81). However, these calculations were made for birds with large gizzards. If knots with a small gizzard differ in their digestive processing rates from knots with a large gizzard, then consumption of soft food may make sense after all.

Their capacity for phenotypic change could now be used in an experiment. By varying the hardness of the food on offer we experimentally manipulated gizzard size in order to study the functional consequences in terms of digestive capacity. A wonderful tool that had been so successful in behavioural ecology, the experimental manipulation of a behavioural choice (see Chapter 6), could

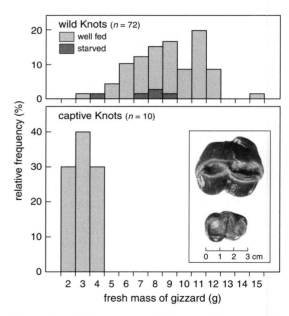

Figure 83. Wild knots (top graph) have much larger gizzards than captive knots (bottom graph), even if the former have starved to death in severe winter weather (darker bars in top graph). Inset in bottom panel shows cross-sectioned gizzards of a 130-g wild (top) and a 125-g captive (bottom) knot.

From Piersma *et al.* (1993b).

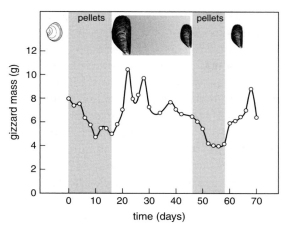

Figure 84. Flexible changes in the gizzard of captive red knots that were offered hard and soft food alternately: coming straight from the field where the birds ate hard-shelled bivalves, they were offered soft trout pellets for the first 2½ weeks of captivity, which halved their average gizzard size. Then they were offered hard-shelled mussels *Mytilus edulis* for more than a month, during which gizzard mass increased back to field dimensions (matching our expectations, gizzards were largest when the birds were offered large mussels, which have lower digestive qualities than the small mussels offered later). This pattern was repeated throughout the rest of the experiment: gizzard atrophy when given pellets and gizzard hypertrophy when back on mussels again. These repeated measurements within the same individuals could only be done with the use of ultrasonography.

From Piersma and Drent (2003).

now be applied to manipulate a 'physiological choice', an individual's gizzard mass. Thus, we manipulated gizzard mass in a few captive knots to obtain two groups: birds with small gizzards and birds with large gizzards. We then offered *ad libitum* hard-shelled food to these groups, one type of prey at a time, but in large, excess quantities. The birds were kept hungry, which motivated them to feed at maximum rates.

The results were very clear cut. Maximum feeding rates were one order of magnitude lower for knots with small gizzards than for knots with large gizzards (Fig. 85A; van Gils *et al.* 2003a). Processing shell material was clearly the bottleneck, as indicated by the fact

that, across different prey types varying in the amount of shell mass per item, the amount of shell mass processed remained constant within each 'gizzard-group' (at 0.24 mg/s in the small gizzards and 2.58 mg/s in the large gizzards; Fig. 85A). The result was upheld when allowing small-gizzard birds to develop a large gizzard and *vice versa*. Between individuals,

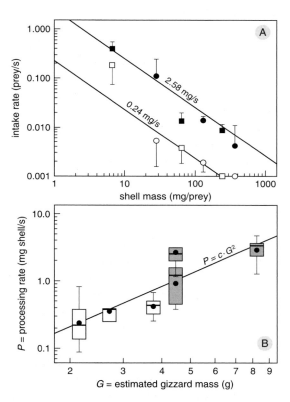

Figure 85. (A) Intake rate is constrained by the rate of processing shell material, with the latter differing tremendously between large and small gizzards. Whereas birds in the large-gizzard group were able to process 2.58 mg of dried shell material per second (filled symbols), birds in the small-gizzard group could only process 0.24 mg per second (open symbols). Prey species and size classes as defined for Fig. 79. (B) Lumping the different prey species and size classes together, but plotting shell-mass processing rate against an individual's gizzard mass, we found a quadratic relationship between the two. Box-and-whisker-plots, as explained in Fig. 80, with grey-filled boxes for birds in the large-gizzard group and non-filled boxes for birds in the small-gizzard group.

From van Gils *et al.* (2003a).

the shell mass processing rate increased quadratically with an increase in gizzard mass (Fig. 85B). With this simple quadratic function to hand, we could now go back into the field and recalculate the intake rates of the different phenotypes in the different food patches.

For each possible gizzard mass, at each food patch, we calculated the maximum intake rate that a feeding knot could attain. When there are multiple prey types within a single patch, this maximum is predicted by the so-called Digestive Rate Model (DRM), which not only takes account of a prey type's abundance, but also of its digestive quality (see Hirakawa 1995 for details and van Gils *et al.* 2005a, 2005b and Quaintenne *et al.* 2010 for applications). Generally speaking, the model predicts that intake rate depends on both food density and

digestive capacity (Fig. 86A). At a given prey density (solid line in Fig. 86A), intake rate is constrained by rates of processing for small gizzards, and thus an increase in gizzard mass leads to an increased intake rate. However, above a certain gizzard mass, processing capacity is so high that it exceeds the rate at which food is found and ingested, and thus it becomes the food density that constrains the intake rate at large gizzard masses (thus obeying Holling's (1959) classical disc equation: Fig. 86A).

Now, when there are two types of patches, those containing low densities of high-quality food (crustacean patches) and those containing high densities of low-quality food (cockle patches), then knots with small gizzards will achieve their maximum intake rate in low-

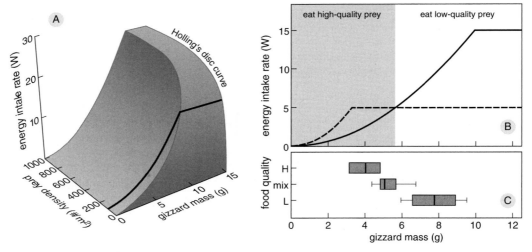

Figure 86. (A) The functional response of red knots potentially facing a digestive constraint. With an extremely large gizzard of 15 g, intake rate increases at a decelerating rate, as given by Holling's disc model. However, for smaller gizzards, the intake rate is only given by the disc model for low food densities. At higher food densities, intake is constrained by the rate of processing shell material, which increases quadratically as a function of gizzard mass. Solid line gives the dependence of intake rate on gizzard size, e.g. food density of 100 items per m². (B) Energy intake rate as a function of gizzard mass in two different patches: a high-density patch, but offering low-quality food (solid line, the same as the solid line in (A)) and a low-density-but-high-quality patch (dashed line). This shows that rate-maximizing knots should select the high-quality food when their gizzard is small (< 6 g), but should switch to low-quality food when their gizzard is large (≥ 6 g). (C) As we have seen before (Fig. 82A), this is exactly what our radio-marked knots did.

Modified from van Gils *et al.* (2005b).

density/high-quality patches, while knots with large gizzards will achieve their maximum intake rate in high-density/low-quality patches (Fig. 86B). Put simply, this is because knots with large gizzards hardly ever face a digestive constraint. For them food quality is unimportant, it is the food quantity that matters. For small-gizzard knots facing a stringent digestive bottleneck, it is the other way around: quality matters, quantity does not (above some threshold minimum density). Thus, although food choice will determine gizzard size, it is now clear, from a functional perspective, that gizzard size also determines food choice (Fig. 86C).

Even though knots with small gizzards attain their maximum intake rate at soft-food patches, their intake rate is still lower than in knots with large gizzards feeding in hard-shelled mollusc patches (Fig. 86B). To compensate for this lower intake rate, knots with small gizzards are predicted to feed longer on a daily basis. And indeed they do, as we found out by

timing the daily departure and arrival from/at the central roost of our radio-tagged knots (Fig. 87A; see Fig. 87B on *how* they do this). Besides the risk of not finding enough high-quality food, such long working days are an obvious cost of having a small gizzard (see Chapter 4 for a discussion of the long-term survival costs of hard work). Therefore, one obvious new question cries out for an answer: if having a small gizzard is so costly, and if gizzard size is a flexible trait that individuals can fine-tune themselves, why do some individuals 'decide' to have a small gizzard?

The answer has to do with the migratory lifestyle of knots. Because the costs of flight increase with body mass (Kvist *et al.* 2001), long intercontinental non-stop flights are likely to be cheaper when unnecessary body weight is 'thrown overboard'. Indeed we found that, in the days before take-off for a long flight, and in the days just after their return, digestive organs in knots are relatively small (Piersma *et al.* 1999a, 1999b). This also occurs in other

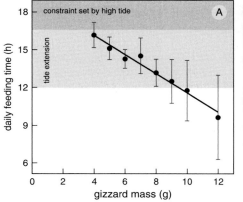

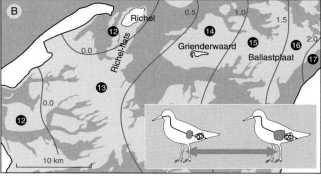

Figure 87. (A) The smaller their gizzard, the longer the radio-marked knots were away from their roost at Richel, feeding on the mudflat instead (means ± standard errors). (B) In the western Dutch Wadden Sea, knots can extend their daily feeding period by 4–5 h beyond the usual 12 h by moving eastwards along with the outgoing tide (numbered dots; also tide-isoclines are indicated, expressed in hours delay per tidal cycle relative to the tide at Richel). Inset shows a knot with a small (left) and a knot with a large (right) gizzard.

Modified from van Gils *et al.* (2005b, 2007a).

migratory shorebirds (Piersma 1998, Piersma and Gill 1998), in waterfowl (van Gils *et al.* 2008), and in songbirds (Bauchinger *et al.* 2009).

Consequently we think that the inter-individual variation in gizzard masses, which we see at any given moment during the late-summer arrival period, is due to different individuals arriving in the Wadden Sea on different dates (Dietz *et al.* 2010). Thus, birds with small gizzards are likely to have just arrived and are in the process of growing a larger gizzard. It takes at least a week (under relaxed laboratory conditions; Dekinga *et al.* 2001) and up to three weeks (under harsher, natural, field conditions; Piersma *et al.* 1999a) to increase gizzard size from small to large. Large-gizzard knots are, therefore, likely to have been around for a couple of weeks already. For several reasons, departure dates from the arctic differ between individuals: males care for the brood, females leave right after hatching, juveniles need some extra weeks to complete their growth, and failed breeders leave before successful breeders (Tomkovich and Soloviev 1996). Thus, we do not think that there are typical small-gizzard knots and typical large-gizzard knots, but rather that timing differences between individuals are causing gizzard mass differences between individuals at any given time.

Optimal gizzards

We have now come full circle with the gizzard-food interplay (the 'chicken-and-egg problem'): gizzard size determines food choice, but food also determines gizzard size. The latter can now be understood functionally: if knots on a daily basis need to assimilate as much energy as they expend, and if there is a cost to carrying and maintaining a large gizzard, then it pays to maintain a gizzard that is just large enough to obtain the daily necessary amount of kiloJoules (a gizzard for which we

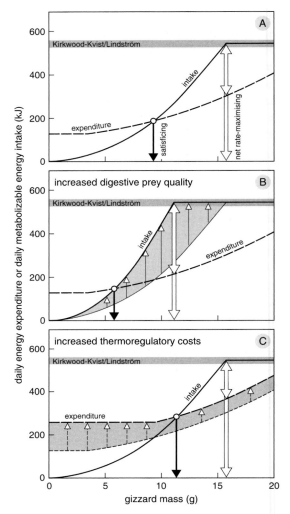

Figure 88. (A) A knot can optimize its gizzard size in two ways. First, by balancing energy income with energy expenditure (which both increase as a function of gizzard mass), leading to a 'satisficing' gizzard. Second, by maximizing the difference between energy income and energy expenditure, leading to a 'net rate-maximizing' gizzard. (B) An increase in the food's digestive quality will reduce the optimal size of both gizzard types. (C) By contrast, an increase in a knot's energy expenditure (e.g. through an increase in thermoregulatory costs) will enlarge the optimal satisficing gizzard, while leaving the rate-maximizing gizzard unaltered.

From van Gils *et al.* (2005c).

coined the term 'satisficing gizzard'; solid arrow in Fig. 88A). Soft, high-quality food, with little shell mass relative to the amount of meat, can be processed rapidly (solid arrow in Fig. 88B) and thus would require a smaller gizzard than hard, low-quality food (solid arrow in Fig. 88A). Thus, offering soft trout-food pellets yields small gizzards, while offering hard-shelled mussels yields large gizzards (Fig. 84; Dekinga *et al.* 2001). A proximate mechanism explained functionally.

This reasoning implies that we can think of gizzards as having an optimal size for each particular ecological context. They can be too small, thus not yielding enough daily energy, or they can be too large, thus costing too much energy to carry and maintain (van Gils *et al.* 2006a). Getting back to the optimality models that had worked so well in behavioural ecology (see Chapter 6), it occurred to us that we may be able to *predict* optimal gizzard sizes for different ecological scenarios, in which both food quality and metabolic requirements vary independently. This would imply that the smallest gizzards are best when food quality is high under low metabolic demands, while large gizzards are required in the opposite case (Fig. 88).

We added one extra dimension to these predictions. As knots are long-distance migrants, they do not always just need to balance energy income with energy expenditure. There are times of the year, when (net) income needs to be maximized, which is when the birds are fuelling up for migration. At these times, they may need a gizzard that is larger than a satisficing gizzard, a gizzard that is so large that it is possible for a knot to reach the upper limit to its daily metabolizable energy intake, a gizzard for which we coined the term a 'rate-maximizing gizzard' (open arrows throughout Fig. 88; this upper limit was determined by Kvist and Lindström (2003) when they offered easily digestible mealworms in excess quantities to captive fuelling red knots; see Fig. 43).

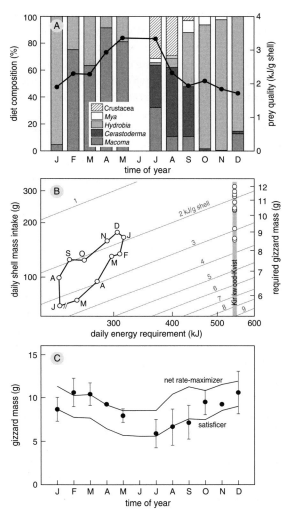

Figure 89. (A) Monthly composition (bars scaled on left axis) and quality (dots scaled on right axis) of the diet of knots wintering in the Dutch Wadden Sea. (B) How diet quality (diagonal lines, having a slope of 1) and daily energy expenditure (horizontal axis) together determine the amount of shell material that needs to be processed daily (left vertical axis), and thus determine the require gizzard mass (right vertical axis) for satisficing knots (monthly dots connected by lines in lower left corner) and for rate-maximizing knots (monthly dots in vertical Kirkwood-Kvist bar). (C) Gizzard masses of *islandica* knots in the Dutch Wadden Sea fit the satisficer prediction throughout most of the year, except during periods of fuelling, when gizzards fit the rate-maximizing prediction (especially in spring, but even during early winter when fuelling for a small midwinter body mass peak).

From van Gils *et al.* (2003a).

Formulating the proper cost and benefit functions (plotted conceptually in Fig. 88), we were now able to generate month-specific, year-round predictions for the optimal gizzard mass of knots wintering or fuelling in the Dutch Wadden Sea (van Gils *et al.* 2003a). The benefit function was determined by the seasonal variation in diet quality, as observed through our analysis of knot-droppings (Fig. 89A and diagonal lines in Fig. 89B); the cost function was given by the seasonal variations in metabolic requirements, which were assessed earlier using heated taxidermic mounts (horizontal axis in Fig. 89B; Wiersma and Piersma 1994 and Chapter 2). We predicted that the largest gizzards would occur in midwinter when demands are high and food quality is low, whereas the smallest gizzards were predicted for late summer, when demands are lowest and food quality is still rather high (Fig. 89B). In the Wadden Sea, knots fuel-up during spring, and thus, in spite of the high-quality food during that time of year, we predicted relatively large, rate-maximizing 'spring gizzards' (Fig. 89B). When we collated all the gizzard data that we had built up over almost 20 years of research (920 individual knots; measured initially by dissection of carcasses, and, in later years, by ultrasonography), we found a fit with the predictions that was even better than we could ever have dreamed. Throughout the year, gizzard size varied in response to demands and food quality, with knots maintaining a rate-maximizing gizzard in spring and a satisficing gizzard for the rest of the year; it even seems that during early winter their gizzards are accommodated to fuel their moderate midwinter mass peak (Fig. 89C).

The gizzard optimality model also worked well when predicting variation in gizzard masses between years during the late-summer arrival period (van Gils *et al.* 2006a, 2006b, 2007a). Again based on analysis of droppings, we found some considerable variation in late-summer food quality across a period of five consecutive years (1998–2002), with 1998 offering the best-quality food and 2002 the worst-quality food. This variation was reflected in the birds' gizzards, with the smallest gizzards found in 1998 and the largest ones in 2002 (Fig. 90). In Chapter 9 we will come back to this gradual decline in food quality over the years, and discuss its causes, and consequences, for red knot survival.

With the gizzard model working so well in our own backyard, the Dutch Wadden Sea, we started thinking globally, and wondered whether we could explain gizzard mass variation between knot-staging sites worldwide. Digging through the literature and scrolling through old data-files on food quality, we were able to predict 'over-

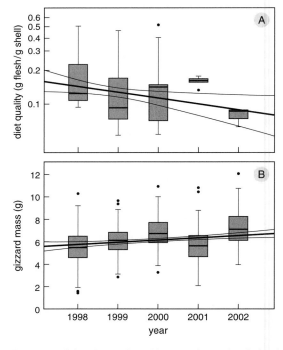

Figure 90. (A) In the Dutch Wadden Sea, diet quality declined between 1998 and 2002, and, as expected, (B) red knots responded by increasing the size of their gizzard. Box-and-whisker-plots as in Fig. 80; lines representing regressions (straight lines) and their 95%-confidence intervals (curved lines).

Modified from van Gils *et al.* (2006b).

wintering gizzard masses' for five out of the six existing knot subspecies, and for three of those we could additionally find food data from their spring stopover sites (van Gils *et al.* 2005c). The latter yielded the surprise. Where we had expected to predict large fuelling gizzards during spring, enabling a rapid build-up of energy stores, we found that the predicted rate-maximizing gizzards at the spring stopover sites were of similar size (or even smaller in the case of the nominate subspecies) than the satisficing gizzards predicted for the wintering sites (Fig. 91). We were surprised by our own model, although the reason is obvious. Food qualities were twice as high at the spring stopover sites as at the wintering sites (Fig. 91). Once again our predictions were reflected in the data: during winter, gizzards varied between sites as expected (from about 6 g in Roebuck Bay, northwest Australia to about 10 g in Mauritania, West Africa), while the 'spring gizzards' hardly showed any variation and averaged out relatively small at about 8 g (Fig. 92).

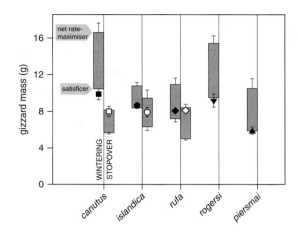

Figure 92. Observed mean gizzard masses (± standard errors) at wintering grounds (filled symbols) and stopover sites (open symbols) match well with predicted gizzard masses (grey bars ± standard errors, with bottom of each bar giving the local satisficer prediction and top of each bar giving the local rate-maximizing prediction).

From van Gils *et al.* (2005c).

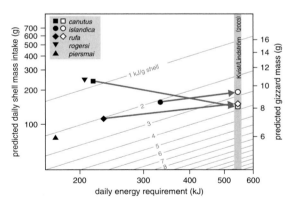

Figure 91. Using the same framework as explained in Fig. 89B, we calculated satisficing gizzard masses for five knot subspecies at their wintering grounds (filled symbols). Additionally, for three of them we could also make predictions on rate-maximizing gizzard masses at their stopovers (open symbols). This shows that, due to an increase in food quality (crossing several food-quality isoclines), little change in gizzard size is required when moving from wintering to stopover site (arrows).

Modified from van Gils *et al.* (2005c).

This led us to infer that knots select their spring stopover sites on the basis of food quality. Selection of such 'hotspots' would speed up migration. This is for two reasons. First, better quality food enables faster fuelling. This is because the rate-maximizing gizzard can then be smaller and is thus energetically cheaper, leading to higher net energy intake rates (compare the lengths of the double-headed open arrows in Figs 88A and B). Second, with hardly any size difference between the satisficing wintering gizzard and the rate-maximizing spring gizzard, knots lose none of the precious time that would otherwise be given over to increasing gizzard mass (Piersma *et al.* 1999a, 1999b). In their site selection, knots may even follow a northward 'wave' in food quality, just as arctic-breeding geese are thought to follow a northward 'green wave' in the quality of their food during spring migration (Drent *et al.* 1978). This is because mollusc prey is usually of best quality during spring, due to the production of nutritious gametes (Honkoop and van der

Meer 1997). Since spring occurs later in the year with increasing latitude, there may be a 'flesh wave' carrying knots to their breeding grounds (also see Piersma *et al.* 1994b). Two observations on spring fuelling rates (published by Piersma *et al.* 2005) are in accordance with this idea: (1) in non-seasonal environments, the tropics, fuelling rates are lower than in seasonal environments; (2) in seasonal environments actually experiencing the austral autumn, Southern Hemisphere stopovers, 'spring' fuelling rates are lower than at Northern Hemisphere stopovers.

Synopsis

Optimal foraging theory has produced elegant models of diet choice and habitat use. Until now such models have assumed rather constant environments and bodies. Based on our own work on shorebirds, we here take these models a step further by including the dynamics of organ size, energy expenditure, and prey quality. Our observations on diet choice, patch use, and stopover-site selection by red knots accord with new versions of foraging models that allow for flexible adjustments of food-processing capacities. Furthermore, not only behaviour is optimized with respect to processing capacity—processing capacity itself is optimized with respect to physiological demand. Red knots fine-tune the size of their digestive organs, notably their gizzard, such that energy income balances with energy expenditure, while, at times of (migratory) fuelling, gizzards are enlarged to such an extent that net energy income is maximized.

Part IV

Towards a fully integrated view

CHAPTER 8

Beyond the physical balance: disease and predation

Running with the Red Queen

Our prime witnesses in this book, red knots, breed in the High Arctic, about as far north as one can get. Specialized molluscivores outside the breeding season, red knots shift to eating surface arthropods during the summer months on the tundra (see Chapter 1). Because they are able to live off intertidal bivalves as well as surface arthropods, there is no obvious reason why red knots should necessarily restrict themselves either to tundra during the breeding season (they ought to be able to find such arthropod prey in temperate meadows as well; Schekkerman *et al.* 2003) or to coastal intertidal habitats during the non-breeding season (they should also be able to use freshwater wetlands with shellfish resources, such as the Niger floodplains in Mali; Zwarts *et al.* 2009).

Expanding this comparison, other High Arctic-breeding shorebird species also winter in marine coastal wetlands, including grey plover *Pluvialis squatarola*, sanderling *Calidris alba*, purple sandpiper *Calidris maritima*, and bar-tailed godwit *Limosa lapponica*. These species provide a striking contrast with their more southerly breeding congeners (respectively Eurasian golden plover *Pluvialis apricaria*, Temmick's stint *Calidris temminckii*, dunlin *Calidris alpina*, and black-tailed godwit *Limosa limosa*) that additionally, or uniquely, rely on freshwater wetlands during the non-breeding season. Indeed, when the breeding and wintering habitats of the 24 different species of the sandpiper subfamily Calidrinae are ranked as 'decreasingly High

Arctic and alpine' or 'decreasingly marine/saline', the covariation shows up clearly (Fig. 93); ruffs *Philomachus pugnax* and red knots occupy the outlying positions. When the interspecific associations were subjected to formal statistical tests, the empirical linkages between the degrees of northerly breeding and the use of marine or coastal non-breeding habitats were confirmed (Piersma 1997); and these associations do not stop at shorebirds. Comparisons among gulls and terns, ducks and passerines all suggest that only the far-northern breeding taxa uniquely use coastal or saline habitats (Piersma 2007). This biogeographic pattern is not linked to any sort of trophic specialization.

In *The Red Queen: sex and the evolution of human nature*, Matt Ridley (1994) explores sex, or rather the reasons why such a seemingly wasteful life-history phenomenon is the rule rather than the exception, and concludes that sex evolved to fend off pathogens and parasites. By continuous intergenerational re-arrangement of the genome—which is what sex is all about (Lane 2009a)—especially the genes coding for proteins that are part of the line of defence against invaders of the body, organisms try to remain one step ahead of their adversaries. This mechanism of an 'evolutionary arms race' (Van Valen 1973), in this case between pathogens and their hosts, was named the 'Red Queen hypothesis', after Lewis Carroll's character from *Through the looking-glass* (Carroll 1960).

In the relevant part of the wonderland story, Alice is running as fast as she possibly can

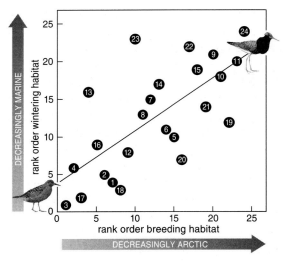

Figure 93. Significant correlation between the extent to which shorebird species belonging to the subfamily Calidrinae (Scolopacidae) breed in decreasingly arctic/alpine, open and climatically extreme habitats (rank order breeding habitat) and the extent to which they winter increasingly in freshwater rather than marine habitats (rank order wintering habitat). The linear regression line is shown to lead the eye. 1: *Aphriza virgata*; 2: *Calidris tenuirostris*; 3: *Calidris alba*; 4: *Calidris pusilla*; 5: *Calidris mauri*; 6: *Calidris mauri*; 7: *Calidris ruficollis*; 8: *Calidris minuta*; 9: *Calidris temminckii*; 10: *Calidris subminuta*; 11: *Calidris minutilla*; 12: *Calidris fuscicollis*; 13: *Calidris bairdii*; 14: *Calidris melanotos*; 15: *Calidris acuminata*; 16: *Calidris ferruginea*; 17: *Calidris maritima*; 18: *Calidris ptilocnemis*; 19: *Calidris alpina*; 20: *Eurynorhynchus pygmeus*; 21: *Limicola falcinellus*; 22: *Micropalama himantopus*; 23: *Tryngites subruficollis*; 24: *Philomachus pugnax*.

From Piersma (2003).

alongside the Red Queen, yet notices that their position does not change. Almost breathless, Alice mentions to the Queen that in *her* country 'you'd generally get to somewhere else—if you ran very fast for a long time, as we've been doing'. 'A slow sort of country!' said the Queen. 'Now, *here*, you see, it takes all the running you can do, to keep in the same place. If you want to get somewhere else, you must run at least twice as fast as that!' If temperate freshwater habitats offer greater disease pressure than High Arctic and marine habitats, shorebirds in freshwater habitats *have to run faster* than shorebirds in High Arctic and marine habitats (Piersma 1997, 2003). Indeed, 'there are ticks and lice, bugs and flies, moths, beetles, grasshoppers, millipedes and more, in all of which males disappear as one moves from the tropics to the poles' (Ridley 1994).[1] This increase in parthenogenetic (male-less) reproduction is consistent with the idea that sex exists to beat pathogens, but also, and this is the point in the present discourse, with the idea that in the High Arctic, pathogen pressure is low.

The patterns also carry another implication. Although shorebirds breeding furthest north and wintering in shoreline habitats cover the greatest migration distances between suitable patches of habitat, they, and not the freshwater species, might be able to pull off such energetically demanding migrations because they do not need to invest as heavily in immune defence (Piersma 1997). Alternatively, because of the low pathogen pressures, 'clean' polar and marine habitat would allow high levels of energy expenditure without negative fitness consequences (see Chapter 4).

It goes without saying that scientists have gone out to test for themselves whether the abundance of disease organisms is actually lower in some environments than others. There is increasing evidence indicating that blood parasites and/or their vectors are indeed relatively sparse in both High Arctic and marine/saline habitats compared with temperate and tropical freshwater habitats (e.g. Bennett *et al.* 1992, Figuerola 1999, Jovani *et al.* 2001, Mendes *et al.*

[1] This citation was incorrectly attributed to Daly and Wilson (1983). In addition, and ironically, it was cited back to front (confirmed by M. Ridley pers. comm.).

2005). We would also expect contrasts in overall immune investments, or aspects of immunity, between High Arctic/marine species and lower latitude/freshwater species—contrasts that would follow logically from the differential exposure to disease pressures during either the egg- or chick- phases or later in life—but these have proven to be difficult to investigate and to confirm (Mendes *et al.* 2006a, 2006b).

The responsive nature of 'constitutive' innate immunity

Do freshwater specialists exposed to a greater density and variety of disease organisms, especially in the tropics, indeed 'run faster' than birds that limit their presence to marine, saline, or High Arctic environments? We must be honest now and let you know that we will not be giving you an answer here. Science has simply not yet conquered much of this territory; the main hurdle being the measurement of 'running speeds' in immunological terms. Assessments of habitat-related differences in immunity between species are hampered by the immense complexity of the immune system (Janeway *et al.* 2004).

In a straightforward and perhaps somewhat simplistic categorization, the immune system can be divided into innate (non-specific) and acquired (specific) arms, and further divided into constitutive (non-induced) and induced branches (Schmid-Hempel and Ebert 2003). Obviously, the acquired and induced branches of the immune system should closely reflect the diseases in the environments in which the organisms have found themselves. To be able to separate organism and environment heuristically, it therefore makes sense that, for a first characterization of the immune system of free-living animals, one should turn to the non-induced (constitutive) and non-specific (innate) parts. Representing circulating levels of immune proteins, rather than responses to

immune challenges, these have the additional benefit of being measurable from samples taken on single capture occasions.

Three techniques are now commonly applied: (1) the killing of standardized microbes by fresh or stored deep-frozen and then thawed blood or plasma (Tieleman *et al.* 2005, Millet *et al.* 2007), (2) concentrations of different leukocyte types (Campbell 1995), and (3) 'complement' and 'natural antibody' assays (Matson *et al.* 2005). Microbial killing of particular strains of the Gram-negative *Escherichia coli* bacterium, the Gram-positive *Staphylococcus aureus* bacterium, and the yeast *Candida albicans* should indicate the overall capacity to limit microbial infections (Millet *et al.* 2007). Leukocyte concentrations provide information on circulating immune cells but have the problem that they are also indicators of health and ongoing infections, and are thus potentially induced (Campbell 1995). However, differential leukocyte counts can be used in multivariate analysis in terms of their relationship to functional measures of immunity, such as microbial killing (see Fig. 94). Heterophils and eosinophils mediate innate immunity against novel pathogens and are important phagocytes, whereas monocytes link innate and acquired defence. Lymphocytes mediate pathogen-specific antibody and cell-mediated responses of the acquired immune system (Campbell 1995). Complement cascade and natural antibodies link innate and acquired immunity, thus providing the first line of defence against spreading infections including viruses (Ochsenbein and Zinkernagel 2000). Are you still with us?

A body of work on immunity in red knots carried out by Debbie Buehler and co-workers has shown that these innate measures of immunity are repeatable within individuals (Buehler *et al.* 2008a). However, even under constant environmental conditions, they nevertheless varied with season (Buehler *et al.* 2008b). The immune measures did not respond to

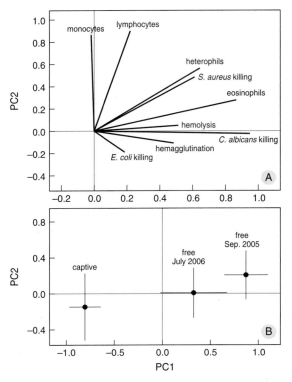

Figure 94. Structure of the measures of constitutive innate immunity in captive and free-living red knots. In panel (A) the axes represent the first two components of a principal component analysis. Vectors are the loadings of each immune measure and the length of the vector indicates how much of the variation in an immune measure is explained by the two axes. The angle between two vectors gives the degree of correlation between them. Adjacent vectors are highly correlated with each other, orthogonal (90°) vectors are uncorrelated, and vectors pointing in opposite directions (180°) are negatively correlated. (B) The groupings of principal component scores (captive red knots, free-living birds caught in September 2005 and free-living birds caught in July 2006). For each group the mean ± SE of PC1 is plotted against the mean ± SE of PC2. Captive and free-living birds are distinguished on PC1 but not on PC2.

Modified from Buehler *et al.* (2008b).

experimental differences in temperature (and thus to differences in the costs of living), except when the birds were brought under stress by limiting feeding hours (Buehler *et al.* 2009b). Under energy stress, red knots cut down on the

immunity measures that are the most costly to maintain. Overall, the story seems to be that even these 'least responsive measures of immunity' are actually quite responsive (Buehler *et al.* 2009a), especially to ambient disease pressures (Buehler *et al.* 2008b). Comparing three groups of red knots (Fig. 94), two of which were sampled immediately after capture in the Wadden Sea and one of which was maintained in a hyper-clean aviary (running seawater to keep the floor and the little mud patch clean, and weekly cage disinfection), showed captive birds to have reduced capacities to kill the bacterium *S. aureus* and the yeast *C. albicans*, as well as lower levels of heterophil and eosinophil leukocytes, when compared to their wild counterparts. In a principal component analysis (Fig. 94), the affected variables fell on a single axis that seemed to reflect reduced phagocytosis and inflammation-based immunity in the aviary birds. Not surprisingly, this correlated with densities of microbial colonies measured per gram of mud, being lower in the regularly disinfected aviaries than in the field.

There is no doubt that the immune system represents one of the most flexible parts of the phenotype. There is also no doubt, and even a little evidence, that the interactions between the pathogens and parasites in an environment and the various branches of immunity of potential hosts occurring there greatly affect the use of these environments. The literature on mammals offers several examples of the pre-emptive presence of deadly pathogens, such as the pandemic of viral rinderpest that swept through East Africa in the late-1800s and left vast areas unusable by a variety of ungulates for decades (Price 1980, Dobson 1995). In the same category falls sleeping sickness, caused by a blood parasite and transmitted by tsetse flies, which makes parts of the African savannah inhospitable for humans and their livestock (McNeill 1976). In addition to the avoidance of habitats in order to prevent a disease, habitat use is also affected

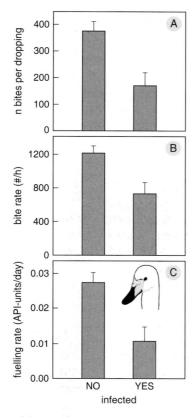

Figure 95. (A) Bewick's swans infected with low-pathogenic avian influenza took fewer bites per faecal dropping than their healthy counterparts, a sign that they experienced poor digestion. (B) Because of this digestive bottleneck, ill swans took fewer bites per hour, (C) leading to slower fuelling rates (expressed as the daily increase in a bird's Abdominal Profile Index, API).

Modified from van Gils *et al.* (2007b).

once it is infected, even if this is by a non-lethal pathogen. GPS-tagged Bewick's swans *Cygnus columbianus bewickii* that were found to be infected by a mild forms of avian influenza (H6N2 and H6N8), stayed longer at their wintering grounds and travelled back to their breeding grounds at a slower pace than uninfected individuals (van Gils *et al.* 2007b). They had hampered digestion (Fig. 95A; consistent with the idea that there is a trade-off between immune function and digestion; Adamo *et al.* 2010), which constrained their intake rate (Fig.

95B). That in turn reduced their rate of accumulating body fat (Fig. 95C). In spite of this, and a few other exceptions (Cornell 1974), we conclude that it has as yet been difficult to build in the presence or absence of more subtly acting disease organisms to explanatory models of habitat use. However, this has been much easier with the more obvious threats to life and limb represented by the predators.

Body-building to defy death

For red knots and other shorebirds, peregrine falcons *Falco peregrinus* are lethal enemies (Fig. 96). Even though peregrines can be held responsible for only a small fraction of the deaths of red knots (van den Hout *et al.* 2008), the knots' use of sites and their flocking behaviour betrays a deep-seated fear of the falcons (Piersma *et al.* 1993c, Dekker 1998, Dekker and Ydenberg 2004, van den Hout *et al.* 2010). Staying away from falcons is one way of coping with the threat they pose (Ydenberg *et al.* 2004, 2007, Leyrer *et al.* 2009), but this is not always possible, especially since they live in such open habitats in which cover is *obstructive* rather than *protective* (as opposed to most passerines; Lima 1993, van den Hout *et al.* 2010).

An interesting situation occurs when such migrating shorebirds have fuelled up to embark on long-distance flights. They are heavy and, at least in red knots, the growth of the pectoral muscle does not keep pace with increases in fat loads (Fig. 97A). The predicted loss of aerodynamic performance (Dietz *et al.* 2007) was demonstrated in a simple experiment. Every week the aviaries at the Royal Netherlands Institute for Sea Research are cleaned, and we use this time to weigh the birds and to examine them for moult and foot infections. Animal caretaker Maarten Brugge trained the red knots to fly from his hand back to their cage. This involved several metres of flight, followed by a 90° turn and a descent

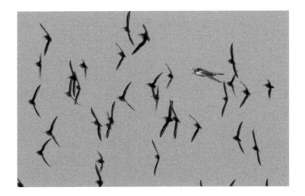

Figure 96. Photographic illustration of fast pure-rotational banks by dunlin *Calidris alpina* chased by a peregrine falcon *Falco peregrinus*.

The photo, from the World Wide Web, was taken by an anonymous photographer.

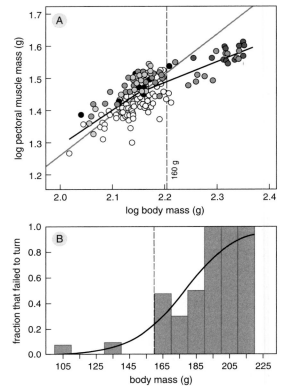

Figure 97. Pectoral muscle mass and the ability to make 90 degree turns in red knots. (A) The allometric relationship between pectoral muscle mass and body mass where the grey line is the theoretical prediction based on aerodynamic principles and the black line the continuous piecewise allometric regression through the data. The dotted vertical line marks 160 g, the body mass level at which red knots start to encounter manoeuvrability problems. Panel (B) shows how the fraction of birds failing to make the 90 degree turn increases steeply above weights over 160 g.

Modified from Dietz *et al.* (2007).

into the aviary. Maarten kept a score of whether or not the birds were able to make this sharp turn. At body masses higher than 160 g, an increasing proportion of birds failed to make it (Fig. 97B), landing somewhere in the corridor and having to walk back and jump over the doorstep into the aviary. Red knots build their pectoral muscles up to a certain point only, and need to accept, or otherwise behaviourally compensate for, the associated risks of decreased flight performance during the final days of fuelling (Piersma *et al.* 2005).

Of course, birds could also assess whether or not predators are around, and then 'decide' whether or not to make bodily adjustments. In an indoor experiment, ruddy turnstones *Arenaria interpres* were exposed to daily but unpredictable disturbances by either a gliding sparrowhawk *Accipiter nisus* model or, alternatively, by a gliding black-headed gull *Larus ridibundus* model (van den Hout *et al.* 2006). During the days of raptor-disturbance, the birds grew larger pectoral muscles (open dots in Fig. 98), while during the days of gull-disturbance their pectoral muscles would shrink. 'Playing' with pectoral muscles in this way changed the birds' flight performance in response to shifts in danger.

As flight performance is expressed as the ratio between pectoral muscle mass and body mass to the power 1.25 (Dietz *et al.* 2007), an alternative way for a bird to boost flight ability is to reduce body mass rather than to boost pectoral muscle size. In fact, 'flight performance' may be too rough a measure to grasp all essential aspects of a bird's escape abilities. Turnstones feed in small flocks on sandy beaches or rocky shores, often in the close

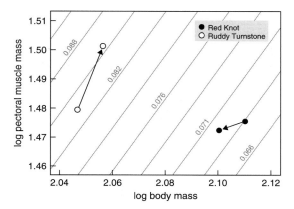

Figure 98. Phase space with lines for equal flight capacity (PMM/BM$^{1.25}$) (Dietz *et al.* 2007), for ruddy turnstones (van den Hout *et al.* 2006) and red knots (van den Hout *et al.* 2010), showing that both species increase flight capacity in response to raptor-model intrusions: red knots by decreasing body mass and ruddy turnstones by increasing pectoral muscle size (all data were log-transformed).

From van den Hout *et al.* 2010.

vicinity of obstructive cover, and raptor attacks are thus usually by surprise. Little time remains for turnstones to escape, and they would thus benefit most if they could rapidly accelerate in order to reach a safe destination, such as water or salt marsh. Acceleration demands the build-up of fast-twitch muscle fibres, and thus turnstones should enlarge their pectoral muscles, rather than reduce overall body mass, in order to boost 'flight performance' and thus increase escape chances. Red knots, however, are typical 'farshore foragers', feeding in huge flocks on wide-open mudflats, often as far from the coastline as possible (Piersma *et al.* 1993c). For them, there is much more time left between predator detection and escape: acceleration capacity may therefore not be their first priority. Red knots will team up, performing united, erratic display flights that confuse the approaching raptor; this requires high turning manoeuvrability of all flock members. Using aerodynamic theory, Piet van den Hout and

co-workers predicted that knots would reduce body mass rather than increase pectoral muscle mass, since being lighter reduces inertia, thus allowing tighter turning angles (van den Hout *et al.* 2010). They repeated the experiment with the stuffed sparrowhawk model, but now substituting red knots for turnstones. Instead of *increasing* pectoral muscle mass, the knots indeed *reduced* their overall body mass, maintaining a more or less constant pectoral muscle mass (solid dots in Fig. 98). This has made us wonder if perhaps new species' names are in order: how about 'stones' and 'turnknots' rather than turnstones and knots?

Although the ability to escape from an approaching predator is an important skill, it may trade off with the ability to process food during times of food scarcity. A starving bird catabolizes its own protein stores, and thus needs to 'decide' between burning down different organs. Will it sacrifice its flight capacities or its food-processing capacities, or both at the same time? Analysing the composition of 103 knot bodies provided the answer (Dietz and Piersma 2007). These birds had all been collected in The Netherlands in winter, and had either died through starvation during severe winter weather or by accident during normal winter conditions (by light-house collisions or during capture). Although all organs considered (gizzard, liver, intestines, pectoral muscles, leg muscles) were smaller in the starved birds than in the non-starved birds, the reductions were much more serious in the pectoral muscles (a decline by *c.* 60%) than in the gizzard (*c.* 21%), especially with respect to the reduction in total body mass (*c.* 33%; Fig. 99). In fact, the starved birds had gizzards of about 8–8.6 g fresh mass, just enough to balance energy income with energy expenditure on a daily basis (the satisficing policy; see Chapter 7). So, the capacity to process food at this stage appears more important than preventing capture and being eaten. After all, impaired escape abilities need not lead to death, as long as there are no successful predatory attacks, but turning off digestive capacity will certainly lead to death.

Such a food-safety trade-off and how it affects the phenotype is also evident in roost selection by knots. Safety at the roost is important, but safe roosts may not always be located near good feeding areas. Therefore, birds may face a trade-off between maximizing roost safety and minimizing daily travel distance from and to the roost. This trade-off is evident in red knots wintering in the western Dutch Wadden Sea, where there are basically two major roosts (Piersma *et al.* 1993c). There is Griend, a small island in the midst of good feeding grounds, but rather dangerous because of a recently elevated dike that enables raptors to perform surprise attacks

on resting shorebirds. The alternative is Richel: a completely flat sandbank, enabling a wide-open unobstructed view of the potentially dangerous sky, but usually rather far from good feeding grounds. Where to rest during high tide: at the risky but energetically cheap Griend, or at the safer but energetically more expensive Richel? The optimal solution may differ between different phenotypes, as revealed by a between-years comparison of roost-use by 26 radio-tagged knots (van Gils *et al.* unpubl.data). In years when almost all roosting took place at Richel, the knots had small pectoral muscles relative to their body mass; while in years of relatively greater roosting at Griend, knots carried relatively large pectoral muscles (Fig. 100). These inter-annual differences in roost-use, and hence in phenotypic appearance, may reflect variation between years in food distribution or raptor abundance. For example, in a typical 'Richel-roosting year', such as 1997, there may have been good food patches near Richel or there may have been relatively many raptors around (the latter increasing predation danger more steeply at the endiked Griend than at the wide-open Richel).

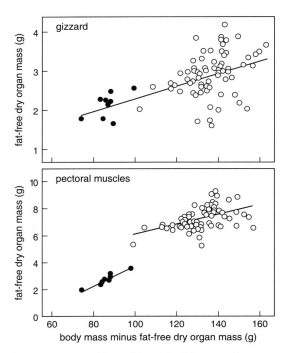

Figure 99. Starved knots (closed dots) have smaller organs than non-starved knots (open dots), but the reductions are much stronger in the pectoral muscles than in the gizzard.
Modified from Dietz and Piersma (2007).

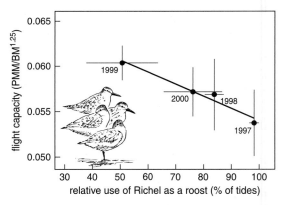

Figure 100. Phenotypic adjustments of flight capacity to roost safety. The more often knots roosted at the relatively safe Richel (relative to the number of times they roosted at Griend), the smaller their flight capacity, i.e. pectoral muscle mass/body mass$^{1.25}$.

Based on J.A. van Gils, A. Dekinga, and T. Piersma *et al.* (unpubl. data.)

Coping with danger

Besides the bodily adjustments just described, animals can respond behaviourally to an increased predation danger. We have already mentioned 'teaming-up' or flocking as an effective way to dilute the risk of being caught among flock mates (Bednekoff and Lima 1998). Selecting a safer habitat and actively looking out for predators (vigilance) are two other behavioural ways to reduce predation risk (note that predation *risk* can be managed by an organism, whereas predation *danger* is a feature of the environment which is uncontrollable and reflects predation *risk* if no anti-predation measures were taken; Lank and Ydenberg 2003).

In most of these cases, the animal pays a price for its safer life. Joel Brown termed this 'the cost of predation' (Brown 1988), and he argued that this price can be expressed in terms of a loss of energy intake. Feeding in dense flocks usually leads to greater interference and reduced food-intake rates (Vahl *et al.* 2005), looking out for potential threats detracts from finding food (Radford *et al.* 2009), and safer habitats may be nutritionally poorer habitats (Cresswell 1994). In the long run, these lower intake rates have knock-on effects on an animal's life history. For example, female elk *Cervus elaphus* give birth to fewer young when wolves *Canis lupus* are abundant (Creel *et al.* 2007). When experimentally exposed to enhanced predation danger, snowshoe hares *Lepus americanus* not only gave birth to fewer young, those young were also smaller and lighter (Sheriff *et al.* 2009).

The discovery of a common—energetic—currency to express 'predation risk' gave rise to a new and flourishing field within animal ecology, the 'ecology of fear' (Brown *et al.* 1999). Patch-use theory played a central role, with the classical marginal value theorem predicting that a forager should give up all its food patches at the same instantaneous intake rate in order to maximize its long-term average intake rate. However, by including fear in this model, an animal trading-off risk of starvation with risk of predation should give up dangerous patches at higher instantaneous intake rates than safe patches. This prediction has been tested and verified many times since Brown's (1988) seminal paper (reviewed by Brown and Kotler 2004). This fundamental principle provides a strong link between the behaviour of fearful prey and the organisation of the species communities of which they are part.

For example, reintroducing wolves into Yellowstone National Park brought back the extensive aspen and willow forests, a typical North-American landscape that had largely disappeared over the last 150 years (Ripple and Beschta 2005). The strong linkage between wolves and plants came about not so much because of wolves killing the major consumers of aspen and willow, elks, but merely due to wolves scaring elks out of potentially dangerous habitats. The slippery slopes of streams, where elks browse aspen branches, are especially risky. The mere avoidance of this riparian zone by elks after the wolves showed up triggered a cascade of ecological events. It 'released' aspen and willow growth along streams, which brought back beavers; their dams then slowed down water flow and streams started to meander more, thereby generating a mosaic of productive biodiverse microhabitats for birds and butterflies (Ripple and Beschta 2004). This is just one example out of many showing the non-lethal fear effects of predators on their prey (Werner and Peacor 2003, Schmitz *et al.* 2004). Fear is a driving ecological force!

Note that for the same *risk* of predation, not every individual experiences the same *cost* of predation. Thus, some individuals showing lower fear-levels will exploit dangerous patches to a greater extent than others. This may seem odd at first, as dying in the claws or jaws of a predator is the end of everyone's fitness. However, we should not forget that the *cost* of predation actually refers to the loss in energetic foraging opportunities due to preventing being killed. Foregone energetic income is most costly to those individuals that are most in need of energy, such as lean individuals with small energy stores. For them, the fitness value of one unit of energy is relatively large. Hence, one unit of predation risk is relatively small on their energetic scale. A good example of lean individuals accepting greater risks of predation than heavier individuals can be found in marine benthic invertebrates, notably bivalves. Being burrowed in the sediment, they protect themselves against predation by shorebirds and fish. The deeper the safer; but this trades off with food intake, as the food needs to be collected from the seafloor surface or the water column using an extendable 'trunk', the inhalent siphon. In this dilemma, shallowly burrowed individuals indeed have significantly lower body masses than deeply burrowed bivalves (Fig. 101; Zwarts and Wanink 1991). Lean individuals appear more willing to give up some of their safety in exchange for a higher energy income.

The same patterns have been found in migratory songbirds stopping over at a small offshore island in the German part of the North Sea. Dierschke (2003) analysed the body weights of birds caught by local cats or raptors and compared them to the body weights of birds trapped for ringing purposes. He found that the lightest individuals had the highest chance of falling victim to predation (Fig. 102). Lank and Ydenberg (2003) correctly pointed out that this result does not imply that lighter individuals are easier to catch (as Dierschke erronously suggested). Presumably it is the other way around, with the *heavier* individuals being easier to catch (think of the heavy red

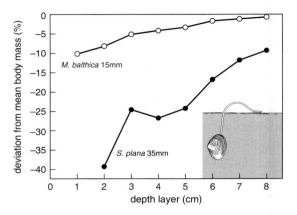

Figure 101. Relative body mass of two marine bivalve species in the upper layer of the sediment (the upper 1 cm, the upper 2 cm, etc.).

Modified from Zwarts and Wanink (1991).

knots failing to turn into their cage; Fig. 97B). However, because for them the fitness value of food is lower than for light individuals, they show less risky behaviour by devoting more time to vigilance and feeding in poorer but safer patches. Because of this hiding behaviour of the most well-nourished prey individuals, hungry predators usually face the most marginal prey in their quest for food.

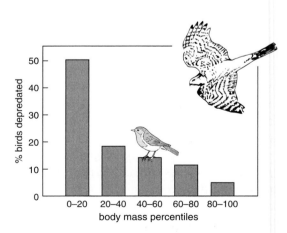

Figure 102. Distribution of body masses of depredated birds. About half of the killed birds belonged to the 20% lightest birds.

Modified from Dierschke (2003).

Although not necessarily lean, migratory birds fuelling for migration are also in great need of energy. This is especially so when there is little time left to fuel, such as in long-distance migrants on their final stopover on the way to the High Arctic; for example, in bar-tailed godwits *Limosa lapponica*. Wintering in West Africa and stopping over in May in the Dutch Wadden Sea, the *taymyrensis*-subspecies faces a tight travel schedule to arrive on time at its Siberian breeding grounds in June (Drent and Piersma 1990, Piersma and Jukema 1990). This greatly contrasts with the *lapponica*-subspecies, which has the whole winter to fuel up for migration as it spends the winter resident in the Wadden Sea. Moreover, this subspecies breeds relatively close by in northern Scandinavia and thus requires less fuel for migration. As a result, in the Dutch Wadden Sea, *taymyrensis* fuels up to 10 times faster than *lapponica* (Fig. 103A; Duijns *et al.* 2009) and the fitness-value of one unit of energy will be much greater in *taymyrensis*. This should reduce the cost of predation, which in turn would make them accept greater predation risks. These subspecific differences are indeed reflected in their feeding distributions on the mudflat, where they face a steep food-safety trade-off: *taymyrensis* is consistently found feeding near shore, where both food abundance but also predation danger are relatively high; *lapponica* prefers to feed in the safer outer-coast mudflats at the expense of finding much lower food densities (Fig. 103B; Duijns *et al.* 2009).

Independent of the short-term value of energy, more long-term fitness prospects, such as the annual survival probability, may also differ between individuals. According to the modern, fear-based foraging theory, better fitness prospects should increase the cost of predation, thus increasing the fear response for a given risk of predation (this principle has been coined the 'asset-protection principle', with the asset being an animal's fitness

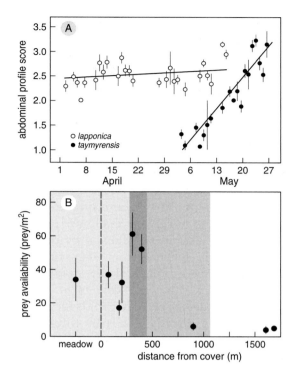

Figure 103. (A) Based on changes in the abdominal profile, the *taymyrensis* subspecies of the bar-tailed godwit fuels up faster and to a greater extent than the *lapponica* subspecies. (B) Furthermore, they feed in higher prey densities in meadows and closer to the coastline than the *lapponica* subspecies. The light grey area on the left gives the 95% confidence interval around the average distance to the coast for *taymyrensis*; the dark grey area on the right does so for *lapponica*; and the darkest grey area in the middle gives the overlap.

Modified from Duijns *et al.* (2009).

prospects in need of protection; Clark 1994). An individual's fitness prospects (equivalent to its reproductive value) usually increases with age, as older individuals have higher survival chances and have greater reproductive capacities. Indeed, in red knots (van den Hout *et al.* in prep), and also in redshanks *Tringa totanus* (Cresswell 1994), juveniles are usually found in the riskier but more productive habitats whereas the adults feed in the safer but effectively poorer habitats. In parrotfish *Scarus* spp. and *Sparisoma* spp., reproductive value also increases with body size and, indeed, the largest individuals showed the greatest fear response towards an approaching scuba diver (Gotanda *et al.* 2009). Getting back to burrowing invertebrate creatures, the degree of risk acceptance in the form of burrowing depth reflected the perceived fitness prospects by knots' favourite prey, the tellinid bivalves *Macoma balthica*. Even though they are tiny little animals with nerve nodes rather than a brain, the accuracy of their perceptions was amazing. Years in which *Macoma* burrowed shallowly (accepting greater risks) were followed by population declines, whereas the years of deep burrowing were followed by population increases (Fig. 104; van Gils *et al.* 2009). Obviously *Macoma*—but this holds for all animals—cannot directly assess its future survival chances, but must rely on some sort of cue, which correlates with these prospects. Food abundance, food quality, or body condition may often be reliable cues.

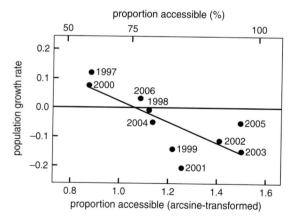

Figure 104. The asset-protection principle at work: the population of *Macoma balthica* increased after individuals burrowed themselves relatively deeply, protecting their asset against predation. After individuals burrowed themselves relatively shallowly, the population would decline. This suggests that *Macoma* is able to assess its own fitness perspectives reasonably well.

From van Gils *et al.* (2009).

Predicting carrying capacity in the light of fear

We have seen that animals give up dangerous patches at higher food densities than safe patches. This notion should be included in models that try to predict how many days animals can live on a given supply of food (van Gils *et al.* 2004). This number of days is usually called an area's 'carrying capacity' (Sutherland and Anderson 1993). We can think of an area as having a carrying capacity of, let's say, 1000 bird-days, which means that there is enough food for a single bird to live for 1000 days, two birds for 500 days, or 1000 birds for a single day. These calculations always take into account the time that animals need to harvest their food, and thus there is a critical food-density below which it will take more time to collect the daily required amount of food than there are hours in a day.

Leaving predation aside, one would expect that, eventually, all patches would be harvested down to a level where gross food-intake rate would just counterbalance an animal's minimal energy requirements. In those rare cases where all patches are equally costly to exploit in terms of energy expenditure, this would lead to similar 'giving-up densities' of food across all patch types (Fig. 105A). This was confirmed recently in an experimental study on bank voles *Myodes glareolus* that were offered identical patches and which experienced virtually no predation threat (Eccard and Liesenjohann 2008). However, the metabolic exploitation costs can frequently differ between patches. Under such circumstances, costly patches should be given up at higher threshold food-densities than cheaper patches (Fig. 105B). This was verified in Bewick's swans feeding on pondweed tubers, which left behind more food in shallow water, where clay bottoms made the tuber-harvest more costly than in sandy patches (Nolet *et al.* 2001).

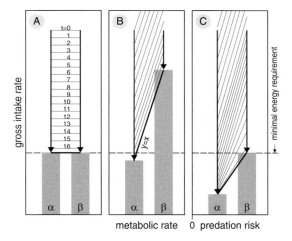

Figure 105. An area's carrying capacity predicted by the original (A) and modified versions (B–C) of Sutherland and Anderson's (1993) depletion model. In these examples, the area consists of two patches, α and β, which initially contain similar prey densities. The parameters of the functional response are kept similar across the figures, which enable direct comparisons between predictions of carrying capacity. Initially, at the onset of exploitation, intake rates are high. Due to depletion, intake rates gradually decline (downward arrows), with the intake rates at time t given by grey lines. Minimal energy requirements are given by dashed horizontal line intersecting the three subfigures.

Modified from van Gils *et al.* (2004).

Consequently, the clay habitat had a lower carrying capacity than the sandy habitat (Nolet and Klaassen 2009).

As predation risk can also be expressed 'energetically' in the form of a predation cost, predation risk can be included in these energy-budget-based carrying-capacity calculations. Initially, when there is still plenty of food throughout the environment, dangerous patches will be avoided by foragers, and only safe patches will be exploited (Fig. 105C). This will change once the difference in intake rate between both types of patches equals the difference in predation cost: from that moment onwards, animals will alternate between both patch types. Once the intake rate in the safer patches has dropped below the minimally required level, the foragers will spend pro-

gressively more time in the dangerous patches. This stops once the intake rate in the dangerous patches has also dropped below the minimally required level. From then on animals either starve or leave the area: carrying capacity has been reached. Foragers thus effectively leave an imprint of their so-called 'landscape of fear' (Laundré *et al.* 2001), a heterogeneous map of variable giving-up food densities that reflects the spatial heterogeneity in predation costs. For example, small mammals often retreat to their burrows upon predator detection, and thus unharvested food densities are usually higher further out from the burrows, yielding physical maps with burrows surrounded by contours of low predation costs (van der Merwe and Brown 2008).

In human-dominated areas, roads often shape the landscape of fear. This is true, for example, of geese and swans that prefer agricultural fields furthest removed from human access. Only when the food in such areas is depleted will they accept greater risks and feed closer to roads (Gill *et al.* 1996; van Gils and Tijsen 2007). Even predators themselves, such as brown bears *Ursus arctos*, live in such road-based landscapes of fear. Moose *Alces alces* sometimes take advantage of this by giving birth on or near roads in the 'safe' shelter of cars scaring off the bears hungry for baby moose (Berger 2007). Having ended here with human-shaped *landscapes of fear*, we will continue in the next chapter with human-wrought *fearsome landscapes*: intertidal mudflats emptied and destroyed by large-scale mechanical fishery activities.

Synopsis

The responses of animals to diseases and predation are diverse and flexible. The immune system is one of the most flexible parts of the phenotype, which can be turned up or down depending on potential pathogenic threat levels. Insightful knowledge on the flexible immune system is still thin on the ground, as immunity is so amazingly complex. What we do know, however, is that shorebird species that breed in the relatively clean High Arctic, winter in relatively clean saline habitats; whereas species that breed at lower latitudes with higher pathogen pressures, winter in pathogen-rich freshwater environments. It is possible that High-Arctic breeders are the only ones able to make super long-distance migrations because they do not need to invest so much in immune defence. We understand much more about phenotypic responses to the threat of predation. By flexibly modifying body composition, prey can reduce their risk of predation. Birds 'play' with pectoral muscle mass and total body mass, depending on the need to escape through acceleration or through manoeuvrability. Obviously, the most immediate response to predation threat is through behaviour: visually scanning the environment, coming together in flocks, or switching habitats. In such ways parasites and predators strongly affect an area's carrying capacity for particular animals, something that is gradually being realized and incorporated in applied contexts and conservation thinking.

CHAPTER 9

Population consequences: conservation and management of flexible phenotypes

We are deep into the book now, and you have read much about phenotypes in interaction with their environments. We have come to understand diet choice and the distributions of animals because we took phenotypic change into account. That is all very interesting in its own right, but is it enough?

The Holy Grail of population biology

The population-oriented ecologists amongst us are usually after even 'holier grails': the understanding of changing numbers of organisms, and of variable population sizes (Kingsland 1995, Hilborn and Mangel 1997, Vandermeer and Goldberg 2003, Ranta *et al.* 2006). Building on what we have learned about specialized molluscivore shorebirds in the previous two chapters, we now want to take the next step by disentangling the reasons behind their changing numbers. Can we understand population dynamics better if we include the notion of phenotypic flexibility? In addition, given that the world's ecology is in constant flux, we would like to establish what differences flexible rather than inflexible phenotypes could make to what animals can do or can tolerate on this rapidly changing planet. This may allow us to predict population consequences somewhat outside the current range of environmental conditions. To set the scene, we will start with a case study on a bird that will not surprise you at all—the red knot.

Dredging out bivalve-rich intertidal flats: a case study on red knots

Tidal flats occur along shallow seas with soft sediment bottoms, in areas where the tidal range is at least a metre or so (Eisma 1998, Lenihan and Micheli 2001, van de Kam *et al.* 2004). The lowest parts of intertidal areas are usually barren except for the occurrence of seagrass meadows and reefs formed by invertebrate 'ecosystem engineers', such as oysters *Ostrea*, mussels *Mytilus*, or tubeworms *Sabellaria*. Intertidal areas are inundated at least once a day. On the landside, flats may be bordered by salt marshes, or, in the warmer regions of the world, by mangrove forests.

The international Wadden Sea, a shared territory of The Netherlands, Germany, and Denmark, is reckoned to be one of the biggest intertidal areas in the world (van de Kam *et al.* 2004, Lotze *et al.* 2005), a distinction shared with the Yellow Sea (see below). Mostly 7 to 10 km wide, it stretches along 450 km of coastline. During spring low tides a total of 490,000 ha of intertidal mud is exposed. Compared with other intertidal areas worldwide, the Wadden Sea harbours particularly large shellfish populations (Piersma *et al.* 1993a). In 2009 it received recognition as a UNESCO World Heritage Site. The Netherlands' 120,000 ha of the Wadden Sea are designated a State Nature Monument, and is protected under various other national and international conventions. Despite this high number of formal conservation accolades, until 2004, three-quarters of the intertidal flats of the Dutch Wadden Sea were open to mechanical dredging for edible cockles *Cerastoderma edule* (Swart and van Andel 2008, Piersma 2009). Most parts are still open to other forms of bottom disturbance with fishing gear of various severity (Watling and Norse 1998).

Suction-dredging of cockles took place from big barges with a draft of only 80 cm (Fig. 106E). The dredges, suspended from either side of the barge, are about a metre wide and have water jets to loosen up the top layer of the sediment. This is then scraped off by a blade at a depth of approximately 5 cm. Within the dredge, a strong water flow ensures that small objects are pushed through a screen. The remaining items, such as large cockles, are sucked on board for further sorting and cleaning. After dredging, the intertidal flats look 'scarred' and can remain so for a whole year (Fig. 106F). In the case of mechanical cockle fishing in the Wadden Sea, GPS loggers were installed to record the position and fishing-activity of each barge (Fig. 106A–D). On the basis of these records, an average 4.3% of the intertidal flats was affected by dredging each year (Kamermans and Smaal 2002). Cumulatively, between 1992 and 2001, 19% of the intertidal area was affected by mechanical cockle-dredging at least once (Kraan *et al.* 2007). If this still seems a small percentage to you, note that cockle-dredgers selected areas that red knots would select too, and that by far the greatest loss of feeding area (Fig. 107C) was accounted for by cockle-dredging (van Gils *et al.* 2006b, Kraan *et al.* 2009).

A direct, immediate effect of dredging was the complete removal of all organisms larger than 19 mm in the top layer. As the dredged sites were usually the most biodiverse (Kraan *et al.* 2007), dredging also affected smaller cockles, other bivalves such as blue mussels *Mytilus edulis*, Baltic tellins *Macoma balthica*, and sandgapers *Mya arenaria*, polychaetes, and crustaceans such as shorecrabs *Carcinus maenas*. More indirectly, and over longer time scales, sediments lost fine silts during dredging events, and changing sediment characteristics may have led to long-term reductions in settlement rates by both cockles and Baltic tellins (Piersma *et al.* 2001, Hiddink 2003).

Population consequences for the molluscivores

Several decades of serious fishing pressure not only led to a stark loss of the shellfish resources, but also resulted in a serious decline of many molluscivore birds in the Wadden Sea. The oystercatchers, most of which are resident in The Netherlands, showed a steep decline, incurring the greatest losses when high food demands coincided with restricted access to partially ice-covered mudflats during severe winters (Camphuysen *et al.* 1996, van Roomen *et al.* 2005). Eider ducks *Somateria mollissima*, consisting of local breeders and a much larger contingent migrating in from the Baltic, also showed food-scarcity related mass starvation and movements out of the Wadden Sea (Camphuysen *et al.* 2002). Oystercatchers and eiders are large birds. Their actual death is easily noticed. This is in marked contrast to the smallest dedicated molluscivore, the red knot, which shows rather cryptic mortality in the form of declining head counts. Following earlier losses (Piersma *et al.* 1993c), red knots declined by 42% between winters in the late-1990s and the early 2000s (Fig. 107A). A decrease in annual local survival from 89% to 82% per year could account for almost half of this decline (Fig. 107B). The remaining red knots moved elsewhere in Western Europe to winter (Kraan *et al.* 2009). Red knots *en route* to wintering areas in West Africa must store fuel for flight and thus have greater energy demands. Their numbers in late summer showed an even

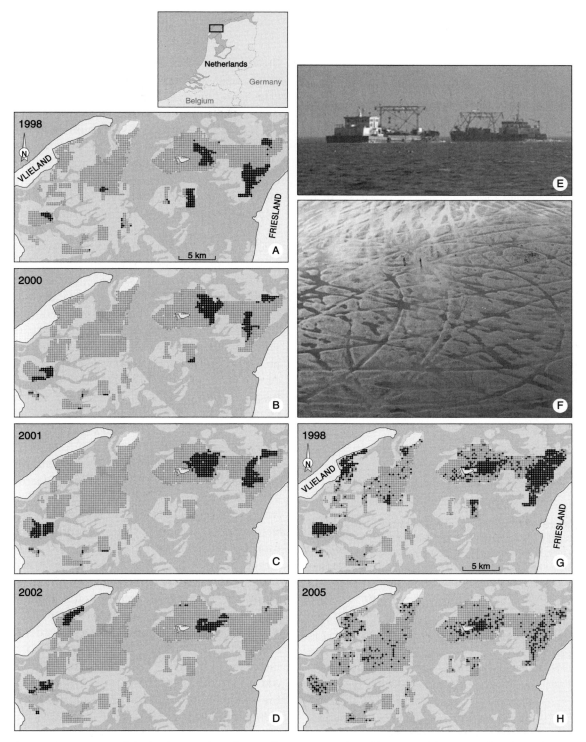

Figure 106. Summary of the benthic sampling efforts in the western Dutch Wadden Sea and the extent of mechanical cockle-dredging in late-1998 (A), 2000 (B), 2001 (C), and 2002 (D). Dredged sampling stations are indicated by a filled dot, sampling stations not affected by mechanical dredging by open circles (1999 not shown as no dredging took place). (E) Cockle-dredging barges in action and (F) an aerial view of the surface scars of cockle dredging six months after the event (for scale, note the two human figures). Distribution of sampling locations where red knots could have achieved the required minimal intake rate of 0.3 mg AFDM/s in 1998 (G) and 2005 (H), based on the measured harvestable food densities and a diet-selection model.

Compiled from Kraan *et al.* (2007, 2009); photographs by M. de Jonge (E) and J. de Vlas (F).

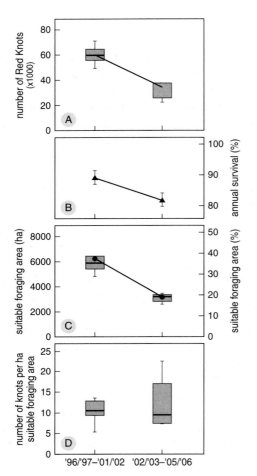

Figure 107. (A) Decline of red knot numbers (box plots) and (B) their annual local survival (means ± SE) between winters in the intervals 1996/1997–2001/2002 and 2002/2003–2005/2006. (C) Shows the decrease of suitable foraging area for red knots in the western Dutch Wadden Sea between the two periods (left axis: suitable foraging area expressed in ha; right axis: suitable foraging area in % of total area). (D) Because the 'slopes' of the decline in numbers (A) and suitable area (C) are similar, the average number of red knots per ha of suitable foraging area remained constant. Box plots give the mean (large dot), median (horizontal line inside the box), interquartile range (box), range (bars) and outliers (asterisk).

Adapted from Kraan et al. (2009).

steeper decline over the same period than that of the over-wintering birds (Kraan et al. 2010).

These patterns could all be explained by changes in the food-base of red knots. Using the empirical relationships between prey densities, sizes and quality, and potential intake rates (van Gils et al. 2006a, 2007a), we can predict intake rates for every place on the mudflats at which data on food abundance and quality were collected. On the basis of our knowledge of energy requirements as a function of weather conditions and storage strategies, we can additionally determine the minimal intake rate necessary for the particular category of birds that we are interested in (Piersma et al. 1995b). On this basis, the disappearance of red knots (Fig. 107A) could be explained by the decline in the extent of suitable intertidal flat (Fig. 106G and H). Because the rates of decline in red knot numbers (Fig. 107A) and suitable foraging area (Fig. 107C) were similar, average densities per unit of suitable mudflat remained unchanged (Fig. 107D). As we have seen in Chapters 6 and 7, in molluscivore birds that ingest their prey whole, energy intake rate is limited by the processing capacity of shell material in the gut (van Gils et al. 2003a, Quaintenne et al. 2010). With all their potential for phenotypic change, in principle red knots could build larger gizzards, extend the range of suitable prey (by including items with less shell per unit shell mass) and thereby almost double the area of suitable mudflat (Fig. 108A). In fact, they did grow larger gizzards (Fig. 108B), but not by much. The measured increase of 0.4 g would have led to an expansion of suitable foraging grounds by only 225 ha, or less than 10% (Fig. 108A).

Why did red knots not increase gizzard size further? Enlargements of gizzards involve increases in a number of cost factors that we did not account for. These cost factors include the energy costs of growth and maintenance of a larger gizzard (van Gils et al. 2003a) and effects of added mass on manoeuvrability during escape from aerial predators (Chapter 8, Dietz et al. 2007, van den Hout et al. 2010). Minimizing the overall rate of energy expendi-

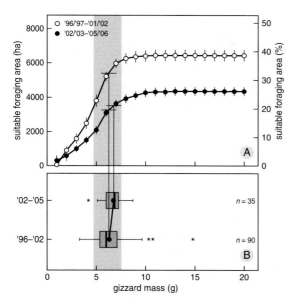

Figure 108. (A) Suitable foraging area in the western Dutch Wadden Sea as a function of gizzard mass during the period 1996/1997–2001/2002 (upper line) and during 2002/2003–2005/2006 (lower line), predicted for a minimally required intake rate of 0.3 mg AFDM/s (left axis: suitable foraging area expressed in ha; right axis: suitable foraging area in % of total area). (B) Distribution of gizzard masses (g) in the two periods (in box plots; see previous figure). Even though the increase in gizzard size between the two periods was small (0.4 g), it was significant (GLM using 125 noninvasively measured gizzards between September and April, and year nested within period; $F_{3,121} = 5.76$; $P = 0.001$). Indicating the range, the grey bar equals the mean ± SD gizzard mass from 1996 to 2005. There are vertical projections of mean gizzard mass per period on to suitable mudflat area.

Adapted from Kraan *et al.* (2009).

ture by carrying the smallest possible gizzard for the energy budget to be in balance, at the expense of the area where they could profitably forage, the birds may have found some kind of optimum. However, red knots may also have failed to increase gizzard size further simply because they were unable to do so. At any particular prey quality on offer, gizzard sizes below the required threshold do not allow the digestive processing of enough shell material for enough energy to be extracted to cover the daily costs. This means that whilst gizzards are

being built up to that size threshold, the birds have to rely on stores that they already carry (van Gils *et al.* 2006b). In the absence of an energy store ('money in the bank'), gizzards cannot be rebuilt as 'the income is too small to pay the carpenter'. Birds will starve locally or, more likely, move away swiftly to try and find places with higher prey quality.

Which individual red knots made it through?

In the Dutch Wadden Sea, from the late-1990s onward, the area of mudflat suitable for foraging red knots declined, not only because the densities of suitable prey declined, but also because the quality, the flesh-to-shell ratios, of these prey declined (van Gils *et al.* 2006b). During these years, in a series of late summers and autumns, we studied red knots returning from the High Arctic breeding grounds in northern Greenland and north-eastern Canada, a time of year when their arthopod foods do not require particularly large gizzards (Piersma *et al.* 1999b, Battley and Piersma 2005). Returning to the Wadden Sea after breeding, red knots would encounter different prey qualities in different years, with a general downward trend. Variations between years in gizzard mass of birds captured during these times reflected prey quality not just qualitatively, but also quantitatively (Fig. 109).

These non-invasively measured gizzard sizes (Dietz *et al.* 1999, Starck *et al.* 2001) of individually colour-marked birds were the basis of an important discovery. Red knots that were subsequently resighted in the study area had significantly larger gizzards at first capture than birds that were not seen again (Fig. 109). Birds that disappeared from view may either have died or moved away permanently. Of course, red knots with small gizzards could have increased gizzard size after capture; in fact, local survival was best quantitatively

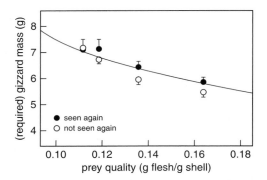

Figure 109. The gizzard mass required to maintain energy balance is predicted to decline with function of prey quality (solid line). Gizzards of individually colour-marked birds seen again within a year after measurement in late summer and autumn fitted the predicted quantitative relationship (grey dots; mean ± SE), whereas gizzards of birds not seen again within a year were significantly smaller (open dots; the two groups almost entirely overlap in the poorest-quality year).

Adapted from van Gils *et al.* (2006b).

explained under the assumption that, after capture, birds increased their gizzards by an average 1 g (van Gils *et al.* 2006b). However, since birds with smaller gizzards also had lower body masses (van Gils *et al.* 2006b), the birds with especially small gizzards would not have been able to make the change. With no 'money in the bank', they would be the ones to move and/or to die.

All this material on red knots in the Wadden Sea goes to show that, by using their capacity to change the size and shell-processing capacity of the gut, red knots have an ability to cope with variations in prey quality between years. It has also become clear that limits to this capacity to change, e.g. by not having the stores to 'pay' for the change, will affect the birds' decision to stay or to move on, and will probably make the difference between life and death. For red knots to capitalize on what the whole world offers them by way of particular, season-specific resources (high quality shellfish at temperate and tropical latitudes when not breeding, safety and abundant arthropods at high latitudes when breeding; Piersma *et al.*

2005, Schekkerman *et al.* 2003), organ flexibility is clearly important. Our latest example also demonstrated how events during one time of year (in this case the building up and use of stores for the migration from arctic breeding grounds to temperate wintering areas) affect the survival chances upon arrival from that migration (in this case affected either by having the stores to build up a gizzard in years with low prey quality, or not). What we have implied, but have not quite shown directly, is that the particular capacity for organ change enhances performance beyond what would be possible with fixed organ sizes. For an illustration, we stay with red knots but move half a world away to the East-Asian–Australasian flyway.

Migrant flexibility and speed of migration

The red knots 'wintering' in the tropics of northwest Australia breed on tundra on the remote New Siberian Islands in the Arctic Ocean and, coincidentally, have been named after one of us by the Russian ornithologist Pavel S. Tomkovich (2001)—*Calidris canutus piersmai*. We were surprised to discover that these '*piersmai*-knots' leave northwest Australia only in early to mid-May (Battley *et al.* 2005), more than a month after the exodus of most other locally wintering migrant shorebirds (Tulp *et al.* 1994). At this time they have less than a month to negotiate 10,400 km of airspace before having to arrive on the tundra in early June. On the basis of resightings of colour-ringed birds, we know that they stage somewhere in the Yellow Sea. With one million hectares of mud, the Yellow Sea mudflats are more extensive than those of the Wadden Sea, but, unlike the Wadden Sea, they are still being reclaimed and transformed with great speed and efficiency (Fig. 110; Barter 2002, van de Kam *et al.* 2008). We now know that during northward migration red knots choose the

Figure 110. That the world is changing fast is illustrated here by newly built motorways crossing mudflats near Incheon, Greater-Seoul, South-Korea. The extent of areas where shorebirds and fishermen can both find their marine resources is steadily becoming smaller.

Photographed on 7 September 2007 by Jan van de Kam.

extensive western mudflats bordering Bohai Bay, in China, a few hundred km east of its capital Beijing (Battley *et al.* 2005, Yang *et al.* 2008).

Depending on the quality of the prey that they find in the Yellow Sea (expressed in terms of units of energy per unit dry shell mass), and depending on the body mass they have to reach before the onward flight to the New Siberian Islands, red knots might well have enough time to refuel in less than the single month available (Fig. 111). However, for birds with inflexible gizzards, prey quality in China has to be five times the measured prey quality in northwest Australia (Fig. 111B); whereas for birds with flexible gizzards, prey quality is not such a constraint (Fig. 111D). Still, at prey qualities somewhat lower than those in Australia, red knots would have to double the size of a 7-g gizzard upon arrival in China (Fig. 111C), for which they would need to bring stores ('money in the bank') and which would take time (about a week under the best of conditions; Dekinga *et al.* 2001). Although the recently measured quality of knot-food in Bohai Bay suggests that red knots in recent years had rather little adjusting to do upon

arrival (H.-Y. Yang *et al.* in prep.), this example emphasizes that flexible migrants will be faster migrants, and certainly will have more options during lean times. It also indicates that changes in prey quality at staging sites easily interfere with the annual cycles of such demanding migrants.

Global change and phenotypic change: plasticity prevails

The reclamation of mudflats currently taking place in the Yellow Sea (Fig. 110) is just the tip of an iceberg of human-wrought change to the planet (e.g. Vitousek *et al.* 1997). The Millenium Ecosystem Assessment (2005), carried out at the initiative of the United Nations, reports that 60% of the world's ecosystem services have been degraded. Many Mediterranean and temperate habitats had already been converted by 1950. In the tropics this process of conversion and degradation of habitats is still in full swing (Fig. 112). With a quarter of the earth's land surface now cultivated, and with much of the seafloor affected by fishing gear (Pauly 2007), the biosphere is heavily affected by humankind. Even the current warming of the atmosphere is attributed in part to the emission of greenhouse gases by an ever expanding human population (e.g. Allen *et al.* 2009). Despite the considerable power of organisms to cope with environmental change, for many, the complete disappearance of (micro-) habitats exceeds their ability to cope. Species extinction rates are now 100–1000 times the background rate (Lawton and May 1995, Millenium Ecosystem Assessment 2005).

As biologists there is thus much for us to deplore, but the good news is that by now our trade has quantified biological phenomena for long enough to look for patterns in what phenotypic changes can teach us. Andrew Hendry and Michael Kinnison (1999) endeavoured to assemble and maintain a compilation

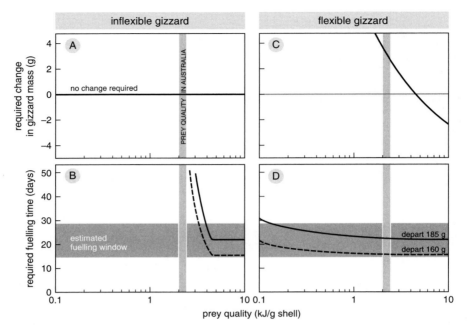

Figure 111. Fuelling schedules of red knots preparing to migrate onward to High Arctic breeding grounds from an intertidal staging area in the Yellow Sea. Left-hand plots (A and B) are for birds that do not adjust their gizzard mass after arrival from northwest Australia in the Yellow Sea staging area (inflexible gizzards), and right-hand plots (C and D) are for birds that do (flexible gizzards). Prey quality in the Yellow Sea is varied along the X-axes. The upper plots show the predicted change in gizzard mass for birds that would fuel as fast as possible (note that in A an invariant gizzard is modelled so no change occurs). The lower plots show the time required to fuel up to departure masses of either 160 g or 185 g as a function of prey quality and, in (D), gizzard mass. The estimated 'fuelling window' based on observed arrival and departure dates is shaded, as is, for reference, the prey quality in northwest Australia.

From Battley *et al.* (2005).

of quantitative changes of animal phenotypes through time (up to 200 generations). They included phenotypic changes at single sites, but also changes taking place when animals were moved, or moved themselves, from one site to another. Phenotypic change can be expressed in two alternative, but correlated, metrics called Darwins and Haldanes (Hendry and Kinnison 1999). The simplest is the Haldane, the absolute change in standard deviations of a trait per generation. In their latest summary of the data, Hendry *et al.* (2008) show that phenotypic change is pervasive in all studied contexts (Fig. 113). Nevertheless, rates of phenotypic change were measurably greater in habitats changed by humans compared with 'natural contexts'. In

fact, phenotypic change due to human-driven disturbances of various kinds can be very abrupt. Of obvious interest in the context of this book is the question whether these phenotypic changes reflect changes in genetic make-up, i.e. demonstrating natural selection in action (Endler 1986), or whether they mainly represent phenotypic plasticity.

On the basis of comparisons between assessments of phenotypic change in free-living organisms, Hendry *et al.* (2008) concluded that 'our analysis suggests a particularly important contribution from phenotypic plasticity'. In an independent meta-analysis for genetic responses to climate change, Gienapp *et al.* (2008) similarly concluded that 'clear-cut evidence indicating a significant role for evolutionary adaptation

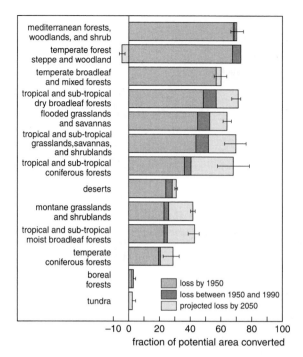

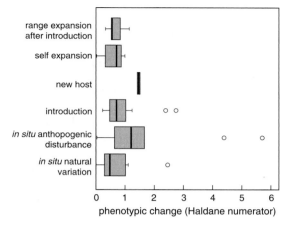

Figure 113. Phenotypic changes associated with each category of environmental change (see text for details). The data summarized here are based on the mean value per system of the Haldane metric for phenotypic change (the absolute change in standard deviations per generation calculated as the difference between two sample means divided by the within-population standard deviation and the number of elapsed generations; Haldane 1949, Gingerich 1993). Boxes include 50% of the data, central horizontal lines represent medians, whiskers contain the remainder of the data excluding outliers (open circles).

Based on a compilation of 2414 Haldane rates.
Modified from Hendry *et al.* (2008).

Figure 112. There are numerous ways to visualize what humans do to planet earth. By way of example, we have chosen a summary of historical, ongoing, and projected conversions of terrestrial biomes worldwide from the United Nations' Millenium Ecosystem Assessment. A biome is the largest unit of ecological classification that can be conveniently recognized across the globe. The potential area of biomes was based on soil and climatic conditions and forms the basis for establishing the extent to which these had been converted by 1950, between 1950 and 1990, and how much will have been converted by 2050. Near-terrestrial biomes, such as mangroves and intertidal flats, were not included because their areas were too small to be accurately assessed. Most of the conversion of biomes is to cultivated systems.

From Millenium Ecosystem Assessment (2005).

to ongoing climate warming is conspicuously rare'. A case in point is a 47-year study on ring-billed gulls *Larus novaehollandiae* in New Zealand (Teplitsky *et al.* 2008). Over time these birds became smaller and lighter, a change that could be correlated with increasing local temperatures. Using advanced quantitative genetic modelling, Céline Teplitsky and co-workers were able to separate the expected effects of genes that individuals would pass on to off-

spring from the effects of the environment, to conclude that there was *no* evidence for *any* genetic change.

Perhaps such conclusions were made rather reluctantly; after all, quantitative geneticists strive to measure just that: quantitative genetic changes. This leads us to think that the evidence from phenotype monitoring in a changing world so far provides abundant evidence that phenotypic responses to environmental change reflect phenotypic plasticity rather than genetic change. Let us now examine some more examples of this in relation to our changing climate.

How flexible phenotypes cope with advancing springs

As we have admitted in Chapter 1, the two authors are both products of the Dutch

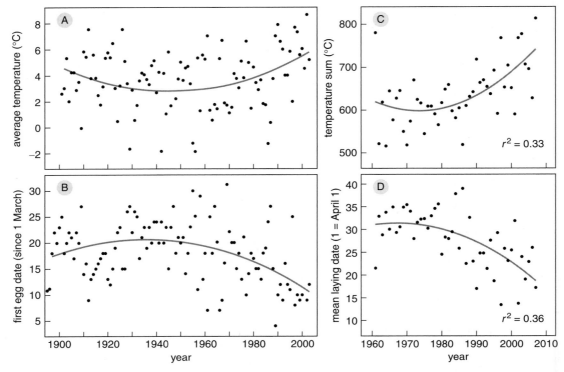

Figure 114. Changes in spring temperatures in The Netherlands (A, average temperature between 16 February and 15 March) and in Oxfordshire, UK (C, sum of daily maximum temperatures between 1 March and 25 April) in relation to the first egg date of northern lapwing in the province of Fryslân (B) and mean laying of great tits in Whytham Wood, Oxford (D).

Compiled from Both *et al.* (2005) and Charmantier *et al.* (2008).

countryside, a totally man-made temperate latitude human habitat. For us northerners, the arrival of spring is a happy event, and any changes in the timing of seasonal temperature rises (Fig. 114A) will become a topic of conversation. In the bad old days, one of the most welcome features of spring was the sudden availability of fresh protein in the form of eggs of the earliest laying bird species. Mothers sent their children out to gather them, and, even though our own mothers did not encourage us, at this point we have to admit that part of our ecological education was learning to find (and then take and eat) the eggs of lapwings *Vanellus vanellus*, a harvesting tradition still clung to in the province of Fryslân (Jensma 2009).

Then as now, finding the first egg of the season brings honour and fame in its wake, including a reception by royalty or their provincial representative, and ensuing newspaper coverage (Jensma 2009). Using these archival services we were able to go back more than a century to plot first egg dates in Fryslân (Fig. 114B). These varied by almost a month. Over this period there have been huge changes in the management of the grasslands where lapwings breed (e.g. Beintema *et al.* 1985, Schekkerman *et al.* 2008, Schroeder *et al.* 2009), but, to our surprise, most of the variation in first egg dates was explained by weather, with little variation left to account for statistically by meadowland-management (Both *et al.* 2005). A decrease in spring temperatures from 1900 to 1940 coincided with later first egg dates. During the period of temperature increase that followed, lapwing eggs were found increasingly early.

Similar patterns of change in the timing of reproduction relative to temperature changes have now been established in several birds, with nest-box breeding species yielding the most detailed information because of ease of study. An increase in spring temperature in the southern UK (Fig. 114C), was correlated with an advance in laying date of the famous great tits of Wytham Woods, Oxford (Fig. 114D). In this analysis (Charmantier *et al.* 2008), and in a similar study of great tits in The Netherlands (Nussey *et al.* 2005), the change in timing is explained by individual adjustment of behaviour, rather than by genetics. Hanging on to their predominantly quantitative genetic paradigm, both studies emphasize that selection would favour highly phenotypically plastic individuals.

Observational studies at single sites where trait changes are correlated with singular climatic change variables, run the risk of confusing correlation with causation. Capitalizing on Europe-wide and largely independent studies on another nest-box breeder, the long-distance migrating pied flycatcher *Ficedula hypoleuca* that winters in sub-Sahelian West Africa, Christiaan Both and 22 co-authors established near-experimental evidence that changing breeding times are actually causally related to climate change (Fig. 115). As we have seen (Fig. 114A and C), over the last half-century, Western Europe has experienced steadily warming springs. Further south and north in Europe this is not the case (Fig. 115). It is impressive to find that pied flycatchers bred increasingly earlier in areas where temperatures had gone up, but delayed egg-laying in areas where temperatures had gone down! Despite the suggestion made by their name, in many woods, pied flycatchers feed their young by relying on a peaked resource of small caterpillars that feast on crops of freshly budded

leaves. Caterpillar-phenology is tightly coupled with temperature (Both *et al.* 2009), such that pied flycatchers that advance egg-laying in warmer springs produce more offspring than flycatchers that do not (Both *et al.* 2006). The adaptiveness of earlier egg-laying in pied flycatchers is thus well-established. Whether this phenotypic change also involves any genetic change remains to be seen (Both and Visser 2005, Both 2010).

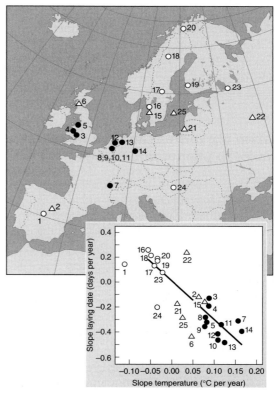

Figure 115. The timing of pied flycatcher breeding has been followed over long periods in many nestbox studies distributed across Europe. At some study sites there was a trend towards colder springs (open circles), at some there was mild warming (open triangles, between 0 and +0.08°C per year), and at the rest there was a clear spring temperature increase (closed dots, more than +0.08°C per year). The inset shows that, depending on the temporal change in spring temperature, pied flycatchers either delayed or advanced egg laying.

Modified from Both *et al.* (2004).

For now we conclude that the phenotypic changes observed in so many taxa and systems in seeming response to climate change (Parmesan and Yohe 2003, Walther *et al.* 2005), do indeed demonstrate the adaptive power of phenotypic plasticity. Arguably, for conservation managers, it becomes most interesting when the limits to such plasticity are reached (for more tit examples, see Visser *et al.* 2003, 2004), or even when the environmentally induced changes to the phenotype become maladaptive (Monaghan 2008). When this happens, conservation-conscious governments that halt the dredging or reclamation of mudflats (Fig. 110), reversing destruction of habitats (Fig. 112), or even stabilizing or reversing atmospheric agents of climate change, could make the greatest difference.

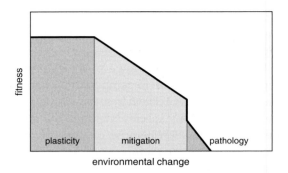

Figure 116. Degrees and means by which organisms can cope with different degrees of environmental change: a three-phased reaction norm. Within a certain range of environmental change, development of an alternative phenotype has no negative fitness consequences. Beyond this range, fitness declines because the development of an optimal phenotype is constrained. As long as further phenotypic adjustments are able to mitigate negative effects on fitness, the decline is slow, but outside this 'mitigation zone', pathologies develop and fitness drops steeply.

From Monaghan (2008).

When organisms can cope no more: limits to phenotypic change

Red knots returning light and with a small gizzard from the tundra were less likely to stay at traditional mudflat sites in The Netherlands and survive (Fig. 109). They illustrate the principle that, when times get tough, organisms may be able to cope locally, may have the option to move, or may even be forced to move. Coping locally or moving would entail small energy costs, so that, over a certain range of environmental changes, fitness cost could be negligible (Fig. 116). At some level of environmental change, however, organisms are no longer able to express the locally optimum phenotype, and in this range, the necessary phenotypic adjustments do entail fitness costs. For an example, recall from Chapter 4 that at some levels of work (energy expenditure) required to survive demanding conditions (including the provisioning of offspring), there are negative downstream consequences, notably on survival. And when the going really gets tough, under severely stressful condi-

tions, initially adaptive adjustments may even turn into pathologies (Sapolsky 2004). This will reduce fitness even further (Fig. 116).

Moving to another habitat patch is a coping strategy that is relevant as long as suitable alternatives remain, and as long as organisms have both the capacity and the resources to make a move. Many organisms cannot make a move—think of barnacles on a rock—and simply have to adjust their phenotype when the environment changes. A category of coping, particularly relevant to humans, is the downstream effect of adjustments made early in life on health and vigour later in life. In a study of the joyful dancing corvids called red-billed choughs *Pyrrhocorax pyrrhocorax* on the Scottish island of Islay, Jane Reid and co-workers (2003) showed that birds raised in poor environments carried this stamp from their childhood throughout life: they survived a little worse and showed lower breeding success, even if they had meanwhile moved to the best chough habitats later in life. In choughs, chicks tend to cope with days of food shortage (and no growth), by faster compensatory growth as soon as the food supply improves (Schew and Ricklefs 1998, Metcalfe and Monaghan 2001). That such growth accelerations may actually come with costs later in life was shown in an elegant experiment with zebra finches. Here the degree of growth compensation was negatively correlated with the speed of learning a simple discriminatory task later in life (Fisher *et al.* 2006).

Similar negative delayed effects of compensatory growth on cognition have been reported for small-born human infants (e.g. Estourgie-van Burk *et al.* 2009). It is the absence of a 'silver spoon' that tragically condemns some human groups to disease and poverty, partly because of the delayed detrimental effects of poor nutrition in early life (Gluckman *et al.* 2009), or even during the pregnancy of the mother (Roseboom *et al.* 2006). What is worse, longitudinal studies of children born in Amsterdam immediately after the Dutch winter famine at the end of the Second World War, showed that pre-natal famine may lead to poorer health, even in the offspring of these post-war children (Painter *et al.* 2008). That health transcends generations suggests that information other than

genes influences phenotypes. Whether there is more to phenotypic evolution than gene selection, and the very role of phenotypic plasticity in evolution, will be the subject of the next and final chapter.

Synopsis

Humans modify habitats at a fast pace. With the disappearance of species-specific 'good' habitat there is no way to avoid individual disappearance and species extinctions. Species that persist tend to show considerable directional phenotypic change across generations. Despite the best attempts to demonstrate that such changes reflect genetic change, firm evidence for such micro-evolutionary processes in multicellular organisms is still quite limited. Instead, plants and animals have great phenotypic plasticity, which, in several well-studied examples from the global 'experiment' of climate change, helps these organisms to adjust and survive. Unless environmental conditions become so bad that coping strategies generate delayed negative effects, or even phenotypic breakdown, the various kinds of phenotypic plasticity certainly help organisms to survive in a world of change.

CHAPTER 10

Evolution in five dimensions: phenotypes first!

... actually, if you say that genetics is relating variation in the genome to variation in phenotype, there is more accessible variation in expression than there is in sequence, and there is more variation in phenotype between cells and tissues and organs than between people.

P. Brown in an interview by J. Gitschier (2009)

And the older identical twins become, the easier it is to tell them apart! Not surprisingly, the shaping roles of development and environment thus leave their marks on bodies that carry identical genomes. Bodies are flexible, they change with age and with environment, but how does this relate to evolution? In this chapter we will elaborate, perhaps in a somewhat partisan manner, the view that there is much more to evolution than selection of genes, which is the single axis of inheritance acknowledged in the 'Modern Synthesis' (e.g. Mayr 1982, 2002, Coyne 2009a). As scientists we seem to be on the verge of a 'post-Modern Synthesis' in which the flexible phenotype, in interaction with its environment, will take centre stage.

Flexible phenotypes and the study of adaptation

This book has presented many examples of organisms responding to environmental challenges, including humans during space-travel and competing in sports, and migrant birds conquering much of the globe in the course of their demanding annual itineraries. We hope to have succeeded in building an ever more powerful case that organisms and their particular environments are inseparable: changes in an individual's shape, size, and capacity will often be direct functions of the ecological demands placed upon them. Bodies express ecology.

What we have been talking about are adaptations, the functionally appropriate adjustment of phenotypes. In assembling our case, we have tried to foster the view that, for evolutionary insight, phenotype–environment interactions need to be studied functionally (Feder and Watt 1992, Watt 2000). Rather than emphasizing simply that the *capacity* for phenotypic change is an adaptation (which it might well be; see Chapter 5 and Pigliucci 2001a, 2001b), inter- and especially intra-individual *trait variation* can be used to evaluate the 'goodness of design' criterion for phenotypic adaptation. As argued by Piersma and Drent (2003), intra-individual phenotypic variations can demonstrate how alternative phenotypic representations of a single genotype represent adaptations. This is particularly appropriate when interspecific comparisons cannot be made. For example, to overcome the absence of matched food-specialists within the same taxon when studying digestive adaptations, Levey and Karasov (1989) compared the flexible phenotypes of birds that were primarily frugivorous in autumn and insectivorous in spring.

Perhaps most importantly, reversible intra-individual phenotypic variation gives us a powerful tool to study phenotypic design *experimentally* (Sinervo and Basolo 1996, Sinervo 1999), almost in the same way that behavioural ecologists have so successfully unravelled the functions of behavioural traits (Krebs and Davies 1987). In one tradition, such 'phenotypic engineering' has made use of hormone-implants, including the male sex hormone testosterone (Sapolsky 1997, Ketterson and Nolan 1999). In the dark-eyed juncos *Junco hyemalis* studied by the extended wife-and-husband team of Ellen D. Ketterson and Val Nolan (Ketterson *et al.* 1991, 1996), testosterone affected an array of phenotypic traits and life-history characteristics. This wide range of (interacting) effects actually limited the power of the hormonal implant approach. One of the examples of how natural and experimental variations of phenotype can enlighten issues of evolutionary adaptation, is spelled out in Chapters 6 and 7. It is concerned with the variably large muscular gizzards of birds eating hard-shelled prey (Piersma *et al.* 1993b). Our work with red knots showed that birds not only increased gizzard size when they shifted feeding from soft to hard food, but also increased their capacity to process that hard food, and *vice versa* (van Gils *et al.* 2003a). As we have seen in Chapter 9, in an ecological context where large gizzards are necessary, individual red knots that managed to build up larger gizzards after returning from the Arctic tundra, were the ones who survived longest (van Gils *et al.* 2006b).

At this point in the argument, the study of appropriate phenotypic adjustment within generations meets the study of cross-generational aspects of phenotypic adjustment—evolution proper. In evolutionary biology, a phenotypic trait can be considered an adaptation only if there is evidence that it has been moulded in specific ways during its evolutionary history to make it more effective for its particular role (Williams 1966b, West-Eberhard

1992). The question of whether this evolutionary moulding is strictly a process of novel genetic instruction, or mostly one of the self-organized nature of pre-existing variations (Newman and Müller 2000, Camazine *et al.* 2001, Turner 2007), cries out for much more investigation than we have the opportunity to indulge in here. Consequently we shall be obliged to leave it to one side and focus on the links between phenotypic plasticity and evolution.

Separating the environment from the organism...

In an insightful essay, Richard Lewontin and Richard Levins (2007) suggest that the most powerful contribution made by Charles Darwin to the development of modern biology was *not* his satisfactory evolutionary mechanism, but his 'rigorous separation of internal and external forces that had, in previous theories, been inseparable'. Before Charles Darwin, during the times of Lamarck and Charles' grandfather Erasmus, many scientists entertained the thought of evolution involving modification by descent. However, by allowing animals to propagate acquired characters in their offspring, they 'totally confounded inner and outer forces in an unanalyzable whole' (Lewontin and Levins 2007). Charles Darwin separated a force (mutation) that was entirely internal to the organism and determined the variation between individuals, from another force (natural selection) that was entirely external and determined which of the variations would survive and reproduce (survival of the fittest). Half a century after Darwin's (1859) *'The origin of species'*, Johannsen (1911) proposed another distinction between the 'internal and external forces' when he considered the *genotype* to be the embellishment of a genome with its genetic instruction, and the *phenotype* to be the physical entity that interacts with the outside world. Evolution happens because variable phenotypes are differentially successful in propagating the genetic instructions that they carry to subsequent generations.

This separation of internal and external forces undoubtedly led to the incredibly successful modern reductionist biology that we witness today. However, the many examples of externally-driven but internally-processed and reversible intra-individual phenotypic responses that have been reviewed in this book illustrate the limitations of a world view that ignores the complex interplay between genotype, body, and environment. According to Lewontin and Levins (2007), the 'separation [between internal and external forces] is bad biology and presents a barrier to further progress'.

...and putting them back together: phenotypes first!

In her book of epic proportions *Developmental plasticity and evolution*, Mary Jane West-Eberhard (2003) sought to give developmental biology back its rightful place in evolutionary biology.[1] Most relevant for our current stream of thought is her five-step scenario for the role of phenotypic plasticity in the generation of novelty, as well as adaptedness, in evolution (this summary is adapted from Jablonka 2006).

Step (1), and we have seen abundant examples of this, is the realization that *organisms are inherently responsive to environmental changes* (italics for emphasis). Step (2) states that such *phenotypic responses, which often are adaptive, represent the 'normal' reaction to new challenges, where 'normal' means that this capacity is built-in.* Step (3) has not been introduced so far. It states that *responses to changed conditions depend on the*

modular organization of the phenotype. Step (4) argues that *organisms that respond in a functional manner to new conditions are selected.* This is followed by step (5), where *genetic changes that make the adaptive phenotypic change more precise and less costly are selected.*

Step (2) represents the process of phenotypic accommodation, the 'self-righting property' of developmental systems that ensures that any intrinsic change will trigger a sequence of regulatory changes to generate an integrated phenotype automatically (Alberch 1989, Blumberg 2009, Coyne 2009b). With respect to step (3), arguments too complicated to be brief about here (but expounded in West-Eberhard 2003, Kirschner and Gerhart 2005), have been developed to suggest that a modular design best enables a developing phenotype to sort itself out adaptively under new environmental conditions (see also Parter *et al.* 2008). Step (5) represents the concepts of genetic assimilation and genetic accommodation (Crispo 2007).

Genotypes accommodating environmental information?

These confusing and related concepts, genetic assimilation and accommodation, originated over 100 year ago (Baldwin 1896, Morgan 1896, Osborn 1897, Waddington 1942). They have generated an interesting theoretical literature (e.g. Pigliucci 2001a, West-Eberhard 2003, Lande 2009), which we shall have to leave to one side in order to concentrate on the main purposes of this book. Following Pigliucci *et al.* (2006), we have chosen to use to use the term *genetic assimilation* for the important, but difficult, concept that, after a shift in environmental conditions, developmental plasticity can produce a partially adapted phenotype, which subsequently becomes fine-tuned to the new conditions by natural selection (Lande 2009).

---

[1] The burgeoning literature on the role of plasticity in evolution is testimony to West-Eberhard's success.

In a hierarchical order, natural selection would represent the central force selecting the fittest phenotypes under all conditions. In changing environments these would be represented by properly adjusted phenotypes from a plasticity range (i.e. because the organism is able to perform the appropriate phenotypic accommodation). If environmental conditions persist, any costs of plasticity (see Table 7) would ensure that novel phenotypes become genetically assimilated.

You will have noticed, of course, that the evolutionary scenario of West-Eberhard (2003) represents a radically different view on how evolution proceeds from that which we have all learned at school. In the 'Modern Synthesis', random mutations in the genome have to generate new phenotypic variants, of which the appropriate versions would then be selected in the ambient environment. Appropriate phenotypic adjustments in new environments have to await appropriate mutations and subsequent selection. This takes time, and many animals have to die for a phenotypic change to take place, generating the 'cost of natural selection' or 'mutational load' identified by Haldane (1957, see Nunney 2003). According to this established view, the natural order of things is that *genes change first* and that *phenotypes follow*. The more comprehensive view, which incorporates considerations from developmental biology and acknowledges the important role of phenotypic plasticity (West-Eberhard 2003, 2005), states exactly the opposite: *phenotypes change first* and *genes follow*.

What West-Eberhard (2003, 2005) has quite successfully engaged (see Gottlieb 1992 for an earlier attempt), is the research agenda 'for the mechanisms that connect induction and retention of within-generation accommodations in evolutionary lineages' (Badyaev 2009). The struggle we face reconciling variability and heredity has succinctly been stated by Alexander Badyaev (2009) as 'adaptation to changing environments requires generation of novel developmental variation, but heritability of such variation should, by definition, limit the range of future variability'. Figure 117A illustrates, in the most general terms, how genetic accommodation could take place. Summarizing many years of fieldwork on the rapidly colonizing house finch *Carpodacus mexicanus* (a species that until the 1850s was limited to the warmer parts of west-central USA and northern Mexico), Badyaev (2009) sketches a scenario of hormone-related induction of novel phenotypes and their 'subsequent directional cross-seasonal transfer of a subset of developmental outcomes favoured by natural selection' (Fig. 117B). This scenario would account for the recent fast and locality-specific changes in phenotype and life-history traits of this very successful colonizer.

Threespine sticklebacks *Gasterosteus aculeatus* live in marine, as well as limnic, environments, and are a celebrated case of adaptive radiation (e.g. Schluter 1993, 2000). After the retreat of the glaciers, over the last 15,000 years, throughout the northern hemisphere sticklebacks have become isolated in lakes and often show a kind of sympatric speciation (with genetic differentiation) towards a bottom-living blunt-headed form that specializes in feeding on bottom-living prey, such as insect larvae, and a mid-water form with a long beak that specializes in feeding on zooplankton. According to the genetic assimilation scenario, one would expect the original marine form to express full developmental plasticity for the appropriate shapes of bottom- or mid-water-living fish and the specialized forms to have lost this capacity. Although there was evidence for some subtle differences in plasticity in the bottom-living form (Wund *et al.* 2008), all populations—despite their genetic divergence—displayed the whole range of functionally appropriate plasticity, but with little evidence for genetic assimilation (Fig. 118).

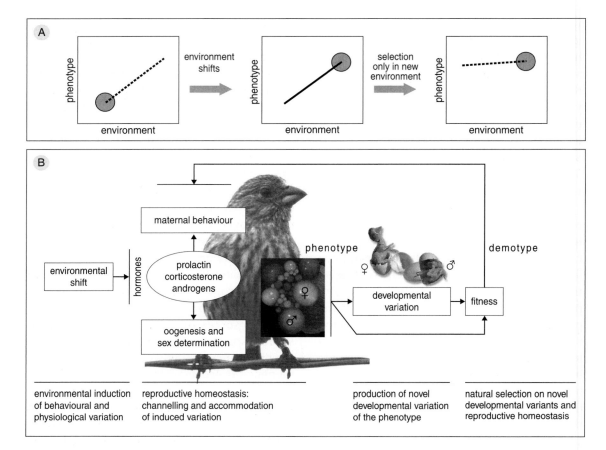

Figure 117. An explanation of how genetic assimilation should work (A), and one of the first empirical examples of how it does work (B). In (A) the population is initially occupying a single environment. When the environment changes, a pre-existing reaction norm allows the population to persist because it produces novel phenotypes with no initial genetic change. If natural selection keeps operating in the new environment, the novel phenotype may become genetically fixed (assimilated), and the original reaction norm may lose plasticity if there are costs associated with maintaining plasticity when it is not favoured. Panel (B) summarizes the origin of novel adaptations during range expansion by house finches *Carpodacus mexicanus*. New environmental conditions induce novel behavioural and physiological variations. The involvement of a single set of hormonal mechanisms in regulating environmental assessment, incubation behaviour, and oogenesis, ensures the channelling and accommodation of the induced variation, resulting in production of novel functional phenotypic variants. Subsequently, natural selection acts on the resulting developmental variants and, if the fitness benefits persist, developmental outcomes of the environmental induction would do too.

Compiled from, respectively, Pigliucci *et al.* (2006) and Badyaev (2009).

Nevertheless, support for the notion of genetic assimilation is now accumulating. In littorinid snails of the rocky shores of Atlantic North America, different populations, which are genetically indistinguishable, have historically been exposed to an invasive predatory crab for variable lengths of time. These time variations enabled Edgell *et al.* (2009) to show that a critical behavioural anti-predation trait (soft tissue withdrawal in the presence of crabs) actually lost its plasticity within a few hundred years. A similar case was made by Aubret and Shine (2009) on the basis of newly established island populations of tiger snakes

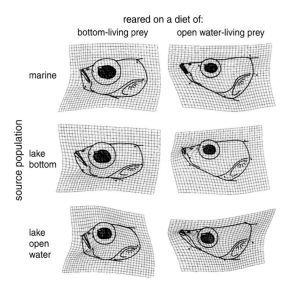

reared on a diet of:
bottom-living prey   open water-living prey

marine

source population

lake bottom

lake open water

Figure 118. Effect of either a diet of bottom-living or open-water living prey on head shape in threespine sticklebacks *Gasterosteus aculeatus* reared from parents from the (evolutionary) source population in the sea (top), from a descendant lake population that lives on bottom-dwelling prey (middle) and a lake population with a mid-water distribution (bottom). The 'deformation' grids are exaggerated four times to highlight shape differences.

Compiled from Wund *et al.* (2008).

Notechis scutatus around the Australian continent. These snakes benefit from large heads in places where large prey are available. Such large heads are achieved by phenotypic plasticity in 30-year young populations but seem fixed by genetic assimilation in 9000-year old populations. Whatever the evidence for genetic assimilation, these examples illustrate once more that genotypes and phenotypes show a fair degree of decoupling.

Enter the tarbutniks, and niche construction

Tarbutniks are hypothetical small rat-like animals with long-whiskered snouts and long tails with two streamers (Fig. 119). They were named after the Hebrew word for culture, i.e. 'tarbut' (Avital and Jablonka 2000, Jablonka and Lamb 2005). Tarbutniks are unique in having perfect DNA-maintenance systems; all these imaginary animals are genetically identical to each other. One would almost say that tarbutniks are unable to evolve. However, because of the downstream effects of chance events during development (we are reminded of the degree to which identical human twins can differ, and see Vogt *et al.* 2008 for a wonderful example on variability in tarbutnik-like, clonal crayfish), tarbutniks do show variations in size and shape and the colour of their fur.

Tarbutniks live in small family groups and join their parents on their foraging trips. In this way they learn about their environments, which foods are good and which foods are better avoided (Fig. 119). Their capacity to learn as young animals, but also as adults, means that new adaptive behaviours accomplished either by accident, or by copying others, will be transmitted to younger generations. During courtship, tarbutniks have the habit of offering nuptial gifts (males) and either accepting these gifts or not (females). Now imagine a tarbutnik female that has been raised in a family that is fond of berries. She flatly refused the attentions of a nut-offering male, but accepts the male that is offering berries (Fig. 119). Everything else being equal, the population would steadily separate itself into a berry- and a nut-loving clan.

The clans would diverge even more rapidly if the berry-loving tarbutniks were to hit upon the invention of protecting and nurturing berry-bearing plants; or if nut-loving tarbutniks were to start cultivating nut-bearing bushes. At this point tarbutniks have 'constructed their niche' (Laland *et al.* 2008), and this niche cannot but reciprocally influence what the tarbutniks will look like. Imagine that berry-bearing plants are small, but that nut-bearing bushes need climbing into. The tarbutniks that love nuts would train their

Figure 119. Socially mediated learning in tarbutniks, long-tailed and long-snouted vertebrates that are genetically identical and have perfect DNA-maintenance systems so that genes never change. Chance events during development ensure that, just like identical twins, they do show small differences in size, shape, fur colour and various aspects of their learned behaviour. In (A) a young tarbutnik is introduced to carrots by its mother, which it consequently eats throughout life (B). In (C) a berry-loving female responds to the advances of a berry-offering male, but (D) shows that she rejects the advances of a male offering nuts.

Illustrations by Anna Zeligowski, from Jablonka and Lamb (2005, p. 157 and p. 160).

young to be agile climbers from a tender age onwards. On the basis of use–disuse and other self-organizing principles that we have encountered in this book, even our untrained eyes would be able to tell a nut-loving tarbutnik from a berry-loving one.

Let us now, for the sake of argument, bring back in genes and genetic variation. Let us turn tarbutniks into something that vaguely resembles them, the short-lived Australian marsupial mice that go under the genus name *Antechinus*, for example. What we would undoubtedly see in antechinal tarbutniks is co-evolution between habits and genes; there would be 'reciprocal causation' (Fig. 120; Laland *et al.* 2008). In the same way that organ-

isms construct their own niches (Odling-Smee *et al.* 1996, 2003, Laland *et al.* 1999, Olff *et al.* 2009), so will the niches 'construct' the organismal phenotype. Here is another reason why, even in the evolutionary sense, organisms and environments are inseparable.

Evolution in four or five dimensions?

In their imaginative book *Evolution in four dimensions*, Eva Jablonka and Marion Lamb (2005) expound the view that, in addition to the single evolutionary dimension that we all are used to, and work with—genetic inheritance—there are three more ways that informational carry-over between generations can generate

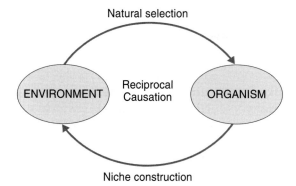

Figure 120. A schematic summary of the two-way interactions between organisms and environment called 'reciprocal causation' combines the forces of natural selection (organisms shaped by the environment) and niche construction (environments shaped by the organism). Another way of describing reciprocal causation is to say that it represents the evolutionary feedbacks between phenotype-induced changes to their ambient environments (niche) and the changed selection pressures by these environments on the phenotype (body).

Diagram designed by F.M. Postma and J. van Gestel.

evolution (Fig. 121). With the tarbutniks, we have just been introduced to what they call 'the behavioural inheritance system' (Avital and Jablonka 2000). This represents the building of the niche; routines can be taught to offspring and thus transferred down the generations. In addition, they distinguish an inheritance route of extra-genetic information through the germline called 'cellular epigenetic inheritance' (Jablonka and Raz 2009), from an inheritance route that includes maternal effects called 'developmental inheritance' (Mousseau and Fox 1998, Bateson *et al.* 2004, Groothuis *et al.* 2005). The final inheritance pathway is one that humans have specialized in, a mechanism using symbols and including the writing of books, we call culture.

There are various ways to cut the cake, of course. Jablonka and Lamb (2005, 2007) somehow decided that their five inheritance path-

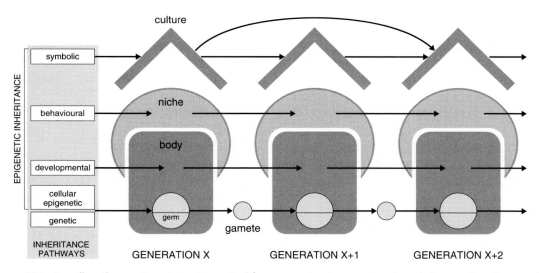

Figure 121. As well as the genetic variation transmitted from generation to generation through the germline, there are four additional 'epigenetic' inheritance pathways. The first of those, 'cellular epigenetic inheritance', transmits acquired information via the gametes. The remaining three are all 'somatic' and include 'developmental inheritance' (e.g. through maternal effects), 'behavioural inheritance' (through niche construction), and 'symbolic inheritance'. One of the reasons that it took biologists so long to distinguish these layers is that they are thoroughly intertwined. There is also 'reciprocal causation', especially between niche and body. Note that symbolic inheritance seems specific to organisms that read books, listen to CDs and watch movies and TV.

Developed from Jablonka and Lamb (2005, 2007), and correspondence with E. Jablonka.

ways (Fig. 121) represent four evolutionary dimensions, hence the name of their book; they merged the cellular-epigenetic and developmental pathways into one and called it 'the epigenetic inheritance system'. Here we stick to the five inheritance pathways identified: genetic, cellular-epigenetic, (somatic) developmental, behavioural, and symbolic. The latter four may also be combined under the liberal banner 'epigenetic inheritance' (Fig. 121).

These five inheritance pathways, combined with the quiet power of reciprocal causation, make phenotypes (and perhaps even genotypes!) respond much faster than what was 'allowed' under the one-dimensional, genetically straightjacketed, 'Modern Synthesis' outlined by Ernst Mayr and others. Phenotypes respond and interact at many different levels. The concept of a 'flexible phenotype' represents the core of post-modern evolutionary thinking.

Context, please! An orchestra in need of a theatre

In *The music of life: Biology beyond the genome*, the renowned physiologist Denis Noble (2006) takes issue with what he regards as the unwarranted 'reductionist causal chain', which sees organisms as simple expressions of their genetic instruction (Fig. 122A). He compares the organs, the systems of the body, to an orchestra. What he calls 'downward causation', the higher level controls on gene expression and the internal developmental feedbacks, is compared to the conductor of that orchestra. As a biomedic, perhaps he should be excused (though perhaps he should not be), but he and many others (e.g. Nüsslein-Volhard 2006) fail to provide their orchestras with somewhere to play, theatres that will co-determine whether an orchestra sounds good, finds a receptive audience and is worth the governmental sub-

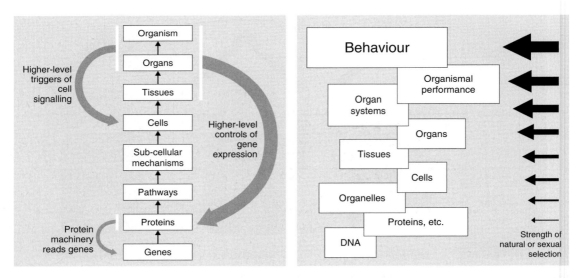

Figure 122. Two visualizations of the types of interaction embodied in phenotypes. In (A), Denis Noble (2006) takes issue with the simplistic reductionist causal chain (small upward arrows), pointing out that at all levels of biological organization there are loops of interacting 'downward and upward' causation, here exemplified by the controls of gene expression by higher organizational levels than the gene. In (B), Garland and Kelly (2006) use the same upward causal chain of increasing complexity to make the point that natural and sexual selection will act most strongly on the highest levels of biological organization. Natural selection typically does not act on genetic variation.

sidies, public donations or philanthropic efforts. The theatre of an organism, of course, is the environment. It is the environment that will determine whether the orchestra will succeed. But is there an audience that determines whether in a particular theatre a particular orchestra sounds good? Not visibly, certainly, but natural and sexual selection can be considered to enact the role of invisible spectators.

As Noble wrestled with downward causation and his musical metaphors, the evolutionary physiologists Theodore Garland and Scott Kelly designed a diagram (Fig. 122B) that embodies some aspects of Noble's representation and some aspects of the diagram with which we started this book (Fig. 1): the positioning of a developing phenotype in the context of its environment. Garland and Kelly (2006) make the point that natural and sexual selection act most strongly on aspects of the phenotype at the highest organizational levels, 'because they are the most strongly correlated with Darwinian fitness', the demotype. Excitingly, it is at the highest organizational layers of an organism, in the phenotype, the ethotype, and the demotype (Fig. 1), where the influence of the environment is most pervasive and where reciprocal causation feedbacks will come to the fore.

Perhaps then, behaviour has primacy in evolution (Bateson 2004, Aubret et al. 2007)? The tarbutniks of Avital, Jablonka, and Lamb show how behavioural change and learning can be the motor of evolution in a genetically invariant organism. Behaviour, however, can also act as a buffer against change. A move to a less stressful environment and a change of diet can shield organisms from strong directional selection (Duckworth 2009). Nevertheless, we have made our point: ecological studies of organisms that ignore the key elements of their physiology, physiological studies with disregard for the organisms' ecology, and behavioural studies carried out in ignorance of the

environmental context, might well all be a waste of time. Drastically paraphrasing Dobzhansky (1973) we believe that nothing in evolution makes sense except in the light of ecology!

Richard Lewontin (2000), on the last pages of his long essay on the interactions between genes, organisms, and environments *The triple helix: gene, organism, and environment*, laments that all too often research efforts focus on the measurable, rather than on the frontiers of insight. The invention of protein-gel electrophoresis to assay genetic diversity led a whole generation of evolutionary geneticists to do just that, assay genetic diversity. The advent of ever more powerful gene-sequencers may now limit biology to the questions that can be answered by DNA-sequences. What has driven the writing of this book is the realization that there is a continuum between organisms and environments that needs much more attention in the complementary realms of physiological, behavioural, ecological, and evolutionary study—the notable complication that even 'small changes in the environment lead to small changes in the organism that, in turn, lead to small changes in the environment' (Lewontin 2000). What we have come to realize as well is that life scientists must not only include the ecological theatre in their thinking and their measurements, but also that in the *biology of the future* there should be equal attention to the two, complementary, roles of the genetic coding of some processes and the self-organizational nature of how these processes are played out in the development and functioning of any organism. The fact that 'organisms are designed not so much because natural selection of particular genes has made them that way, but because agents of homeostasis build them that way' (Turner 2007), may well be the most important reason that organisms appear so wonderfully adapted to their environments and their environments to them.

Synopsis

In this book we have offered a progression of ideas that support the view that, although bodies and environments are recognizable entities, they really are inseparable. In this final chapter we extend this view to evolution—the inheritance of, and selection for, randomly variable phenotypic traits across generations. What we find is that evolutionary change needs systems of inheritance, but we also find that the genetic system that we all work with is just one of five such possible inheritance systems. Since organisms not only provide the beginnings and nurture of their offspring, but also build the environments in which they and their offspring live, there are very tight feedbacks of reciprocal causation, both in the development of organisms and in the relationships between the developing organism and their environments. Bodies are earth, and we would do well to acknowledge that in the ways that we study them.

References

Abdalla, E.B., Kotby, E.A., and Johnson, H.D. (1993). Physiological responses to heat-induced hyperthermia of pregnant and lactating ewes. *Small Ruminant Research* **11**, 125–34.

Adamo, S.A., Barlett, A., Le, J., Spencer, N., and Sullivan, K. (2010). Illness-induced anorexia may reduce trade-offs between digestion and immune function. *Animal Behaviour* **79**, 3–10.

Adams, G.R., Caiozzo, V.J., and Baldwin, K.M. (2003). Skeletal unweighting: spaceflight and ground-based models. *Journal of Applied Physiology* **95**, 2185–2201.

Alberch, P. (1989). The logic of monsters: Evidence for internal constraint in development and evolution. *Geobios* **19**, 21–57.

Alerstam, T. (2003). Bird migration speed. In *Avian migration*, eds P. Berthold, E. Gwinner, and E. Sonnenschein, pp. 253–67. Springer-Verlag, Berlin.

Alerstam, T. and Jönsson, P.E. (1999). Ecology of tundra birds: Patterns of distribution, breeding and migration along the Northeast Passage. *Ambio* **28**, 212–24.

Alexander, R.McN. (1984). Optimum strengths for bones liable to fatigue and accidental fracture. *Journal of Theoretical Biology* **109**, 247–52.

Alexander, R.McN. (1997). A theory of mixed chains applied to safety factors in biological systems. *Journal of Theoretical Biology* **184**, 621–36.

Alexander, R.McN. (1998). Symmorphosis and safety factors. In *Principles of animal design: The optimization and symmorphosis debate*, eds E.R. Weibel, C.R. Taylor, and L. Bolis, pp. 28–35. Cambridge University Press, Cambridge.

Allen, M.R., Frame, D.J., Huntingford, C., Jones, C.D., Lowe, J.A., Meinshausen, M., and Meinshausen, N. (2009). Warming caused by cumulative carbon emissions towards the trillionth tonne. *Nature* **458**, 363–68.

Alonso-Alvarez, C., Bertrand, S., Devevey, G.L., Prost, J., Faivre, B., Sorci, G., and Loeschcke, V. (2004). Increased susceptibility to oxidative stress as a proximate cost of reproduction. *Ecology Letters* **7**, 363–68.

Amann, M., Eldridge, M.W., Lovering, A.T., Stickland, M.K., Pegelow, D.F., and Dempsey, J.A. (2006). Arterial oxygenation influences central motor output and exercise performance via effects on peripheral locomotor muscle fatigue in humans. *Journal of Physiology* **575**, 937–52.

Anderson, K.J. and Jetz, W. (2005). The broad-scale ecology of energy expenditure of endotherms. *Ecology Letters* **8**, 310–18.

Anderson, T.R., Hessen, D.O., Elser, J.J., and Urabe, J. (2005). Metabolic stoichiometry and the fate of excess carbon and nutrients in consumers. *American Naturalist* **165**, 1–15.

Arad, Z., Midtgard, U., and Bernstein, M.H. (1989). Thermoregulation in Turkey vultures: Vascular anatomy, arteriovenous heat exchange, and behaviour. *Condor* **91**, 505–14.

Arngrimsson, S.A., Petitt, D.S., Borrani, F., Skinner, K.A., and Cureton, K.J. (2004). Hyperthermia and maximal oxygen uptake in men and women. *European Journal of Applied Physiology* **92**, 524–32.

Arsenault, D.J., Marchinko, K.B., and Palmer, A.R. (2001). Precise tuning of barnacle leg length to coastal wave action. *Proceedings of the Royal Society, London, B* **268**, 2149–54.

Aschoff, J. (1954). Zeitgeber der tierischen Tagesperiodik. *Naturwissenschaften* **41** 49–56.

Aschoff, J. and Pohl, H. (1970). Der Ruheumsatz von Vögeln als Funktion der Tageszeit und des Körpergrösse. *Journal für Ornithologie* **111**, 38–47.

Ashton, K.G. (2002). Patterns of within-species body size variation of birds: strong evidence for Bergmann's rule. *Global Ecology and Biogeography* **11**, 505–23.

Ashton, K.G., Tracy, M.C., and de Queiroz, A. (2000). Is Bergmann's rule valid for mammals? *American Naturalist* **156**, 390–415.

Aubret, F. and Shine, R. (2009). Genetic assimilation and the postcolonization erosion of phenotypic plasticity in island tiger snakes. *Current Biology* **19**, 1932–36.

Aubret, F., Bonnet, X., and Shine, R. (2007). The role of adaptive plasticity in a major evolutionary transition: Early aquatic experience affects locomotor performance of terrestrial snakes. *Functional Ecology* **21**, 1154–61.

Austad, S.N. (1997). *Why we age: What science is discovering about the body's journey through life*. Wiley, New York.

Avital, E. and Jablonka, E. (2000). *Animal traditions: behavioural inheritance in evolution*. Cambridge University Press, Cambridge.

Bacigalupe, L.D. and Bozinovic, F. (2002). Design limitations and sustained metabolic rate: Lessons from small mammals. *Journal of Experimental Biology* **205**, 2963–70.

Badyaev, A.V. (2009). Evolutionary significance of phenotypic accommodation in novel environments: An empirical test of the Baldwin effect. *Philosophical Transactions of the Royal Society, London, B* **364**, 1125–41.

Bailey, E. (1998). Odds on the FAST gene. *Genome Research* **8**, 569–71.

Baker, A.J., Piersma, T., and Greenslade, A.D. (1999). Molecular versus phenotypic sexing in red knots. *Condor* **101**, 887–93.

Baker, A.J., González, P.M., Piersma, T., Niles, L.J., de Lima Serrano do Nascimento, I., Atkinson, P.W., Clark, N.A., Minton, C.D.T., Peck, M.K., and Aarts, G. (2004). Rapid population decline in red knots: Fitness consequences of decreased refuelling rates and late arrival in Delaware Bay. *Proceedings of the Royal Society, London, B* **271**, 875–82.

Bakken, G.S. (1976). A heat transfer analysis of animals: Unifying concepts and the application of metabolism chamber data to field ecology. *Journal of Theoretical Biology* **60**, 337–84.

Bakken, G.S. and Gates, D.M. (1974). Linearized heat transfer relations in biology. *Science* **183**, 976–77.

Baldwin, J.M. (1896). A new factor in evolution. *American Naturalist* **30**, 354–451.

Banavar, J.R., Damuth, J., Maritan, A., and Rinaldo, A. (2003). Allometric cascades. *Nature* **421**, 713–14.

Barja, G., Cadenas, S., Rojas, C., Perez-Campo, R., and López-Torres, M. (1994). Low mitochondrial free radical production per unit O_2 consumption can explain the simultaneous presence of high longevity and high aerobic metabolic rate in birds. *Free Radical Research* **21**, 317–28.

Barnes, M. (1992). The reproductive periods and condition of the penis in several species of common Cirripedes. *Oceanography and Marine Biology: an Annual Review*, **30** 483–525.

Barter, M. (2002). Shorebirds of the Yellow Sea: importance, threats and conservation. *International Wader Studies* **12**, 1–104.

Bartholomew, G.A. (1958). The role of physiology in the distribution of terrestrial vertebrates. In *Zoogeography*, ed. C.L. Hubbs, pp. 81–95. Publ. No. 51, American Association for the Advancement of Science, Washington, DC.

Bartholomew, G.A. and Cade, T.J. (1963). The water economy of land birds. *Auk* **80**, 504–39.

Bateson, P. (2004). The active role of behaviour in evolution. *Biology and Philosophy* **19**, 283–98.

Bateson, P., Barker, D., Clutton-Brock, T., Deb, D., D'Udine, B., Foley, R.A., Gluckman, P., Godfrey, K., Kirkwood, T., Lahr, M.M., McNamara, J., Metcalfe, N.B., Monaghan, P., Spencer, H.G., and Sultan, S.E. (2004). Developmental plasticity and human health. *Nature* **430**, 419–21.

Battley, P.F. and Piersma, T. (2005). Adaptive interplay between feeding ecology and features of the digestive tract in birds. In *Physiological and ecological adaptations to feeding in vertebrates*, eds J.M. Starck and T. Wang, pp. 201–28. Science Publishers, Enfield.

Battley, P.F., Piersma, T., Dietz, M.W., Tang, S., Dekinga, A., and Hulsman, K. (2000). Empirical evidence for differential organ reductions during trans-oceanic bird flight. *Proceedings of the Royal Society, London, B* **267**, 191–96.

Battley, P.F., Dietz, M.W., Piersma, T., Dekinga, A., Tang, S., and Hulsman, K. (2001). Is long-distance bird flight equivalent to a high-energy fast? Body composition changes in freely migrating and captive fasting great knots. *Physiological and Biochemical Zoology* **74**, 435–49.

Battley, P.F., Rogers, D.I., Piersma, T., and Koolhaas, A. (2003). Behavioural evidence for heat-load problems in great knots in tropical Australia fuelling for long-distance flight. *Emu* **103**, 97–103.

Battley, P.F., Rogers, D.I., van Gils, J.A., Piersma, T., Hassell, C.J., Boyle, A., and Yang, H-Y. (2005). How do red knots *Calidris canutus* leave Northwest Australia in May and reach the breeding grounds in June? Predictions of stopover times, fuelling rates, and prey quality in the Yellow Sea. *Journal of Avian Biology* **36**, 494–500.

Bauchinger, U. and McWilliams, S. (2009). Carbon turnover in tissues of a passerine bird: Allometry, isotopic clocks, and phenotypic flexibility of organ size. *Physiological and Biochemical Zoology* **82**, 787–97.

Bauchinger, U., Kolb, H., Afik, D., Pinshow, B., and Biebach, H. (2009). Blackcap warblers maintain digestive efficiency by increasing digesta retention time on the first day of migratory stopover. *Physiological and Biochemical Zoology* **82**, 541–48.

Bech, C., Langseth, I., and Gabrielsen, G.W. (1999). Repeatability of basal metabolism in breeding female kittiwakes *Rissa tridactyla*. *Proceedings of the Royal Society, London, B* **266**, 2161–67.

Beck, C. (2004). Personal portrait Bart Kempenaers—Ornithology. *Max Planck Research* **2**/2004, 64–67.

Bednekoff, P.A. and Lima, S.L. (1998). Re-examining safety in numbers: Interactions between risk dilution and collective detection depend upon predator targeting behaviour. *Proceedings of the Royal Society, London, B* **265**, 2021–26.

Beintema, A.J., Beintema-Hietbrink, R.J., and Müskens, G.J.D.M. (1985). A shift in the timing of breeding in meadow birds. *Ardea* **73**, 83–89.

Bennett, A.F. and Ruben, J.A. (1979). Endothermy and activity in vertebrates. *Science* **206**, 649–54.

Bennett, G.F., Montgomerie, R., and Seutin, G. (1992). Scarcity of haematozoa in birds breeding on the arctic tundra of North America. *Condor* **94**, 289–92.

Berger, J. (2007). Fear, human shields and the redistribution of prey and predators in protected areas. *Biology Letters* **3**, 620–23.

Bergmann, C. (1847). Über die Verhältnisse der Wärmeökonomie der Thiere zu ihrer Grösse. *Göttinger Studien* **3**, 595–708.

Bergmann, O., Bhardwaj, R.D., Bernard, S., Zdunek, S., Barnabé-Heider, F., Walsh, S., Zupinich, J., Alkass, K., Druid, H., Jivinge, S., and Frisén, J. (2009). Evidence for cariomycete renewal in humans. *Science* **325**, 98–102.

Berry, H. (1990). *From LA to New York, from New York to LA.* Self-published, Chorley, UK.

Bibby, R., Cleall-Harding, P., Rundle, S., Widdicombe, S., and Spicer, J. (2007). Ocean acidification disrupts induced defences in the intertidal gastropod *Littorina littorea*. *Biology Letters* **3**, 699–701.

Biebach, H. (1990). Strategies of trans-Sahara migrants. In *Bird migration*, ed. E. Gwinner, pp. 352–67. Springer-Verlag, Berlin.

Biebach, H., Biebach, I., Friedrich, W., Heine, G., Partecke, J., and Schmidl, D. (2000). Strategies of passerine migration across the Mediterranean Sea and the Sahara desert: a radar study. *Ibis* **142**, 623–34.

Biernaskie, J.M., Walker, S.C., and Gegear, R.J. (2009). Bumblebees learn to forage like Bayesians. *American Naturalist* **174**, 413–23.

Biggerstaff, S. and Mann, M. (1992). Consummatory behaviors and weight regulation in pregnant, lactating, and pregnant-lactating mice. *Physiology and Behavior* **52**, 485–91.

Bigland-Ritchie, B. and Vollestadt, N. (1988). Hypoxia and fatigue: how are they related? In *Hypoxia: the tolerable limits*, eds J.R. Sutton, C.S. Houston, and G. Coates, pp. 315–25. Benchmark, Indianapolis, Ill.

Bijleveld, A.I., Egas, M., van Gils, J.A., and Piersma, T. (2010). Beyond the information centre hypothesis: Communal roosting for information on food, predators, travel companions and mates? *Oikos* **119**, 277–85.

Bilz, S., Ninnis, R., and Keller, U. (1999). Effects of hypoosmolality on whole-body lipolysis in man. *Metabolism - Clinical and Experimental* **48**, 472–76.

Bishop, C.M. (1999). The maximum oxygen consumption and aerobic scope of birds and mammals: Getting to the heart of the matter. *Proceedings of the Royal Society, London, B* **266**, 2275–81.

Black, A.E., Coward, W.A., Cole, T.J., and Prentice, A.M. (1996). Human energy expenditure in affluent societies: An analysis of 574 doubly-labelled water measurements. *European Journal of Clinical Nutrition* **50**, 72–92.

Black, J.L., Mullan, B.P., Lorschy, M.L., and Giles, L.R. (1993). Lactation in the sow during heat-stress. *Livestock Production Science* **35**, 153–70.

Blaxter, K. (1989). *Energy metabolism in animals and man.* Cambridge University Press, Cambridge.

Blumberg, M.S. (2009). *Freaks of nature: What anomalies tell us about development and evolution.* Oxford University Press, New York.

Bohning-Gaese, K., Halbe, B., Lemoine, N., and Oberrath, R. (2000). Factors influencing clutch size, number of broods and annual fecundity of North American and European land birds. *Evolutionary Ecology Research* **2**, 8230839.

Bonner, J.T. (2006). *Why size matters: From bacteria to blue whales.* Princeton University Press, Princeton, NJ.

Bonnet, X., Bradshaw, A.D., and Shine, R. (1998). Capital versus income breeding: An ectothermic perspective. *Oikos* **83**, 333–42.

Bonnie, K.E. and Earley, R.L. (2007). Expanding the scope for social information use. *Animal Behaviour* **74**, 171–81.

Bonser, R.H.C. (1996). The mechanical properties of feather keratins. *Journal of Zoology* **239**, 477–84.

Both, C. (2010). Flexibility of timing of avian migration to climate change masked by environmental conditions en route. *Current Biology* **20**, 243–48.

Both, C. and Visser, M.E. (2005). The effect of climate change on the correlation between life-history traits. *Global Change Biology* **11**, 1606–13.

Both, C., Artemyev, V.A., Blaauw, B., Cowie, R.J., Dekhuijzen, A.J., Eeva, T., Enemar, A., Gustafsson, L., Ivankina, E.V., Järvinen, A., Metcalfe, N.B., Nyholm, N.E.I., Potti, J., Ravussin, P-A., Sanz, J.J., Silverin, B., Slater, F.M., Sokolov, L.V., Török, J., Winkel, W., Wright, J., Zang, H., and Visser, M.E. (2004). Large-scale geographical variation confirms that climate change causes birds to lay earlier. *Proceedings of the Royal Society, London, B* **271**, 1657–62.

Both, C., Piersma, T., and Roodbergen, S. P. (2005). Climatic change explains much of the 20th century advance in laying date of northern lapwing *Vanellus vanellus* in The Netherlands. *Ardea* **93**, 79–88.

Both, C., Bouwhuis, S., Lessells, C.M., and Visser, C.M. (2006). Climate change and population declines in a long-distance migratory bird. *Nature* **441**, 81–83.

Both, C., van Asch, M., Bijlsma, R.G., van den Burg, A.B., and Visser, M.E. (2009). Climate change and unequal phenological changes across four trophic levels: Constraints or adaptations? *Journal of Animal Ecology* **78**, 73–83.

Bouwhuis, S., Sheldon, B.C., Verhulst, S., and Charmantier, A. (2009). Great tits growing old: Selective disappearance and the partitioning of senescence to stages within the breeding cycle. *Proceedings of the Royal Society, London, B* **276**, 2769–77.

Boyd, I.L. (2000). State-dependent fertility in pinnipeds: Contrasting capital and income breeders. *Functional Ecology* **14**, 623–30.

Brakefield, P.M. and Reitsma, N. (1991). Phenotypic plasticity, seasonal climate and the population biology of *Bicyclus* butterflies (Satyridae) in Malawi. *Ecological Entomology* **16**, 291–303.

Brakefield, P.M., Pijpe, J., and Zwaan, B.J. (2007). Developmental plasticity and acclimation both contribute to adaptive responses to alternating seasons of plenty and of stress in *Bicyclus* butterflies. *Journal of Biosciences* **32**, 465–75.

Briffa, M. and Sneddon, L.U. (2007). Physiological constraints on contest behaviour. *Functional Ecology* **21**, 627–37.

Brisbin, I.L. (1969). Bioenergetics of the breeding cycle of the ring dove. *Auk* **86**, 54–74.

Brody, S. (1945). *Bioenergetics and growth*. Reinhold (reprinted Hafner Publishing Company), New York.

Broggi, J., Hohtola, E., Orell, M., and Nilsson, J-Å. (2005). Local adaptation to winter conditions in a passerine spreading north: A common-garden approach. *Evolution* **59**, 1600–1603.

Brown, J.S. (1988). Patch use as an indicator of habitat preference, predation risk and competition. *Behavioral Ecology and Sociobiology* **22**, 37–48.

Brown, J.S. and Kotler, B.P. (2004). Hazardous duty pay and the foraging cost of predation. *Ecology Letters* **7**, 999–1014.

Brown, J.S., Laundré, J.W., and Gurung, M. (1999). The ecology of fear: Optimal foraging, game theory, and trophic interactions. *Journal of Mammalogy* **80**, 385–99.

Bruinzeel, L.W., Piersma, T., and Kersten, M. (1999). Low costs of terrestrial locomotion in waders. *Ardea* **87**, 199–205.

Brunner, D., Kacelnik, A., and Gibbon, J. (1992). Optimal foraging and timing processes in the starling, *Sturnus vulgaris*: Effect of inter-capture interval. *Animal Behaviour* **44**, 597–613.

Brunner, D., Kacelnik, A., and Gibbon, J. (1996). Memory for inter-reinforcement interval variability and patch departure decisions in the starling, *Sturnus vulgaris*. *Animal Behaviour* **51**, 1025–45.

Bryant, D.M. and Tatner, P. (1991). Intraspecies variation in energy expenditure: Correlates and constraints. *Ibis* **133**, 236–45.

Bryant, D.M. and Westerterp, K.R. (1980). The energy budget of the house martin (*Delichon urbica*). *Ardea* **68**, 91–102.

Bubenik, G.A. and Bubenik, A.B. eds (1990). *Horns, pronghorns, and antlers: Evolution, morphology, physiology, and social significance*. Springer-Verlag, New York.

Buehler, D.M. and Baker, A.J. (2005). Population divergence times and historical demography in red knots and dunlins. Condor **107**, 497–513.

Buehler, D.M. and Piersma, T. (2008). Travelling on a budget: Predictions and ecological evidence for bottlenecks in the annual cycle of long-distance migrants. *Philosophical Transactions of the Royal Society, London, B* **363**, 247–66.

Buehler, D.M., Baker, A.J., and Piersma, T. (2006). Reconstructing palaeoflyways of the late Pleistocene and early Holocene red knot (*Calidris canutus*). *Ardea* **94**, 485–98.

Buehler, D.M., Matson, K.D., Piersma, T., and Tieleman, B.I. (2008a). Seasonal redistribution of immune function in a shorebird: Annual-cycle effects override adjustments to thermal regime. *American Naturalist* **172**, 783–96.

Buehler, D.M., Piersma, T., and Tieleman, B.I. (2008b). Captive and free-living red knots *Calidris canutus* exhibit differences in non-induced immunity that suggest different immune strategies in different environments. *Journal of Avian Biology* **39**, 560–66.

Buehler, D.M., Tieleman, B.I., and Piersma, T. (2009a). Age and environment affect constitutive immune function in red knots (*Calidris canutus*). *Journal of Ornithology* **150**, 815–25.

Buehler, D.M., Encinas-Viso, F., Petit, M., Vézina, F., Tieleman, B.I., and Piersma T. (2009b). Limited access to food and physiological trade-offs in a long-distance migrant shorebird. II. Constitutive immune function and the acute-phase response. *Physiological and Biochemical Zoology* **82**, 561–71.

Buehler, D.M., Koolhaas, A., Van't Hof, T.J., Schwabl, I., Dekinga, A., Piersma, T., and Tieleman, B.I. (2009c). No evidence for melatonin-linked immunoenhancement over the annual cycle of an avian species. *Journal of Comparative Physiology A* **195**, 445–51.

Bundle, M.W., Hoppeler, H., Vock, R., Tester, J.M., and Weyand, P.G. (1999). High metabolic rates in running birds. *Nature* **397**, 31–32.

Burness, G.P. (2002). Elephants, mice and red herrings. *Science* **296**, 1245–46.

Buttemer, W.A., Battam, H., and Hulbert, A.H. (2008). Fowl play and the price of petrel: Long-living Procellariiformes have peroxidation-resistant membrane composition compared with short-living Galliformes. *Biology Letters* **4**, 351–54.

Cadée, N., Piersma, T., and Daan, S. (1996). Endogenous circannual rhythmicity in a non-passerine migrant, the knot *Calidris canutus*. *Ardea* **84**, 75–84.

Calder, W.A., III (1984). *Size, function, and life history*. Harvard University Press, Cambridge, Mass.

Calder, W.A., III and King, J.R. (1974). Thermal and caloric relations of birds. In *Avian biology*, eds D. S. Farner and J. R. King, *Vol. 4*, pp. 260–415. Academic Press, New York.

Camazine, S., Deneubourg, J-L., Franks, N.R., Sneyd, J., Theraulaz, G., and Bonabeau, E. (2001). *Self-organization in biological systems*. Princeton University Press, Princeton, NJ.

Campbell, T.W. (1995). *Avian hematology and cytology*. Iowa State University Press, Ames.

Campbell, I.T. and Donaldson, J. (1981). Energy requirements of Antarctic sledge dogs. *British Journal of Nutrition* **45**, 95–98.

Camphuysen, C.J., Ens, B.J., Heg, D., Hulscher, J.B., van der Meer, J., and Smit, C.J. (1996). Oystercatcher *Haematopus ostralegus* winter mortality in The Netherlands: the effect of severe weather and food supply. *Ardea* **84A**, 469–92.

Camphuysen, C.J., Berrevoets, C.M., Cremers, H.J.W.M., Dekinga, A., Dekker, R., Ens, B.J., van der Have, T.M., Kats, R.K.H., Kuiken, T., Leopold, M.F., van der Meer, J., and Piersma, T. (2002). Mass mortality of Common Eiders (*Somateria mollissima*) in the Dutch Wadden Sea, winter 1999/2000: starvation in a commercially exploited wetland of international importance. *Biological Conservation* **106**, 303–17.

Cannon, B. and Nedergaard, J. (2004). Brown adipose tissue: Function and physiological significance. *Physiological Reviews* **84**, 277–359.

Caramujo, M-J. and Boavida, M.J. (2000). Induction and cost of tail spine elongation in *Daphnia hyalina* x *galeata*: Reduction of susceptibility to copepod predation. *Freshwater Biology* **45**, 413–23.

Carmi, N., Pinshow, B., Porter, W.P., and Jaeger, J. (1992). Water and energy limitations on flight duration in small migrating birds. *Auk* **109**, 268–76.

Carroll, L. (1960). *The annotated Alice: Alice's adventures in Wonderland and through the looking-glass*. New American Library, New York.

Cartar, R.V. and Morrison, R.I.G. (2005). Metabolic correlates of leg length in breeding arctic shorebirds: The cost of getting high. *Journal of Biogeography* **32**, 377–82.

Castro, G., Myers, J.P., and Ricklefs, R.E. (1992). Ecology and energetics of sanderlings migrating to four latitudes. *Ecology* **73**, 833–44.

Chambers, W.H. (1952). Max Rubner (June 2, 1854–April 27, 1932). *Journal of Nutrition* **48**, 1–12.

Chappell, M.A., Zuk, M., and Johnsen, T.S. (1996). Repeatability of aerobic performance in red junglefowl: Effects of ontogeny and nematode infection. *Functional Ecology* **10**, 578–85.

Chappell, M.A., Bech, C., and Buttemer, W.A. (1999). The relationship of central and peripheral organ masses to aerobic performance variation in house sparrows. *Journal of Experimental Biology* **202**, 2269–79.

Chappell, M.A., Rezende, E.L., and Hammond, K.A. (2003). Age and aerobic performance of deer mice. *Journal of Experimental Biology* **206**, 1221–31.

Chappell, M.A., Garland, T.Jr, Robertson, G.F., and Saltzman, W. (2007). Relationships among running performance, aerobic physiology and organ mass in male Mongolian gerbils. *Journal of Experimental Biology* **210**, 4179–97.

Charles, J.D. and Bejan, A. (2009). The evolution of speed, size and shape in modern athletics. *Journal of Experimental Biology* **212**, 2419–25.

Charmantier, A., McCleery, R.H., Cole, L.R., Perrins, C., Kruuk, L.E.B., and Sheldon, B.C. (2008). Adaptive phenotypic plasticity in response to climate change in a wild bird population. *Science* **320**, 800–803.

Charnov, E.L. (1976a). Optimal foraging: The marginal value theorem. *Theoretical Population Biology* **9**, 129–36.

Charnov, E.L. (1976b). Optimal foraging: attack strategy of a mantid. *American Naturalist* **110**, 141–51.

Charnov, E.L. (1993). *Life history invariants: Some explorations of symmetry in evolutionary ecology*. Oxford University Press, Oxford.

Charnov, E.L. and Orians, G.H. (1973). Optimal foraging: some theoretical explorations. Unpublished manuscript, http://hdl.handle.net/1928/1649 (last accessed 15 June 2010).

Cherel, Y., Robin, J-P., and Le Maho, Y. (1988). Physiology and biochemistry of long-term fasting in birds. *Canadian Journal of Zoology* **66**, 159–66.

Chew, R.M. (1951). The water exchange of some small mammals. *Ecological Monographs* **21**, 215–25.

Chew, R.M. (1961). Water metabolism of desert-inhabiting vertebrates. *Biological Reviews* **36**, 1–31.

Clark, C.W. (1994). Antipredation and behavior and the asset-protection principle. *Behavioral Ecology* **5**, 159–70.

Cohen, A.A., McGraw, K.J., Wiersma, P., Williams, J.B., Robinson, W.D., Robinson, T.R., Brawn, J.D., and Ricklefs, R.E. (2008). Interspecific associations between circulating antioxidant levels and life-history variation in birds. *American Naturalist* **172**, 178–93.

Constanzo, K. and Monteiro, A. (2007). The use of chemical and visual cues in female choice in the butterfly *Bicyclus anyana*. *Proceedings of the Royal Society, London, B* **274**, 845–51.

Convertino, V.A. (1997). Cardiovascular consequences of bed rest: effect on maximal oxygen uptake. *Medicine and Science in Sports and Exercise* 29, 191–96.

Cook, S.A. and Johnson, M.P. (1968). Adaptation to heterogenous environments. I. Variation in heterophylly in *Ranunculus flammula* L. *Evolution* 22, 496–516.

Cooper, S.J. (1999). The thermal and energetic significance of cavity roosting in mountain chickadees and juniper titmice. *Condor* 101, 863–66.

Cooper, H.M., Herbin, M., and Nevo, E. (1993). Visual system of a naturally microphthalmic mammal: the blind mole rat, *Spalax eherenbergi. Journal of Comparative Neurology* 328, 313–50.

Cooper, W.E., Perez-Mellado, V., and Vitt, L.J. (2004). Ease and effectiveness of costly autotomy vary with predation intensity among lizard populations. *Journal of Zoology* 262, 243–55.

Cooper, C.B., Hochachka, W.M., Butcher, G., and Dhondt, A.A. (2005). Seasonal and latitudinal trends in clutch size: thermal constraints during laying and incubation. *Ecology* 86, 2018–31.

Cornell, H. (1974). Parasitism and distributional gaps between allopatric species. *American Naturalist* 108, 880–83.

Corning, W.R. and Biewener, A.A. (1998). *In vivo* strains in pigeon flight feather shafts: Implications for structural design. *Journal of Experimental Biology* 201, 3057–66.

Costantini, D. (2008). Oxidative stress in ecology and evolution: lessons from avian studies. *Ecology Letters* 11, 1238–51.

Costantini, D., Dell'Ariccia, G., and Lipp, H-P. (2008). Long flights and age affect oxidative status of homing pigeons (*Columba livia*). *Journal of Experimental Biology* 211, 377–81.

Cott, H.B. (1957). *Adaptive coloration in animals.* Methuen, London.

Cox, D.R. (1972). Regression models and life tables. *Journal of the Royal Statistical Society, B* 34, 187–220.

Coyne, J.A. (2009a). *Why evolution is true.* Viking, New York.

Coyne, J.A. (2009b). Evolution's challenge to genetics. *Nature* 457, 382–83.

Creel, S., Christianson, D., Liley, S., and Winnie, J.A. (2007). Predation risk affects reproductive physiology and demography of elk. *Science* 315, 960.

Cresswell, W. (1994). Age-dependent choice of redshank (*Tringa totanus*) feeding location: Profitability or risk? *Journal of Animal Ecology* 63, 589–600.

Crispo, E. (2007). The Baldwin effect and genetic assimilation: Revisiting two mechanisms of evolutionary change mediated by phenotypic plasticity. *Evolution* 61, 2469–79.

Cronin, H. (1991). *The ant and the peacock: Altruism and sexual selection from Darwin to today.* Cambridge University Press, Cambridge.

Croze, H. (1970). Searching image in carrion crows: hunting strategy in a predator and some anti-predator devices in camouflaged prey. *Zeitschrift für Tierpsychologie* Suppl. 5, 1–86.

Crucian, B.E., Stowe, R.P., Pierson, D.L., and Sams, C.F. (2008). Immune system dysregulation following short- vs long-duration spaceflight. *Aviation Space and Environmental Medicine* 79, 835–43.

Currey, J.D. ed. (1984). *The mechanical adaptations of bones.* Princeton University Press, Princeton, NJ.

Cuthill, I. and Kacelnik, A. (1990). Central place foraging: a reappraisal of the 'loading effect'. *Animal Behaviour* 40, 1087–1101.

Cuthill, I.C., Haccou, P., and Kacelnik, A. (1994). Starlings (*Sturnus vulgaris*) exploiting patches: response to long-term changes in travel time. *Behavioral Ecology* 5, 81–90.

Daan, S. and Tinbergen, J.M. (1997). Adaptation in life histories. In *Behavioural ecology: An evolutionary approach*, eds J.R. Krebs and N.B. Davies, 4th edition, pp. 311–33. Blackwell Science, Oxford.

Daan, S., Dijkstra, C., and Tinbergen, J.M. (1990a). Family planning in the kestrel (*Falco tinnunculus*): the ultimate control of covariation of laying date and clutch size. *Behaviour* 114, 83–116.

Daan, S., Masman, D., and Groenewold, A. (1990b). Avian basal metabolic rates: their association with body composition and energy expenditure in nature. *American Journal of Physiology* 259, R333–40.

Daan, S., Deeerenberg, C., and Dijkstra, C. (1996). Increased daily work precipitates death in the kestrel. *Journal of Animal Ecology* 65, 539–44.

Dall, S.R.X., Giraldeau, L-A., Olsson, O., McNamara, J.M., and Stephens, D.W. (2005). Information and its use by animals in evolutionary ecology. *Trends in Ecology and Evolution* 20, 187–93.

Daly, M. and Wilson, M. (1983). *Sex, evolution and behavior*, 2nd edition. Wadsworth, Belmont, CA.

Danchin, É., Giraldeau, L-A., Valone, T.J., and Wagner, R.H. (2004). Public information: From nosy neighbors to cultural evolution. *Science* 305, 487–91.

Danchin, É., Giraldeau, L-A., and Cézilly, F. (2008). *Behavioural ecology.* Oxford University Press, Oxford.

Danks, H.V. (1999). Life cycles in polar arthropods - flexible or programmed? *European Journal of Entomology* 96, 83–103.

Darveau, C-A., Suarez, R.K., Andrews, R.D., and Hochachka, P.W. (2002). Allometric cascade as a unifying principle of body mass effects on metabolism. *Nature* 417, 166–70.

Darveau, C-A., Suarez, R.K., Andrews, R.D., and Hochachka, P.W. (2003). Allometric cascades: reply to West *et al.*, and Banavar *et al. Nature* **421**, 714.

Darwin, C. (1854). *The balanidae (or sessile cirripedes); the verrucidae, etc: A monograph of the sub-class cirripedia, with figures on all the species.* The Ray Society, London.

Darwin, C. (1859). *The origin of species.* John Murray, London.

Dasgupta, S., Singh, R.P., and Kafatos, M. (2009). Comparison of global chlorophyll concentrations using MODIS data. *Advances in Space Research* **43**, 1090–1100.

Davis, M.S., Willard, M.D., Williamson, K.K., Steiner, J.M., and Williams, D.A. (2005). Sustained strenuous exercise increases intestinal permeability in racing Alaskan sled dogs. *Journal of Veterinary Internal Medicine* **19**, 34–39.

Davis, M.S., Willard, M.D., Williamson, K.K., Royer, C., Payton, M., Steiner, J.M., Hinchcliff, K., McKenzie, E., and Nelson, S. Jr (2006). Temporal relationship between gastrointestinal protein loss, gastric ulceration or erosion, and strenuous exercise in racing Alaskan sled dogs. *Journal of Veterinary Internal Medicine* **20**, 835–39.

Dawkins, R. (1995). *River out of Eden: A Darwinian view of life.* Weidenfeld and Nicolson, London.

Dawkins, R. (1996). *Climbing Mount Improbable.* Viking, London.

Dawkins, R. (2005). *The ancestor's tale: A pilgrimage to the dawn of evolution.* Marine Books, New York.

Dawson, W.R. and Bartholomew, G.A. (1968). Temperature regulation and water economy of desert birds. In *Desert biology*, ed. G.W. Brown, pp. 357–94. Academic Press, New York.

Dawson, W.R. and Marsh, R.L. (1989). Metabolic adaptations to cold and season in birds. In *Physiology of cold adaptation in birds*, eds C. Bech and R.E. Reinertsen, pp. 83–94. Plenum Press, New York.

Dawson, W.R., Carey, C., Adkisson, C.S., and Ohmart, R.D. (1979). Responses of Brewer's and chipping sparrows to water restriction. *Physiological Zoology* **42**, 529–41.

Dayton, G.H., Saenz, D., Baum, K.A., Langerhans, R.B., and DeWitt, T.J. (2005). Body shape, burst speed and escape behavior of larval anurans. *Oikos* **111**, 582–91.

Deerenberg, C., Pen, I., Dijkstra, C., Arkies, B.J., Visser, G.H., and Daan, S. (1995). Parental energy expenditure in relation to manipulated brood size in the European kestrel *Falco tinnunculus. Zoology* **99**, 39–48.

Dekinga, A., Dietz, M.W., Koolhaas, A., and Piersma, T. (2001). Time course and reversibility of changes in the gizzards of red knots alternately eating hard and soft food. *Journal of Experimental Biology* **204**, 2167–73.

Dekker, D. (1998). Over-ocean flocking by dunlins, *Calidris alpina*, and the effect of raptor predation at Boundary Bay, British Columbia. *Canadian Field-Naturalist* **112**, 694–97.

Dekker, D. and Ydenberg, R. (2004). Raptor predation on wintering dunlins in relation to the tidal cycle. *Condor* **106**, 415–19.

de Nie, G.J. (2002). Schrikrui bij wespendieven *Pernis apivorus. De Takkeling* **10**, 107–16.

Dennett, D.C. (1995). *Darwin's dangerous idea: Evolution and the meanings of life.* Allen Lane/Penguin Press, London.

Denny, M.W. (2008). Limits to running speeds in dogs, horses and humans. *Journal of Experimental Biology* **211**, 3836–49.

DeWitt, T.J. and Scheiner, S.M. eds (2004) *Phenotypic plasticity: Functional and conceptual approaches.* Oxford University Press, New York.

DeWitt, T.J., Sih, A., and Wilson, D.S. (1998). Costs and limits to phenotypic plasticity. *Trends in Ecology and Evolution* **13**, 77–81.

Dial, K.P., Greene, E., and Irschick, D.J. (2008). Allometry of behaviour. *Trends in Ecology and Evolution* **23**, 394–401.

Diamond, J.M.(1993). Evolutionary physiology. In *The logic of life: The challenge of integrative physiology*, eds C.A.R. Boyd and D. Noble, pp. 89–111. Oxford University Press, Oxford.

Diamond, J.M. (1998). Evolution of biological safety factors: a cost/benefit analysis. In *Principles of animal design: The optimization and symmorphosis debate*, eds E.R. Weibel, C.R. Taylor, and L. Bolis, pp. 21–27. Cambridge University Press, Cambridge.

Diamond, J.M. (2002). Quantitative evolutionary design. *Journal of Physiology* **542**, 337–45.

Diamond, J.M. and Hammond, K.A. (1992). The matches, achieved by natural selection, between biological capacities and their natural loads. *Experientia* **48**, 551–57.

Dierschke, V. (2003). Predation hazard during migratory stopover: are light or heavy birds under risk? *Journal of Avian Biology* **34**, 24–29.

Dietz, M.W. and Piersma, T. (2007). Red knots give up flight capacity and defend food processing capacity during winter starvation. *Functional Ecology* **21**, 899–904.

Dietz, M.W., Dekinga, A., Piersma, T., and Verhulst, S. (1999). Estimating organ size in small migrating shorebirds with ultrasonography: an intercalibration exercise. *Physiological and Biochemical Zoology* **72**, 28–37.

Dietz, M.W., Piersma, T., Hedenström, A., and Brugge, M. (2007). Intraspecific variation in avian pectoral muscle mass: Constraints on maintaining manoeuvrability with increasing body mass. *Functional Ecology* **21**, 317–26.

Dietz, M.W., Spaans, B., Dekinga, A., Klaassen, M., Korthals, H., van Leeuwen, C., and Piersma, T. (2010).

Do red knots (*Calidris canutus islandica*) routinely skip Iceland during southward migration? *Condor* **112**, 48–55.

DiMagno, E.P., Go, G.L.W., and Summerskill, W.H.J. (1973). Relations between pancreatic enzyme outputs and malabsorption in severe pancreatic insufficiency. *New England Journal of Medicine* **288**, 813–15.

di Pamprero, P.E. (1985). Metabolic and circulatory limitations to VO_{2max} at the whole animal level. *Journal of Experimental Biology* **115**, 319–31.

Dobson, A. (1995). The ecology and epidemiology of rinderpest virus in Serengeti and Ngorongoro conservation area. In *Serengeti II: dynamics, management, and conservation of an ecosystem*, eds A.R.E. Sinclair and P. Arcese, pp. 485–505. University of Chicago Press, Chicago.

Dobzhansky, T. (1973). Nothing in biology makes sense except in the light of evolution. *American Biology Teacher* **35**, 125–29.

Doherty, P.A., Wassersug, R.J., and Lee, J.M. (1998). Mechnical properties of the tadpole tail fin. *Journal of Experimental Biology* **201**, 2691–99.

Dorfman, T.A., Levine, B.D., Tillery, T., Peschock, R.M., Hastings, J.L., Schneider, S.M., Macias, B.R., Biolo, G., and Hargens, A.R. (2007). Cardiac atrophy in women following bed rest. *Journal of Applied Physiology* **103**, 8–16.

Dowling, D.K. and Simmons, L.W. (2009). Reactive oxygen species as universal constraints in life-history evolution. *Proceedings of the Royal Society, London, B* **276**, 1737–45.

Drent, J. (2004). *Life history variation of a marine bivalve (Macoma balthica) in a changing world*. PhD thesis, University of Groningen, The Netherlands.

Drent, R.H. and Daan, S. (1980). The prudent parent: Energetic adjustments in avian breeding. *Ardea* **68**, 225–52.

Drent, R. and Piersma, T. (1990). An exploration of the energetics of leap-frog migration in arctic breeding waders. In *Bird migration: Physiology and Ecophysiology*, ed. E. Gwinner, pp. 399–412. Springer-Verlag, Berlin.

Drent, R., Ebbinge, B., and Weijand, B. (1978). Balancing the energy budgets of arctic-breeding geese throughout the annual cycle: a progress report. *Verhandlungen der Ornithologischen Gesellschaft in Bayern* **23**, 239–64.

Drent, R., Both, C., Green, M., Madsen, J., and Piersma, T. (2003). Pay-offs and penalties of competing migratory schedules. *Oikos* **103**, 274–92.

Du Bois, E.F. (1936). *Basal metabolism in health and disease*. Lea & Febiger, Philadelphia.

Duckworth, R.A. (2009). The role of behavior in evolution: a search for mechanism. *Evolutionary Ecology* **23**, 513–31.

Dudley, R. and Gans, C. (1991). A critique of symmorphosis and optimality models in physiology. *Physiological Zoology* **64**, 627–37.

Duijns, S., van Dijk, J.G.B., Spaans, B., Jukema, J., de Boer, W.F., and Piersma, T. (2009). Foraging site selection of two subspecies of bar-tailed godwit *Limosa lapponica*: Time minimizers accept greater predation danger than energy minimizers. *Ardea* **97**, 51–59.

Dykhuizen, D. (1978). Selection for tryptophan auxotrophs of *Escherichia coli* in glucose-limited chemostat as a test of the energy conservation hypothesis of evolution. *Evolution* **32**, 627–37.

Dzialowski, E.M. (2005). Use of operative temperature and standard operative temperature models in thermal biology. *Journal of Thermal Biology* **30**, 317–34.

Eccard, J.A. and Liesenjohann, T. (2008). Foraging decisions in risk-uniform landscapes. *Public Library of Science One* **3**, e3438.

Eckert, R., Randall, D., and Augustine, G. (1988). *Animal physiology: Mechanisms and Adaptations*. W.H. Freeman, New York.

Edelman, G.M. (2004). *Wider than the sky. The phenomenal gift of consciousness*. Yale University Press, Newhaven.

Edgell, T.C., Lynch, B.R., Trussell, G.C., and Palmer, A.R. (2009). Experimental evidence for the rapid evolution of behavioral canalization in natural populations. *American Naturalist* **174**, 434–40.

Eisma, D. (1998). *Intertidal deposits: River mouths, tidal flats, and coastal lagoons*. CRC Press, Boca Raton.

Elgar, M.A. and Harvey, P.H. (1987). Basal metabolic rates in mammals: allometry, phylogeny and ecology. *Functional Ecology* **1**, 25–36.

Elser, J.J. and Hamilton, A. (2007). Stoichiometry and the new biology: the future is now. *Public Library of Science—Biology* **5**, 1403–1405.

Elser, J.J., Dobberfuhl, D.R., MacKay, N.A., and Schampel, J.H. (1996). Organism size, life history, and N:P stoichiometry. *Bioscience* **46**, 674–84.

Emlen, D.J. (2008). The evolution of animal weapons. *Annual Review of Ecology, Evolution, and Systematics* **39**, 387–413.

Emlen, D.J., Lavine, L.C., and Ewen-Campen, B. (2007). On the origin and evolutionary diversification of beetle horns. *Proceedings of the National Academy of Sciences, USA* **104** (Suppl. 1), 8661–68.

Endler, J.A. (1986). *Natural selection in the wild*. Princeton University Press, Princeton.

Engel, S., Biebach, H., and Visser, G.H. (2006). Metabolic costs of avian flight in relation to flight velocity: a study in rose coloured starlings (*Sturnus roseus*, Linnaeus). *Journal of Comparative Physiology B* **176**, 415–27.

Estourgie-van Burk, G.F., Bartels, M., Hoekstra, R.A., Polderman, T.J.C., Delemarre-van der Wal, H.A., and Boomsma, D.I. (2009). A twin study of cognitive costs of low birth weight and catch-up growth. *Journal of Pediatrics* **154**, 29–32.

Feder, M.E. and Watt, W.B. (1992). Functional biology of adaptation. In *Genes in ecology*, eds R.J. Berry, T.J. Crawford, and G.M. Hewitt, pp. 365–92. Blackwell Scientific Publications, Oxford.

Fiennes, R. (2003). *Captain Scott*. Hodder and Stoughton, London.

Figuerola, J. (1999). Effects of salinity on rates of infestation of waterbirds by haematozoa. *Ecography* **22**, 681–85.

Fisher, M.O., Nager, R.G., and Monaghan, P. (2006). Compensatory growth impairs adult cognitive function. *Public Library of Science—Biology* **4**, 1462–66.

Fitts, R.H., Riley, D.R., and Widrick, J.J. (2001) Functional and structural adaptations of skeletal muscle to microgravity. *Journal of Experimental Biology* **204**, 3201–3208.

Forbes-Ewan, C.H., Morrissey, B.L.L., Gregg, G.C., and Waters, D.R. (1989). Use of the doubly labelled water technique in soldiers training for jungle warfare. *Journal of Applied Physiology* **67**, 14–18.

Ford, H. and Crowther, S. (1922). *My life and work*. Electronic edition: www.gutenberg.org/etext/7213 (last accessed 16 June 2010).

Foster, C., Hoyos, J. Earnest, C., and Lucia, A. (2005). Regulation of energy expenditure during prolonged athletic competition. *Medicine and Science in Sports and Exercise* **37**, 670–75.

Freckleton, R.P., Harvey, P.H., and Pagel, M. (2003). Bergmann's rule and body size in mammals. *American Naturalist* **161**, 821–25.

Fretwell, S.D. and Lucas, H.L. (1970). On territorial behaviour and other factors influencing habitat distribution in birds. *Acta Biotheoretica* **19**, 16–36.

Froehle, A.W. and Schoeninger, M.J. (2006). Intraspecies variation in BMR does not affect estimates of early hominin total daily energy expenditure. *American Journal of Physical Anthropology* **131**, 552–59.

Fry, C.H., Ferguson, I.J., and Dowsett, R.J. (1972). Flight muscle hypertrophy and ecophysiological variation of yellow wagtail *Motacilla flava* races at Lake Chad. *Journal of Zoology* **167**, 293–306.

Fuglesteg, B.N, Haga, Ø.E., Folkow, L.P., Fuglei, E., and Schytte Blix, A. (2006). Seasonal variations in basal metabolic rate, lower critical temperature and responses to temporary starvation in the arctic fox (*Alopex lagopus*) from Svalbard. *Polar Biology* **29**, 308–19.

Fuller, A., Carter, R.N., and Mitchell, D. (1998). Brain and abdominal temperatures at fatigue in rats exercising in the heat. *Journal of Applied Physiology* **84**, 877–83.

Furness, L.J. and Speakman, J.R. (2008). Energetics and longevity in birds. *Age* **30**, 75–87.

Gaffney, B. and Cunningham, E.P. (1988). Estimation of genetic trend in racing performance of thoroughbred horses. *Nature* **332**, 722–24.

Galilei, G. (1637). *Dialogues concerning two new sciences* (translated by H. Crew and A. De Salvio). Macmillan, New York, 1914.

Garland, T. Jr (1998). Conceptual and methodological issues in testing predictions of symmorphosis. In *Principles of animal design: The optimization and symmorphosis debate*, eds E.R. Weibel, C.R. Taylor, and L. Bolis, pp. 40–47. Cambridge University Press, Cambridge.

Garland, T. Jr and Huey, R.B. (1987). Testing symmorphosis: does structure match functional requirements? *Evolution* **41**, 1404–1409.

Garland, T. Jr. and Kelly, S.A. (2006). Phenotypic plasticity and experimental evolution. *Journal of Experimental Biology* **209**, 2344–61.

Geen, R.G. (1995). *Human motivation: A social psychological approach*. Wadsworth Publishing, New York.

Geist, V. (1998). *Deer of the world*. Stackpole Books, Mechanicsburg, PA.

Gerth, N., Sum, S., Jackson, S., and Starck, J.M. (2009). Muscle plasticity of Inuit sled dogs in Greenland. *Journal of Experimental Biology* **212**, 1131–39.

Gervasi, S.S. and Foufopoulos, J. (2008). Costs of plasticity: responses to dessication decrease post-metamorphic immune function in a pond-breeding amphibian. *Functional Ecology* **22**, 100–108.

Ghalambor, C.K., McKay, C.K., Carroll, S.P., and Reznick, D.N. (2007). Adaptive versus non-adaptive plasticity and the potential for contemporary adaptation in new environments. *Functional Ecology* **21**, 394–407.

Gibbon, J. (1977). Scalar expectancy theory and Weber's law in animal timing. *Psychological Review* **84**, 279–325.

Gienapp, P., Teplitsky, C., Alho, J.S., Mills, J.A., and Merilä, J. (2008). Climate change and evolution: disentangling environmental and genetic responses. *Molecular Ecology* **17**, 167–78.

Gilbert, S.F. and Epel, D. (2009). *Ecological developmental biology: Integrating epigenetics, medicine, and ecolution*. Sinauer, Sunderland, Mass.

Gill, J., Sutherland, W.J., Watkinson, A.R. (1996). A method to quantify the effects of human disturbance on animal populations. *Journal of Applied Ecology* **33**, 786–92.

Gill, R.E. Jr, Piersma, T., Hufford, G., Servranckx, R., and Riegen, A. (2005). Crossing the ultimate ecological barrier: evidence for an 11 000-km-long nonstop flight from Alaska to New Zealand and eastern Australia by bar-tailed godwits. *Condor* **107**, 1–20.

Gill, R.E. Jr, Tibbitts, T.L., Douglas, D.C., Mulcahy, D.M., Handel, C.M., Gottschalck, J.C., Warnock, N., McCaffery, B.J., Battley, P.F., and Piersma, T. (2009). Extreme endurance flights by landbirds crossing the Pacific Ocean: Ecological corridor rather than barrier? *Proceedings of the Royal Society, London, B* **276**, 447–57.

Gillooly, J.F. and Allen, A.O. (2007). Changes in body temperature influence the scaling of VO_{2max} and aerobic scope in mammals. *Biology Letters* **3**, 99–102.

Gillooly, J.F., Brown, J.H., West, G.B., Savage, V.M., and Charnov, E.L. (2001). Effects of size and temperature on metabolic rate. *Science* **293**, 2248–51.

Gingerich, P.D. (1993). Quantification and comparison of evolutionary rates. *American Journal of Science* **293A**, 453–78.

Gitschier, J. (2009). You say you want a revolution: an interview with Pat Brown. *Public Library of Science - Genetics* **4**, e1000560.

Gloutney, M.L. and Clark, R.G. (1997). Nest site selection by mallards and blue winged teal in relation to microclimate. *Auk* **114**, 381–95.

Gluckman, P.D., Hanson, M.A., Bateson, P., Beedle, A.S., Law, C.M., Bhutta, Z.A., Anokhin, K.V., Bougnères, P., Chandak, G.T., Dasgupta, P., Davey Smith, G., Ellison, P.T., Forrester, T.E., Gilbert, S.F., Jablonka, E., Kaplan, H., Prentice, A.M., Simpson, S.J., Uauy, R., and West-Eberhard, M.J. (2009). Towards a new developmental synthesis: adaptive developmental plasticity and human disease. *Lancet* **373**, 1654–57.

González-Alonso, J., Teller, C., Amdersen, S.L., Jensen, F.B., Hyldig, T and Nielsen, B. (1999). Influence of body temperature on the development of fatigue during prolonged exercise in the heat. *Journal of Applied Physiology* **86**, 1032–39.

Gordon, J.E. (1978). *Structures, or why things don't fall down*. Plenum Press, New York.

Gordon, M.S. (1998). Evolution of optimal systems: overview. In *Principles of animal design: The optimization and symmorphosis debate*, eds E.R. Weibel, C.R. Taylor, and L. Bolis, pp. 37–39. Cambridge University Press, Cambridge.

Gosler, A.G. (1987). Pattern and process in the bill morphology of the great tit *Parus major*. *Ibis* **129**, 451–76.

Goss, R.J. (1983). *Deer antlers: Regeneration, function, and evolution*. Academic Press, New York.

Goss, R.J. (1995). Future-directions in antler research. *Anatomical Record* **241**, 291–302.

Gotanda, K., Turgeon, K., and Kramer, D. (2009). Body size and reserve protection affect flight initiation distance in parrotfish. *Behavioral Ecology and Sociobiology* **63**, 1563–72.

Gottlieb, G. (1992). *Individual development and evolution: The genesis of novel behavior*. Oxford University Press, New York.

Gould, S.J. (1977). *Ontogeny and phylogeny*. Harvard University Press, Cambridge, Mass.

Gould, S.J. and Lewontin, R.C. (1979). The spandrels of San Marcos and the Panglossian paradigm: a critique of the adaptationist program. *Proceedings of the Royal Society, London, B* **205**, 581–98.

Grant, P.R. (1986). *Ecology and evolution of Darwin's finches*. Princeton University Press, Princeton, NJ.

Grant, P.R. and Grant, B.R. (2008). *How and why species multiply: The radiation of Darwin's finches*. Princeton University Press, Princeton, NJ.

Graveland, J., van der Wal, R., van Balen, J.H., and van Noordwijk, A.J. (1994). Poor reproduction in forest passerines from decline of snail abundance on acidified soils. *Nature* **368**, 446–48.

Green, R.F. (1984). Stopping rules for optimal foragers. *American Naturalist* **123**, 30–43.

Green, R.F. (2006). A simpler, more general method of finding the optimal foraging strategy for Bayesian birds. *Oikos* **112**, 274–84.

Green, M., Alerstam, T., Clausen, P., Drent, R., and Ebbinge, B.S. (2002). Dark-bellied brent geese *Branta bernicla bernicla*, as recorded by satellite telemetry, do not minimize flight distance during spring migration. *Ibis* **144**, 106–21.

Greene, H.W. (1997). *Snakes: The evolution of mystery in nature*. University of California Press, Berkeley.

Grimshaw, H.M., Ovington, J.D., Betts, M.M., and Gibb, J.A. (1958). The mineral content of birds and insects in plantations of *Pinus sylvestris* L. *Oikos* **9**, 26–34.

Groothuis, T.G.G., Muller, W., von Engelhardt, N., Carere, C., and Eising, C. (2005). Maternal hormones as a tool to adjust offspring phenotype in avian species. *Neuroscience and Biobehavioral Reviews* **29**, 329–52.

Groscolas, R. (1986). Changes in body mass, body temperature and plasma fuel levels during the natural breeding fast in male and female emperor penguins *Aptenodytes forsteri*. *Journal of Comparative Physiology B* **156**, 521–27.

Gudmundsson, G.A., Lindström, Å., and Alerstam, T. (1991). Optimal fat loads and long-distance flights by migrating knots, *Calidris canutus*, sanderlings, *C. alba*, and turnstones, *Arenaria interpres*. *Ibis* **133**, 140–52.

Guezennec, C.Y., Satabin, P., Legrand, H., and Bigard, A.X. (1994). Physical performance and metabolic changes induced by combined prolonged exercise and different energy intakes in humans. *European Journal of Applied Physiology* **68**, 525–30.

Guglielmo, C.G. and Williams, T.D. (2003). Phenotypic flexibility of body composition in relation to migratory state, age, and sex in the western sandpiper (*Calidris mauri*). *Physiological and Biochemical Zoology* **76**, 84–98.

Guglielmo, C.G., Piersma, T., and Williams, T.D. (2001). A sport-physiological perspective on bird migration: Evidence for flight-induced muscle damage. *Journal of Experimental Biology* **204**, 2683–90.

Gwinner, E. (1986). *Circannual rhythms: Endogenous annual clocks in the organization of seasonal processes.* Springer-Verlag, Berlin.

Hacklander, K., Tataruch, F., and Ruf, T. (2002). The effect of dietary fat content on lactation energetics in the European hare (*Lepus europaeus*). *Physiological and Biochemical Zoology* **75**, 19–28.

Hails, C.J. (1983). The metabolic rate of tropical birds. *Condor* **85**, 61–65.

Haldane, J.B.S. (1949). Suggestions as to quantitative measurement of rates of evolution. *Evolution* **3**, 51–56.

Haldane, J.B.S. (1957). The cost of natural selection. *Journal of Genetics* **55**, 511–24.

Hammill, E., Rogers, A., and Beckerman, A.P. (2008). Costs, benefits and the evolution of inducible defences: a case study with *Daphnia pulex. Journal of Evolutionary Biology* **21**, 705–15.

Hammond, K.A. and Diamond, J.M. (1992). An experimental test for a ceiling on sustained metabolic rate in lactating mice. *Physiological Zoology* **65**, 952–77.

Hammond, K.A. and Diamond, J.M. (1994). Limits to dietary nutrient intake and intestinal nutrient uptake in lactating mice. *Physiological Zoology* **67**, 282–303.

Hammond, K.A. and Diamond, J.M. (1997). Maximal sustained energy budgets in humans and animals. *Nature* **386**, 457–62.1

Hammond, K.A. and Kristan, D.M. (2000). Responses to lactation and cold exposure by deer mice (*Peromyscus maniculatus*). *Physiological and Biochemical Zoology* **73**, 547–56.

Hammond, K.A., Konarzewski, M., Torres, R., and Diamond, J.M. (1994). Metabolic ceilings under a combination of peak energy demands. *Physiological Zoology* **67**, 1479–1506.

Hammond, K.A., Lam, M., Lloyd, K.C.K., and Diamond, J.M. (1996a). Simultaneous manipulation of intestinal capacities and nutrient loads in mice. *American Journal of Physiology* **271**, G969–79.

Hammond, K.A., Lloyd, K.C.K and Diamond, J.M. (1996b). Is mammary output capacity limiting to lactational performance in mice? *Journal of Experimental Biology* **199**, 337–49.

Hammond, K.A., Chappell, M.A., Cardullo, R.A., Lin, R-S., and Johnsen, T.S. (2000). The mechanistic basis of aerobic performance in red junglefowl. *Journal of Experimental Biology* **203**, 2053–64.

Harman, D. (1956). Aging: a theory based on free radical radiation chemistry. *Journal of Gerontology* **11**, 298–300.

Harwood, J. (1996). Weimar culture and biological theory: a study of Richard Woltereck (1877–1944). *History of Science* **34**, 347–77.

Hatch, D.E. (1970). Energy conserving and heat dissipating mechanisms of Turkey vulture. *Auk* **87**, 111–24.

Haugen, M., Williams, J.B., Wertz, P., and Tieleman, B.I. (2003a). Lipids of the stratum corneum vary with cutaneous water loss among larks along a temperature-moisture gradient. *Physiological and Biochemical Zoology* **76**, 907–17.

Haugen, M., Tieleman, B.I., and Williams, J.B. (2003b). Phenotypic flexibility in cutaneous water loss and lipids of the stratum corneum. *Journal of Experimental Biology* **206**, 3581–88.

Hawking, S. (1988). *A brief history of time: From the big bang to black holes.* Bantam Books, London.

Hayes, J.P. and Garland, T. Jr. (1995). The evolution of endothermy: testing the aerobic capacity model. *Evolution* **49**, 836–47.

Hayes, J.P., Bible, C.A., and Boone, J.D. (1998). Repeatability of mammalian physiology: evaporative water loss and oxygen consumption of *Dipodomys merriami. Journal of Mammalogy* **79**, 475–85.

Hedenström, A. and Alerstam, T. (1995). Optimal flight speeds of birds. *Philosophical Transactions of the Royal Society, London, B* **348**, 471–87.

Heinrich, B. (1997). *The trees in my forest.* HarperCollins, New York.

Heinrich, B. (2001). *Racing the antelope: What animals can teach us about running and life.* HarperCollins, New York.

Hendry, A.P. and Kinnison, M.T. (1999). The pace of modern life: measuring of contemporary microevolution. *Evolution* **53**, 1637–53.

Hendry, A.P., Farrugia, T.J., and Kinnison, M.T. (2008). Human influences on rates of phenotypic change in wild animal populations. *Molecular Ecology* **17**, 20–29.

Henry, C.J.K. (2005). Basal metabolic rate studies in humans: Measurement and development of new equations. *Public Health Nutrition* **8**, 1133–52.

Hessen, D.O. and Anderson, T.R. (2008). Excess carbon in aquatic organisms and ecosystems: physiological, ecological and evolutionary implications. *Limnology and Oceanography,* **53**, 1685–96.

Hiddink, J.G. (2003). Effects of suction-dredging for cockles on non-target fauna in the Wadden Sea. *Journal of Sea Research* **50**, 315–23.

Higashi, M., Abe, T., and Burns, T.P. (1992). Carbon-nitrogen balance and termite ecology. *Proceedings of the Royal Society, London, B* **249**, 303–308.

Hilborn, R. and Mangel, M. (1997). *The ecological detective: Confronting models with data.* Princeton University Press, Princeton, NJ.

Hill, J.A. and Olson, E.N. (2008). Cardiac plasticity. *New England Journal of Medicine* **358**, 1370–80.

Hill, R.W., Wyse, G.A., and Anderson, M. (2004). *Animal physiology*. Sinauer Associates, Sunderland, MA.

Hilton, G.M., Houston, D.C., Barton, N.W.H., and Furness, R.W. (1999). Digestion strategies of meat- and fish-eating birds. In *Proceedings of the 22nd International Ornithological Congress*, eds N. J. Adams and R.J. Slotow, pp. 2184–97. BirdLife South Africa, Johannesburg.

Hinchcliff, K.W., Reinhart, G.A., Burr, J.R., Schreier, C.J., and Swenson, R.A. (1997). Metabolizable energy intake and sustained energy expenditure of Alaskan sled dogs during heavy exertion in the cold. *American Journal of Veterinary Research* **58**, 1457–62.

Hinchcliff, K.W., Constable, P.D., and DiSilvestro, R.A. (2004). Muscle injury and antioxidant status in sled dogs competing in a long-distance sled dog race. *Equine and Comparative Exercise Physiology* **1**, 81–85.

Hinds, D.S., Baudinette, R.V., MacMillen, R.E., and Halpern, E.A. (1993). Maximum metabolism and the aerobic factorial scope of endotherms. *Journal of Experimental Biology* **182**, 41–56.

Hirakawa, H. (1995). Diet optimization with a nutrient or toxin constraint. *Theoretical Population Biology* **47**, 331–46.

Ho, C-K., Pennings, S.C., and Carefoot, T.H. (2010). Is diet quality an overlooked mechanism for Bergmann's rule? *American Naturalist* **175**, 269–76.

Hoch, J.M. (2008). Variation in penis morphology and mating ability in the acorn barnacle, *Semibalanus balanoides*. *Journal of Experimental Marine Biology and Ecology* **359**, 126–30.

Hochachka, P.W. and Beatty, C.L. (2003). Patterns of control of maximum metabolic rate in humans. *Comparative Biochemistry and Physiology A* **136**, 215–25.

Hochachka, P.W., Beatty, C.L., Burelle, Y., Trump, M.E., McKenzie, D.C., and Matheson, G.O. (2002). The lactate paradox in human high altitude physiological performance. *News in Physiological Sciences* **17**, 122–26.

Hochachka, P.W., Darveau, C-A., Andrews, R.D., and Suarez, R.K. (2003). Allometric cascade: A model for resolving body mass effects on metabolism. *Comparative Biochemistry and Physiology A* **134**, 675–91.

Hodgson, W.R., McCutcheon, J., Byrd, S.K., Brown, W.S., Bayly, W.M., Brengelmann, G.L., and Gollnick, P.D. (1993). Dissipation of metaboic heat in the horse during exercise. *Journal of Applied Physiology* **74**, 1161–70.

Hoffman, R.J. (1978). Environmental uncertainty and evolution of physiological adaptation in *Colias* butterflies. *American Naturalist* **112**, 999–1015.

Hofman, M.A. and Swaab, D.F. (1992). Seasonal changes in the suprachiasmatic nucleus of man. *Neuroscience Letters* **139**, 257–60.

Holick, M.F. (1998). Perspectives on the impact of weightlessness on calcium and bone metabolism. *Bone* **22**, 105S–111S.

Holick, M.F. (2000). Microgravity–induced bone loss—will it limit human space exploration? *Lancet* **355**, 1569–70.

Holling, C.S. (1959). Some characteristics of simple types of predation and parasitism. *Canadian Entomologist* **91**, 385–98.

Holloszy, J.O. and Smith, E.K. (1986). Longevity of cold-exposed rats: A reevaluation of the 'rate-of-living theory'. *Journal of Applied Physiology* **61**, 1656–60.

Holmes, D.J. and Austad, S.N. (1995a). Birds as animal models for the comparative biology of aging: A prospectus. *Journal of Gerontology A* **50**, B59–B66.

Holmes, D.J. and Austad, S.N. (1995b). The evolution of avian senescence patterns: Implications for understanding primary aging processes. *American Zoologist* **35**, 307–17.

Holmes, D.J. and Martin, K. (2009). A bird's-eye view of aging: What's in it for ornithologists? *Auk* **126**, 1–23.

Holmgren, N.M.A. and Olsson, O. (2000). A three-neuron model of information processing during Bayesian foraging. In *Artificial neural networks in medicine and biology: Proceedings of the ANNIMAB-1 conference, Göteborg, Sweden, 13–16 May 2000*, eds H. Malmgren, M. Borga, and L. Niklasson, pp. 265–70. Springer-Verlag, London.

Honkoop, P.J.C. and van der Meer, J. (1997). Reproductive output of *Macoma balthica* populations in relation to winter-temperature and intertidal-height mediated changes of body mass. *Marine Ecology Progress Series* **149**, 155–62.

Hoppeler, H. and Flück, M. (2002). Normal mammalian skeletal muscle and its phenotypic plasticity. *Journal of Experimental Biology* **205**, 2143–52.

Hoppeler, H., Altpeter, E., Wagner, M., Turner, D.L., Hokanson, J., König, M., Stalder-Navarro, V.P., and Weibel, E.R. (1995). Cold acclimation and endurance training in guinea pigs: changes in lung, muscle and brown fat tissue. *Respiratory Physiology* **101**, 189–98.

Hoppema, M. and Goeyens, L. (1999). Redfield behavior of carbon, nitrogen and phosphorus depletions in Antarctic surface water. *Limnology and Oceanography* **44**, 220–24.

Horak, P., Saks, L., Ots, I., and Kollist, H. (2002). Repeatability of condition indices in captive greenfinches (*Carduelis chloris*). *Canadian Journal of Zoology* **80**, 636–43.

Houston, A.I. (1995). Energetic constraints and foraging efficiency. *Behavioral Ecology* **6**, 393–96.

Houston, A.I., Stevens, M., and Cuthill, I.C. (2007). Animal camouflage: compromise or specialize in a 2 patch-type environment? *Behavioral Ecology* **18**, 769–75.

Hoverman, J.T. and Relyea, R.A. (2007). How flexible is phenotypic plasticity? Developmental windows for trait induction and reversal. *Ecology* 88, 693–705.

Hoyt, D.F. and Taylor, C.R. (1981). Gait and the energetics of locomotion in horses. *Nature* 292, 239–40.

Hoyt, R.W., Jones, T.E., Stein, T.P., McAnich, G.W., Lieberman, H.R., Askew, E.W., and Cymerman, A. (1991). Doubly labelled water measurement of human energy expenditure during strenuous exercise. *Journal of Applied Physiology* 71, 16–22.

Hoyt, R.W., Opstad, P.K., Haugen, A–H., DeLany, J.P., Cymerman, A., and Friedl, K.E. (2006). Negative energy balance in male and female rangers: Effects of 7 days of sustained exercise and food deprivation. *American Journal of Clinical Nutrition* 83, 1068–75.

Hughes, M.R., Roberts, J.R., and Thomas, B.R. (1987). Total body water and its turnover in free-living nestling glaucous-winged gulls with a comparison of body water and water flux in avian species with and without salt glands. *Physiological Zoology* 60, 481–91.

Hulscher, J.B. (1985). Growth and abrasion of the oyster-actcher bill in relation to dietary switches. *Netherlands Journal of Zoology* 35, 124–54.

Humphrey, N.K. (1983). *Consciousness regained*. Oxford University Press, Oxford.

Hunte, W., Meyers, R.A., and Doyle, R.W. (1985). Bayesian mating decisions in an amphipod, *Gammarus lawrencianus* Bousfield. *Animal Behaviour* 33, 366–72.

Huntford, R. (1985). *The last place on Earth*. Pan Books, London.

Iwasa, Y., Highashi, M., and Yamamura, N. (1981). Prey distribution as a factor determining the choice of optimal foraging strategy. *American Naturalist* 117, 710–23.

Jablonka, E. (2006). Genes as followers in evolution—a post-synthesis synthesis? *Biology and Philosophy* 21, 143–54.

Jablonka, E. and Lamb, M.J. (2005). *Evolution in four dimensions: Genetic, epigenetic, behavioral and symbolic variation in the history of life*. MIT Press, Cambridge, Mass.

Jablonka, E. and Lamb, M.J. (2007). Precis of *Evolution in four dimensions*. *Behavioral and Brain Sciences* 30, 353–65.

Jablonka, E. and Raz, G. (2009). Transgenerational epigenetic inheritance: prevalence, mechanisms, and implications for the study of heredity and evolution. *Quarterly Review of Biology* 84, 131–76.

Jackson, S. and Diamond, J.M. (1996). Metabolic and digestive responses to artificial selection in chickens. *Evolution* 50, 1638–50.

Jacobs, J.D. and Wingfield, J.C. (2000). Endocrine control of lifecycle stages: a constraint on response to the environment? *Condor* 102, 35–51.

Janeway, C.A., Travers, P., Walport, M., and Shlomchik, M. (2004). *Immunobiology: The immune system in health and disease*, 6th edition. Garland Publishing, New York.

Jarrett, J.N. (2008). Inter-population variation in shell morphology of the barnacle *Chthamalus fissus*. *Journal of Crustacean Biology* 28, 16–20.

Jenni, L. and Jenni-Eiermann, S. (1998). Fuel supply and metabolic constraints in migrating birds. *Journal of Avian Biology* 29, 521–28.

Jenni-Eiermann, S., Jenni, L., Kvist, A., Lindström, Å., Piersma, T., and Visser, G.H. (2002). Fuel use and metabolic response to endurance exercise: a wind tunnel study of a long-distance migrant shorebird. *Journal of Experimental Biology* 205, 2453–60.

Jensma, G. (2009). Van professie naar traditie. Het 'ljipaai-sykjen' als uitgevonden traditie. In *Friese sport: Tussen traditie en professie,* eds G. Jensma and P. Breuker, pp. 18–62. Bornmeer, Gorredijk.

Jeschke, J.M. (2007). When carnivores are 'full and lazy'. *Oecologia* 152, 357–64.

Jeschke, J.M., Kopp, M., and Tollrian, R. (2002). Predator functional responses: discriminating between handling and digesting prey. *Ecological Monographs* 72, 95–112.

Johannsen, W. (1911). The genotype conception of heredity. *American Naturalist* 45, 129–59.

Johnson, M.S. and Speakman, J.R. (2001). Limits to sustained intake. V. Effect of cold-exposure during lactation in *Mus musculus*. *Journal of Experimental Biology* 204, 1967–77.

Johnson, M.S., Thomson, S.C., and Speakman, J.R. (2001a). Limits to sustained intake. I. Lactation in the laboratory mouse *Mus musculus*. *Journal of Experimental Biology* 204, 1925–35.

Johnson, M.S., Thomson, S.C., and Speakman, J.R. (2001b). Limits to sustained intake. II. Interrelationships between resting metabolic rate, life-history traits and morphology in *Mus musculus*. *Journal of Experimental Biology* 204, 1937–46.

Johnson, M.S., Thomson, S.C., and Speakman, J.R. (2001c). Limits to sustained intake. III. Effects of concurrent pregnancy and lactation *Mus musculus*. *Journal of Experimental Biology* 204, 1947–56.

Johnson, J.B., Burt, D.B., and DeWitt, T.J. (2008). Form, function, and fitness: pathways to survival. *Evolution* 62, 1243–51.

Johnstone, A.M., Rance, K.A., Murison, S.D., Duncan, J.S., and Speakman, J.R. (2006). Additional anthropometric measures may improve the predictability of basal metabolic rate in adult subjects. *European Journal of Clinical Nutrition* 60, 1437–44.

Jones, J.H. (1998a). Symmorphosis and the mammalian respiratory system: what is optimal design and does it exist?

In *Principles of animal design: The optimization and symmorphosis debate*, eds E.R. Weibel, C.R. Taylor, and L. Bolis, pp. 241–48. Cambridge University Press, Cambridge.

Jones, J.H. (1998b). Optimization of the mammalian respiratory system: symmorphosis versus single species adaptation. *Comparative Biochemistry and Physiology B* **120**, 125–38.

Jones, P.H., Jacobs, I., Morris, A., and Ducharme, M.B. (1993). Adequacy of food rations in soldiers during an arctic exercise measured by doubly labelled water. *Journal of Applied Physiology* **75**, 1790–97.

Jonsson, K.I. (1997). Capital and income breeding as alternative tactics of resource use in reproduction. *Oikos* **78**, 57–66.

Jovani, R., Tella, J.L., Forero, M.G., Bertelotti, M., Blanco, G., Ceballos, O., and Donázar, J.A. (2001). Apparent absence of blood parasites in the Patagonian seabird community: is it related to the marine environment? *Waterbirds* **24**, 430–33.

Juanes, F. and Hartwick, E.B. (1990). Prey size selection in Dungeness crabs: the effect of claw damage. *Ecology* **71**, 744–58.

Kaandorp, J.A. (1999). Morphological analysis of growth forms of branching marine sessile organisms along environmental gradients. *Marine Biology* **134**, 295–306.

Kacelnik, A. (1984). Central-place foraging in starlings (*Sturnus vulgaris*). I. Patch residence time. *Journal of Animal Ecology* **53**, 283–99.

Kahl, P.M. Jr. (1963). Thermoregulation in the wood stork, with special reference to the role of the legs. *Physiological Zoology* **36**, 141–51.

Kamermans, P. and Smaal, A.C. (2002). Mussel culture and cockle fisheries in The Netherlands: Finding a balance between economy and ecology. *Journal of Shellfish Research* **21**, 509–17.

Karasov, W.H. (1986). Energetics, physiology and vertebrate ecology. *Trends in Ecology and Evolution* **4**, 101–104.

Karasov, W.H. and Martínez del Rio, C. (2007). *Physiological ecology: How animals process energy, nutrients, and toxins*. Princeton University Press, Princeton, NJ.

Kendeigh, S.C. (1970). Energy requirements for existence in relation to size of bird. *Condor* **72**, 60–65.

Kersten, M. and Piersma, T. (1987). High levels of energy expenditure in shorebirds: Metabolic adaptations to an energetically expensive way of life. *Ardea* **75**, 175–87.

Kersten, M. and Visser, W. (1996). The rate of food provisioning in the oystercatcher: Food intake and energy expenditure constrained by a digestive bottleneck. *Functional Ecology* **10**, 440–48.

Kersten, M., Bruinzeel, L.W., Wiersma, P., and Piersma, T. (1998). Reduced basal metabolic rate of migratory waders wintering in coastal Africa. *Ardea* **86**, 71–80.

Ketterson, E.D. and Nolan, V. Jr (1999). Adaptation, exaptation, and constraint: A hormonal perspective. *American Naturalist*, **154**, S4–S25.

Ketterson, E.D., Nolan, V. Jr, Wolf, L., Ziegenfus, C., Dufty, A.M., Ball, G.F., and Johnsen, T.S. (1991). Testosterone and avian life histories: the effect of experimentally elevated testosterone on corticosterone and body-mass in dark-eyed juncos. *Hormones and Behavior* **25**, 489–503.

Ketterson, E.D., Nolan, V. Jr, Cawthorn, M.J., Parker, P.G., and Ziegenfus, C. (1996). Phenotypic engineering: using hormones to explore the mechanistic bases of phenotype variation in nature. *Ibis* **138**, 70–86.

Kierdorf, U., Kierdorf, H., and Szuwart, T. (2007). Deer antler regeneration: cells, concepts and controversies. *Journal of Morphology* **268**, 726–38.

King, J.M., Parsons, D.J., Turnpenny, J.R., Nyangaga, J., Bakari, P., and Wathes, C.M. (2006). Modelling energy metabolism of Friesians in Kenya smallholdings shows how heat stress and energy deficit constrain milk yield and cow replacement rate. *Animal Science* **82**, 705–16.

King, J.R. (1964). Oxygen consumption and body temperature in relation to ambient temperature in the white-crowned sparrow. *Comparative Biochemistry and Physiology* **12**, 13–24.

King, J.R (1974). Seasonal allocation of time and energy resources in birds. In *Avian energetics*, ed. R.A. Paynter Jr, pp. 4–85. Nuttall Onithological Club, Cambridge, Mass.

Kingsland, S.E. (1995). *Modeling nature: Episodes in the history of population ecology*, 2nd edition. University of Chicago Press, Chicago.

Kingsolver, J.G. (1988). Thermoregulation, flight and the evolution of wing pattern in pierid butterflies: the topography of adaptive landscapes. *American Zoologist* **28**, 899–912.

Kingsolver, J.G. (1995a). Viability selection on seasonally polyphenic traits: wing melanin patterns in western white butterflies. *Evolution* **49**, 932–41.

Kingsolver, J.G. (1995b). Fitness consequences of seasonal polyphenism in western white butterflies. *Evolution* **49**, 942–54.

Kirkwood, J.K. (1983). A limit to metabolisable energy intake in mammals and birds. *Comparative Biochemistry and Physiology A* **75**, 1–3.

Kirschner, M.W. and Gerhart, J.C. (2005). *The plausibility of life: Resolving Darwin's dilemma*. Yale University Press, New Haven.

Kishida, O. and Nishimura, K. (2006). Flexible architecture of inducible morphological plasticity. *Journal of Animal Ecology* **75**, 705–12.

Kishida, O., Mizuta, Y., and Nishimura, K. (2006). Reciprocal phenotypic plasticity in predator-prey interaction between larval amphibians. *Ecology* **87**, 1599–1604.

Kishida, O, Trussell, G.C., and Nishimura, K. (2009). Top-down effects on antagonistic inducible defense and offense. *Ecology* **90**, 1217–26.

Klaassen, M. (1995). Water and energy limitations on flight range. *Auk* **112**, 260–62.

Klaassen, M. (1996). Metabolic constraints on long-distance migration in birds. *Journal of Experimental Biology* **199**, 57–64.

Klaassen, M. (2004). May dehydration risk govern long-distance migratory behaviour? *Journal of Avian Biology* **35**, 4–6.

Klaassen, M. and Ens, B.J. (1990). Is salt stress a problem for waders wintering on the Banc d'Arguin, Mauritania? *Ardea* **78**, 67–74.

Klaassen, M. and Nolet, B.A. (2008). Stoichiometry of endothermy: shifting the quest from nitrogen to carbon. *Ecology Letters* **11**, 785–92.

Klaassen, M., Kvist, A., and Lindström, Å. (2000). Flight costs and fuel composition of a bird migrating in a wind tunnel. *Condor* **102**, 444–51.

Klasing K.C. (1998). *Comparative avian nutrition*. CAB International, Wallingford.

Kleiber, M. (1961). *The fire of life*. John Wiley, New York.

Klepal, W. (1990). The fundamentals of insemination in Cirripedes. *Oceanography and Marine Biology: an Annual Review* **28**, 353–79.

Klepal, W., Barnes, H., and Munn, E.A. (1972). The morphology and histology of the cirripede penis. *Journal of Experimental Marine Biology and Ecology* **10**, 243–65.

Knaus, W. (2009). Dairy cows trapped between performance demands and adaptability. *Journal of the Science of Food and Agriculture* **89**, 1107–14.

Kodandaramaiah, U., Vallin, A., and Wiklund, C. (2009). Fixed eyespot display in a butterfly thwarts attacking birds. *Animal Behaviour* **77**, 1415–19.

Koiter, T.R., Moes, H., Valkhof, N., and Wijkstra, S. (1999). Interaction of late pregnancy and lactation in rats. *Journal of Reproduction and Fertility* **115**, 341–47.

Konarzewski, M. and Diamond, J.M. (1995). Evolution of basal metabolic rate and organ masses in laboratory mice. *Evolution* **49**, 1239–48.

Kooi, R.E. and Brakefield, P.M. (1999). The critical period for wing pattern induction in the polyphenic tropical butterfly *Bicyclus anynana* (Satyrinae). *Journal of Insect Physiology* **45**, 201–12.

Kopp, M. and Tollrian, R. (2003). Trophic size polyphenism in *Lembadion bullinum*: costs and benefits of an inducible offense. *Ecology* **84**, 641–51.

Koteja, P. (1991). On the relation between basal and field metabolic rates in birds and mammals. *Functional Ecology* **5**, 56–64.

Koteja, P. (1995). Maximum cold-induced energy assimilation in a rodent, *Apodemus flavicollis*. *Comparative Biochemistry and Physiology A* **112**, 479–85.

Koteja, P. (1996). Limits to the energy budget in a rodent, *Peromyscus maniculatus*: Does gut capacity set the limit? *Physiological Zoology* **69**, 994–1020.

Koteja, P., Swallow, J.G., Carter, P.A., and Garland, T. (2001). Maximum cold-induced food consumption in mice selected for high locomotor activity: implications for the evolution of endotherm energy budgets. *Journal of Experimental Biology* **204**, 1177–90.

Kraan, C., Piersma, T., Dekinga, A., Koolhaas, A., and Van der Meer, J. (2007). Dredging for edible cockles *Cerastoderma edule* on intertidal flats: short-term consequences of fishermen's patch-choice decisions for target and non-target benthic fauna. *ICES Journal of Marine Science* **64**, 1735–42.

Kraan, C., van Gils, J.A., Spaans, B., Dekinga, A., Bijleveld, A.I., van Roomen, M., Kleefstra, R., and Piersma, T. (2009). Landscape-scale experiment demonstrates that Wadden Sea intertidal flats are used to capacity by molluscivore migrant shorebirds. *Journal of Animal Ecology* **78**, 1259–68.

Kraan, C., van Gils, J.A., Spaans, B., Dekinga, A., and Piersma, T. (2010). Why Afro-Siberian red knots *Calidris canutus canutus* have stopped staging in the western Dutch Wadden Sea during southward migration. *Ardea* **97**, in press.

Kramer, D.L. and Nowell, W. (1980). Central place foraging in the eastern chipmunk, *Tamias striatus*. *Animal Behaviour* **28**, 772–78.

Krebs, H.A. (1950). Body size and tissue respiration. *Biochimica et Biophysica Acta* **4**, 249–69.

Krebs, J.R. and Davies, N.B. (1978, 1984, 1991, 1997). *Behavioural ecology: An evolutionary approach*. Blackwell Science, Oxford.

Krebs, J.R. and Davies, N.B. (1987). *An introduction to behavioural ecology*. Blackwell Science, Oxford.

Krebs, J.R., Ryan, J.C., and Charnov, E.L. (1974). Hunting by expectation or optimal foraging? A study of patch use by chickadees. *Animal Behaviour* **22**, 953–64.

Krebs, J.R., Erichsen, J.T., Webber, M.I., and Charnov, E.L. (1977). Optimal prey selection in the great tit (*Parus major*). *Animal Behaviour* **25**, 30–38.

Krogh, A. (1916). *The respiratory exchange of animals and man*. Longmans-Green, London.

Król, E. and Speakman, J.R. (2003a). Limits to sustained intake. VI. Energetics of lactation in laboratory mice at thermoneutrality. *Journal of Experimental Biology* **206**, 4255–66.

Król, E. and Speakman, J.R. (2003b). Limits to sustained intake. VII. Milk energy ouput in laboratory mice at thermoneutrality. *Journal of Experimental Biology* **206**, 4267–81.

Król, E., Johnson, M.S., and Speakman, J.R. (2003). Limits to sustained intake. VIII. Resting metabolic rate and organ morphology of laboratory mice lactating at thermoneutrality. *Journal of Experimental Biology* **206**, 4283–91.

Król, E., Murphy, M., and Speakman, J.R. (2007). Limits to sustained intake. X. Effects of fur removal on reproductive performance in laboratory mice. *Journal of Experimental Biology* **210**, 4233–43.

Książek, A., Konarzewski, M., and Łapo, I.B. (2004). Anatomic and energetic correlates of divergent selection for basal metabolic rate in laboratory mice. *Physiological and Biochemical Zoology* **77**, 890–99.

Książek, A., Czerniecki, J., and Konarzewski, M. (2009). Phenotypic flexibility of traits related to energy acquisition in mice divergently selected for basal metabolic rate (BMR). *Journal of Experimental Biology* **212**, 808–14.

Künkele, J. (2000). Effects of litter size on the energetics of reproduction in a highly precocial rodent, the guinea pig. *Journal of Mammalogy* **81**, 691–700.

Kvist, A. and Lindström, Å. (2000). Maximum daily energy intake: It takes time to lift the metabolic ceiling. *Physiological and Biochemical Zoology* **73**, 30–36.

Kvist, A. and Lindström, Å. (2001). Basal metabolic rate in migratory waders: Intra-individual, intraspecific, interspecific and seasonal variation. *Functional Ecology* **15**, 465–73.

Kvist, A. and Lindström, Å. (2003). Gluttony in migratory waders – unprecedented energy assimilation rates in vertebrates. *Oikos* **103**, 397–402.

Kvist, A., Lindström, Å., Green, M., Piersma, T., and Visser, G.H. (2001). Carrying large fuel loads during sustained bird flight is cheaper than expected. *Nature* **413**, 730–32.

Labocha, M.K., Sadowska, E.T., Baliga, K., Semer, A.K., and Koteja, P. (2004). Individual variation and repeatability of basal metabolism in the bank vole, *Clethrionomys glareolus*. *Proceedings of the Royal Society,London, B* **271**, 367–72.

Lack, D. (1947). *Darwin's finches*. Cambridge University Press, Cambridge.

Lackner, J.R., and DiZio, P. (2000). Artificial gravity as a countermeasure in long-duration space flight. *Journal of Neuroscience Research* **62**, 169–76.

Laland, K.N., Odling-Smee, F.J., and Feldman, M.W. (1999). Evolutionary consequences of niche construction and their implications for ecology. *Proceedings of the National Academy of Sciences, USA* **96**, 10242–47.

Laland, K.N., Odling-Smee, J., and Gilbert, S.F. (2008). EvoDevo and niche construction: Building bridges. *Journal of Experimental Zoology B* **310B**, 549–66.

Lam, M., O'Connor, T.P., and Diamond, J.M. (2002). Loads, capacities, and safety factors of maltase and the glucose transporter SGLT1 in mouse intestinal brush border. *Journal of Physiology* **542**, 493–500.

Lampert, W., Tollrian, R., and Stibor, H. (1994). Chemische Induktion von Verteidigungsmechanismen bei Süßwassertieren. *Naturwissenschaften* **81**, 375–82.

Lande, R. (2009). Adaptation to an extraordinary environment by evolution of phenotypic plasticity and genetic assimilation. *Journal of Evolutionary Biology* **22**, 1435–46.

Landys, M.M., Piersma, T., Visser, G.H., Jukema, J., and Wijker, A. (2000). Water balance during real and simulated long-distance migratory flight in the bar-tailed godwit. *Condor* **102**, 645–52.

Landys-Ciannelli, M.M., Jukema, J., and Piersma, T. (2002). Blood parameter changes during stopover in a long-distance migratory shorebird, the bar-tailed godwit *Limosa lapponica taymyrensis*. *Journal of Avian Biology* **33**, 451–55.

Lane, N. (2005). *Power, sex, suicide: Mitochondria and the meaning of life*. Oxford University Press, Oxford.

Lane, N. (2009a). *Life ascending: the ten great inventions of evolution*. Norton, New York.

Lane, N. (2009b). The furnace within. *New Scientist* **2694**, 42–45.

Lank, D.B. and Ydenberg, R.C. (2003). Death and danger at migratory stopovers: Problems with 'predation risk'. *Journal of Avian Biology* **34**, 225–28.

Lasiewski, R.C. and Dawson, W.R. (1967). A re-examination of the relation between standard metabolic rate and body weight in birds. *Condor* **69**, 13–23.

Laundré, J.W., Hernández, L., and Altendorf, K.B. (2001). Wolves, elk, and bison: Reestablishing the 'landscape of fear' in Yellowstone National Park, USA. *Canadian Journal of Zoology* **79**, 1401–1409.

Laurien-Kehnen, C. and Trillmich, F. (2003). Lactation performance of guinea pigs (*Cavia porcellus*) does not respond to experimental manipulation of pup demands. *Behavioral Ecology and Sociobiology* **53**, 145–52.

Lawton, J.H. and May, R.M. eds (1995). *Extinction rates*. Oxford University Press, Oxford.

Le Maho, Y., Vu Van Kha, H., Koubi, H., Dewasmes, G., Girard, J., Ferre, P., and Cagnard, M. (1981). Body composition, energy expenditure, and plasma metabolites in long-term fasting geese. *American Journal of Physiology* **241**, E342–54.

Lenihan, H.S. and Micheli, F. (2001). Soft-sediment communities. In, *Marine community ecology*, eds M.D. Bertness and S.D. Gaines, pp. 253–88. Sinauer Publishers, Sunderland.

Lessells, C.M. (1991). The evolution of life histories. In *Behavioural ecology: An evolutionary approach*, eds J.R.

Krebs and N.B. Davies, 3rd edition, pp. 32–68. Blackwell Science, Oxford.

Lessells, C.M. (2007). Neuroendocrine control of life histories: what do we need to know to understand the evolution of phenotypic plasticity? *Philosophical Transactions of the Royal Society, London, B* **363**, 1589–98.

Levey, D.J. and Karasov, W.H. (1989). Digestive responses of temperate birds switched to fruit or insect diets. *Auk* **106**, 675–86.

Levins, R. (1968). *Evolution in changing environments.* Princeton University Press, Princeton, NJ.

Levitan, D.R. (1989). Density-dependent size regulation in *Diadema antillarum*: Effects on fecundity and survivorship. *Ecology* **70**, 1414–24.

Levitan, D.R. (1991). Skeletal changes in the test and jaws of the sea urchin *Diadema antillarum* in response to food limitation. *Marine Biology* **111**, 431–35.

Lewontin, R.C. (2000). *The triple helix: Gene, organism, and environment.* Harvard University Press, Cambridge, Mass.

Lewontin, R.C. and Levins, R. (2007). *Biology under influence: Dialectical essays on ecology, agriculture, and health.* Monthly Review Press, New York.

Leyrer J., Pruiksma S., and Piersma T. (2009). On 4 June 2008 Siberian red knots at Elbe Mouth kissed the canonical evening migration departure rule goodbye. *Ardea* **97**, 71–79.

Li, N.K. and Denny, M.W. (2004). Limits to phenotypic plasticity: flow effects on barnacle feeding appendages. *Biological Bulletin* **206**, 121–24.

Lignot, J-H., Helmstetter, C., and Secor, S.M. (2005). Postprandial morphological response of the epithelium of the Burmese python (*Python molurus*). *Comparative Biochemsitry and Physiology A* **141**, 280–91.

Lima, S.L. (1993). Ecological and evolutionary perspectives on escape from predatory attack – a survey of North-American birds. *Wilson Bulletin* **105**, 1–47.

Lincoln, G.A. (1992). Biology of antlers. *Journal of Zoology* **226**, 517–28.

Lind, M.I. and Johansson, F. (2009). Costs and limits of phenotypic plasticity in island populations of the common frog *Rana temporaria* under divergent selection pressures. *Evolution* **63**, 1508–18.

Lindstedt, S.L. and Conley, K.E. (2001). Human aerobic performance: too much ado about limits to VO_{2max}. *Journal of Experimental Biology* **204**, 3195–99.

Lindstedt, S.L. and Jones, J.H. (1987). Symmorphosis and the concept of optimal design. In *New directions in ecological physiology*, eds M.E. Feder, A.F. Bennett, W.W. Burggren, and R.B. Huey, pp. 289–309. Cambridge University Press, Cambridge.

Lindström, Å. (1991). Maximum fat deposition rates in migrating birds. *Ornis Scandinavica* **22**, 12–19.

Lindström, Å. (2003). Fuel deposition rates in migrating birds: causes, constraints and consequences. In *Avian migration*, eds P. Berthold, E. Gwinner, and E. Sonnenschein, pp. 307–20. Springer-Verlag, Berlin.

Lindström, Å. and Kvist, A. (1995). Maximum energy intake rate is proportional to basal metabolic rate in passerine birds. *Proceedings of the Royal Society, London, B* **261**, 337–43.

Lindström, Å. and Nilsson, J.Å. (1988). Birds doing it the octopus way—fright moulting and distraction of predators. *Ornis Scandinavica* **19**, 165–66.

Lindström, Å., Klaassen, M., and Kvist, A. (1999). Variation in energy intake and basal metabolic rate of a bird migrating in a wind-tunnel. *Functional Ecology* **13**, 352–59.

Liu, Z., Jaitner, J., Reinhardt, F., Pasman, E., Rensing, S., and Reents, R. (2008). Genetic evaluation of fertility traits of dairy cattle using a multiple-trait animal model. *Journal of Dairy Science* **91**, 4333–43.

Lively, C.M. (1986). Predator-induced shell dimorphism in the acorn barnacle *Chthamalus fissus*. *Evolution* **40**, 232–42.

Lively, C.M. (1999). Developmental strategies in spatially variable environments: Barnacle shell dimorphism and strategic models of selection. In *The ecology and evolution of inducible defenses*, eds R. Tollrian and C.D. Harvell, pp. 245–58. Princeton University Press, Princeton, NJ.

Lively, C.M., Hazel, W.N., Schellenberger, M.J., and Michelson, K.S. (2000). Predator-induced defense: Variation for inducibility in an intertidal barnacle. *Ecology* **81**, 1240–47.

Lotze, H.K., Reise, K., Worm, B., Van Beusekom, J., Busch, M., Ehlers, A., Heinrich, D., Hoffmann, R.C., Holm, P., Jensen, C., Knottnerus, O.S., Langhanki, N., Prummel, W., Vollmer, M., and Wolff, W.J. (2005) Human transformations of the Wadden Sea ecosystem through time: a synthesis. *Helgolander Marine Research* **59**, 84–95.

Lovegrove, B.G. (2000). The zoogeography of mammalian basal metabolic rate. *American Naturalist* **156**, 201–19.

Lucia, A., Hoyos, J., Santalla, A., Earnest, C., and Chicharro, J.L. (2003). Tour de France versus Vuelta a España: which is harder? *Medicine and Science in Sports and Exercise* **35**, 872–78.

Lucy, M.C. (2001). Reproductive loss in high-producing dairy cattle: where will it end? *Journal of Dairy Science* **84**, 1277–93.

Luttbeg, B. and Warner, R.R. (1999). Reproductive decision-making by female peacock wrasses: flexible versus fixed behavioural rules in a variable environment. *Behavioral Ecology* **10**, 666–74.

Luttikhuizen, P.C., Honkoop, P.J.C., Drent, J., and van der Meer, J. (2004). A general solution for optimal egg size during external fertilization, extended scope for intermediate optimal egg size and the introduction of Don Ottavio 'tango'. *Journal of Theoretical Biology* **231**, 333–43.

MacArthur, R.H. and Pianka, E.R. (1966). On optimal use of a patchy environment. *American Naturalist* **100**, 603–609.

Machado, A. and Keen, R. (1999). Learning to time (LeT) or scalar expectancy theory (SET)? A critical test of two models of timing. *Psychological Science* **10**, 285–90.

MacLean, S.F. Jr. (1974). Lemming bones as a source of calcium for Arctic sandpipers (*Calidris* spp.). *Ibis* **116**, 552–57.

Maclean, G. (1996). *Ecophysiology of desert birds*. Springer-Verlag, Berlin.

MacLeod, R., Barnett, P., Clark, J.A., and Cresswell, W. (2005). Body mass change strategies in blackbirds *Turdus merula*: the starvation–predation trade-off. *Journal of Animal Ecology* **74**, 292–302.

Malapani, C. and Fairhurst, S. (2002). Scalar timing in animals and humans. *Learning and Motivation* **33**, 156–76.

Marchinko, K.B. (2003). Dramatic phenotypic plasticity in barnacle legs (*Balanus glandula* Darwin): magnitude, age dependence, and speed of response. *Evolution* **57**, 1281–90.

Marchinko, K.B. (2007). Feeding behavior reveals the adaptive nature of plasticity in barnacle feeding limbs. *Biological Bulletin*, **213**, 12–15.

Marchinko, K.B. and Palmer, A.R. (2003). Feeding in flow extremes: dependence of cirrus form on wave-exposure in four barnacle species. *Zoology* **106**, 127–41.

Masman, D., Daan, S., and Dijkstra, C. (1988). Time allocation in the kestrel (*Falco tinnunculus*), and the principle of energy minimization. *Journal of Animal Ecology*, **57**, 411–32.

Matson, K.D., Ricklefs, R.E., and Klasing, K.C. (2005). A hemolysis-hemagglutination assay for characterizing constitutive innate humoral immunity in wild and domestic birds. *Developmental and Comparative Immunology* **29**, 275–86.

Mauroy, B., Filoche, M., Weibel, E.R., and Sapoval, B. (2004). An optimal bronchial tree may be dangerous. *Nature* **427**, 633–36.

Mayr, E. (1963). *Animal species and evolution*. Belknap Press of Harvard University Press, Cambridge, Mass.

Mayr, E. (1982). *The growth of biological thought: Diversity, evolution, and inheritance*. Harvard University Press, Cambridge, Mass.

Mayr, E. (2002). *What evolution is*. Weidenfeld and Nicholson, London.

McEwen, B.S. and J.C. Wingfield (2003). The concept of allostasis in biology and biomedicine. *Hormones and Behavior* **43**, 2–15.

McFarland, D.J. (1977). Decision making in animals. *Nature* **269**, 15–21.

McKechnie, A.E., Freckleton, R.O., and Jetz, W. (2006). Phenotypic plasticity in the scaling of avian basal metabolic rate. *Proceedings of the Royal Society, London, B* **273**, 931–37.

McLandress, M.R. and Raveling, D.G. (1981). Changes in diet and body composition in Canada geese before spring migration. *Auk* **98**, 65–79.

McLaughlin, R.L. and Montgomerie, D.R. (1990). Flight speeds of parent birds feeding nestlings: maximization of foraging efficiency or food delivery rate? *Canadian Journal of Zoology* **68**, 2269–74.

McNab, B.K. (1986). The influence of food habits on the energetics of eutherian mammals. *Ecological Monographs* **56**, 1–19.

McNab, B.K. (1988). Food habits and the basal rate of metabolism in birds. *Oecologia* **77**, 343–49.

McNab, B.K. (1994). Energy conservation and the evolution of flightlessness in birds. *American Naturalist* **144**, 628–42.

McNab, B.K. (1996). Metabolism and temperature regulation of kiwis (Apterygidae). *Auk* **113**, 687–92.

McNab, B.K. (2002). *The physiological ecology of vertebrates: A view from energetics*. Cornell University Press, Ithaca.

McNair, J.N. (1982). Optimal giving-up times and the marginal value theorem. *American Naturalist* **119**, 511–29.

McNamara, J.M. (1982). Optimal patch use in a stochastic environment. *Theoretical Population Biology* **21**, 269–88.

McNamara, J.M. and Houston, A.I. (1997). Currencies for foraging based on energetic gain. *American Naturalist* **150**, 603–17.

McNamara, J.M., Green, R.F., and Olsson, O. (2006). Bayes' theorem and its application in animal behaviour. *Oikos* **112**, 243–51.

McNamara, J.M., Barta, Z., Wikelsi, M., and Houston, A.I. (2008). A theoretical investigation of the effect of latitude on avian life histories. *American Naturalist* **172**, 331–45.

McNeill, W.H. (1976). *Plagues and people*. Doubleday, New York.

McWhirter, N. (1980). *Guinness book of world records*, 41st edition. Sterling Publishing Co., New York.

Meck, J.V., Reyes, C.J., Perez, S.A., Goldberger, A.L., and Ziegler, M.G. (2001). Marked exacerbation of orthostatic intolerance after long- vs. short-duration spaceflight in veteran astronauts. *Psychosomatic Medicine* **63**, 865–73.

Meijer, T. and Drent, R. (1999). Re-examination of the capital and income dichotomy in breeding birds. *Ibis* **141**, 399–414.

Meiri, S. and Dayan, T. (2003). On the validity of Bergmann's rule. *Journal of Biogeography* **30**, 331–51.

Mendes, L., Piersma, T., Lecoq, M., Spaans, B., and Ricklefs, R.E. (2005). Disease-limited distributions? Contrasts in the prevalence of avian malaria in shorebird species using marine and freshwater habitats. *Oikos* **109**, 396–404.

Mendes, L., Piersma, T., Hasselquist, D., Matson, K.D., and Ricklefs, R.E. (2006a). Variation in the innate and acquired arms of the immune system among five shorebird species. *Journal of Experimental Biology* **209**, 284–91.

Mendes, L., Piersma, T., and Hasselquist, D. (2006b). Two estimates of the metabolic costs of antibody production in migratory shorebirds: low costs, internal reallocation, or both? *Journal of Ornithology* **147**, 274–300.

Messner, R. (1979). *Everest: Expedition to the ultimate*. Kay and Ward, London.

Metcalfe, N.B. and Monaghan, P. (2001). Compensation for a bad start: grow now, pay later? *Trends in Ecology and Evolution* **16**, 254–60.

Meyers, L.A. and Bull, J.J. (2002). Fighting change with change: adaptive variation in an uncertain world. *Trends in Ecology and Evolution* **17**, 551–57.

Michimae, H. (2006). Differentiated phenotypic plasticity in larvae of the cannibalistic salamander *Hynobius retardatus*. *Behavioral Ecology and Sociobiology* **60**, 205–11.

Michimae, H., Nishimura, K., and Wakahara, M. (2005). Mechanical vibrations from tadpoles' flapping tails transform salamander's carnivorous morphology. *Biology Letters* **1**, 75–77.

Midgley, J.J., Midgley, G., and Bond, W.J. (2002). Why were dinosaurs so large? A food quality hypothesis. *Evolutionary Ecology Research* **4**, 1093–95.

Millenium Ecosystem Assessment (2005). *Ecosystems and human well-being: Synthesis*. Island Press, Washington, DC.

Miller, G. (2001). *The mating mind: How sexual choice shaped the evolution of human nature*. Vintage, London.

Miller, L.P. (2007). Feeding in extreme flows: behavior compensates for mechanical constraints in barnacle cirri. *Marine Ecology Progress Series* **349**, 227–34.

Millet, S., Bennet, J., Lee, K.A., Hau, M., and Klasing, K.C. (2007). Quantifying and comparing constitutive immunity across avian species. *Developmental and Comparative Immunology* **31**, 188–201.

Miner, B.G., Sultan, S.E., Morgan, S.G., Padilla, D.K., and Relyea, R.A. (2005). Ecological consequences of phenotypic plasticity. *Trends in Ecology and Evolution* **20**, 685–91.

Mitchell, A.C. (1901). On the convection of heat by air currents. *Transactions of the Royal Society of Edinburgh* **40**, 39.

Møller, A.P., Nielsen, J.T., and Erritzoe, J. (2006). Losing the last feather: feather loss as an antipredator adaptation in birds. *Behavioral Ecology* **17**, 1046–56.

Monaghan, P. (2008). Early growth conditions, phenotypic development and environmental change. *Philosophical Transactions of the Royal Society, London, B* **363**, 1635–45.

Monaghan, P., Charmantier, A., Nussey, D.H., and Ricklefs, R.E. (2008). The evolutionary ecology of senescence. *Functional Ecology* **22**, 371–78.

Monaghan, P., Metcalfe, N.B., and Torres, R. (2009). Oxidative stress as a mediator of life history trade-offs: Mechanisms, measurements and interpretation. *Ecology Letters* **12**, 75–92.

Montgomerie, R., Lyon, B., and Holder, K. (2001). Dirty ptarmigan: behavioral modification of conspicuous male plumage. *Behavioral Ecology* **12**, 429–38.

Moran, N.A. (1992). The evolutionary maintenance of alternative phenotypes. *American Naturalist* **139**, 971–89.

Morgan, C. (1896). Of modification and variation. *Science* **4**, 733–40.

Morrison, R.I.G., Davidson, N.C., and Piersma, T. (2005). Transformation at high latitudes: Why do red knots bring body stores to the breeding grounds? *Condor* **107**, 449–57.

Morrison, S.F., Nakamura, K., and Madden, C.J. (2008). Central control of thermogenesis in mammals. *Experimental Physiology* **93**, 773–97.

Mousseau, T.A. and Fox, C.W. (1998). The adaptive significance of maternal effects. *Trends in Ecology and Evolution* **13**, 403–407.

Munday, P.L., Buston, P.M., and Warner, R.R. (2005). Diversity and flexibility of sex-change strategies in animals. *Trends in Ecology and Evolution* **21**, 89–95.

Munday, P.L., White, J.W., and Warner, R.R. (2006). A social basis for the development of primary males in a sex-changing fish. *Proceedings of the Royal Society, London, B* **283**, 2845–51.

Murata, A., Imafuku, M., and Abe, M. (2001). Copulation by the barnacle *Tetraclita japonica* under natural conditions. *Journal of Zoology* **253**, 275–80.

Murton, R.K. and Westwood, N.J. (1977). *Avian breeding cycles*. Clarendon Press, Oxford.

Nager, R.G., Keller, L.F., and van Noordwijk, A.J. (2000). Understanding natural selection on traits that are influenced by environmental conditions. In *Adaptive genetic variation in the wild*, eds T.A. Mousseau, B. Sinervo, and J.A. Endler, pp. 95–115. Oxford University Press, New York.

Nassuna-Musoke, G.M., Kabasa, J.D., and King, M.J. (2007). Microclimate dynamics in smallholder Friesian dairies of the tropical warm-humid central Uganda. *Journal of Animal and Veterinary Advances* **6**, 907–11.

Navarro, A., Gomez, C., Lopez-Cepero, J.M., and Boveris, A. (2004). Beneficial effects of moderate exercise on mice aging: Survival, behavior, oxidative stress, and mitochondrial electron transfer. *American Journal of Physiology* **286**, R505–11.

Naya, D.E., Karasov, W.H., and Bozinovic, F. (2007). Phenotypic plasticity in laboratory mice and rats: a meta-analysis of current ideas on gut size flexibility. *Evolutionary Ecology Research* **9**, 1363–74.

Naya, D.E., Veloso, C., Dabat, P., and Bozinovic, F. (2009). The effect of short- and long-term fasting on digestive and metabolic flexibility in the Andean toad, *Bufo spinulosus*. *Journal of Experimental Biology* **212**, 2167–75.

Neufeld, C.J. and Palmer, A.R. (2008). Precisely proportioned: Intertidal barnacles alter penis form to suit coastal wave action. *Proceedings of the Royal Society, London, B* **275**, 1081–87.

Nevill, A.M. and Whyte, G. (2005). Are there limits to running world records? *Medical Science Sports and Exercise* **37**, 1785–88.

Newman, R.A. (1992). Adaptive plasticity in amphibian metamorphosis. *BioScience* **32**, 671–78.

Newman, S.A. and Müller, G.B. (2000). Epigenetic mechanisms of character origination. *Journal of Experimental Zoology B*, **288B**, 304–17.

Newton, I. (1701). Scala graduum caloris: Calorum descriptiones & figna. *Philosophical Transactions of the Royal Society, London* **22**, 824–29.

Newton, I. (2008). *The migration ecology of birds.* Elsevier/Academic Press, London.

Nijhout, H.F. (1999). Control mechanisms of polyphenic development in insects. *BioScience* **49**, 181–92.

Noakes, T.D. (1992). *Lore of running: Discover the science and the spirit of running.* 3rd Ed. Leisure Press, Champaign, Ill.

Noakes, T.D. (2006). The limits of endurance exercise. *Basic Research in Cardiology* **101**, 408–17.

Noakes, T.D. (2007). The limits of human endurance: what is the greatest endurance performance of all time? Which factors regulate performance at extreme altitude? In *Hypoxia and the circulation*, ed. R.C. Roach, pp. 259–80. Springer, New York.

Noakes, T.D., Peltonen, J.E., and Rusko, H.K. (2001). Evidence that a central governor regulates exercise performance during acute hypoxia and hyperoxia. *Journal of Experimental Biology* **201**, 3225–34.

Noakes, T.D., St Clair Gibson, A., and Lambert, E.V. (2004). From catastrophe to complexity: a novel model of integrative central neural regulation of effort and fatigue during exercise in humans: Summary and conclusions. *British Journal of Sports Medicine* **39**, 120–24.

Noble, D. (2006). *The music of life: Biology beyond the genome.* Oxford University Press, Oxford.

Nolet, B.A. and Klaassen, M. (2009). Retrodicting patch use by foraging swans in a heterogeneous environment using a set of functional responses. *Oikos* **118**, 431–39.

Nolet, B.A., Langevoord, O., Bevan, R.M., Engelaar, K.R., Klaassen, M., Mulder, R.J.W., and van Dijk, S. (2001). Spatial variation in tuber depletion by swans explained by differences in net intake rates. *Ecology* **82**, 1655–67.

Nottebohm, F. (1981). A brain for all seasons: cyclical anatomical changes in song control. *Science* **214**, 1368–70.

Nunney, L. (2003). The cost of natural selection revisited. *Annales Zoologici Fennici* **40**, 185–94.

Nussey, D.H., Postma, E., Gienapp, P., and Visser, M.E. (2005). Selection on heritable phenotypic plasticity in a wild bird population. *Science* **310**, 304–306.

Nüsslein-Volhard, C. (2006). *Coming to life: How genes drive development.* Kales Press, Carlsbad, CA.

Oaten, A. (1977). Optimal foraging in patches: a case for stochasticity. *Theoretical Population Biology* **12**, 263–85.

Ochsenbein, A.F. and Zinkernagel, R.M. (2000). Natural antibodies and complement link innate and acquired immunity. *Immunology Today* **21**, 624–30.

O'Connor, T. and Diamond, J.M. (1999). Ontogeny of intestinal safety factors: Lactase capacities and lactose loads. *American Journal of Physiology* **276**, R753–65.

O'Connor, T., Lam, M., and Diamond, J.M. (1999). The magnitude of functional adaptation after intestinal resection. *American Journal of Physiology* **276**, R1265–75.

Odling-Smee, F.J., Laland, K.N., and Feldman, M.W. (1996). Niche construction. *American Naturalist* **147**, 641–48.

Odling-Smee, F.J., Laland, K.N., and Feldman, M.W. (2003). *Niche construction: The neglected process in evolution.* Princeton University Press, Princeton, NJ.

Odum, E.P., Rogers, D.T., and Hicks, D.L. (1964). Homeostasis of the non-fat components of migrating birds. *Science* **143**, 1037–39.

Olff, H., Alonso, D., Berg, M.P., Eriksson, B.K., Loreau, M., Piersma, T., and Rooney, N. (2009). Parallel ecological networks in ecosystems. *Philosophical Transactions of the Royal Society, London, B* **364**, 1755–79.

Oliver, J.C., Robertson, K.A., and Monteiro, A. (2009). Accomodating natural and sexual selection in butterfly wing pattern evolution. *Proceedings of the Royal Society, London, B* **276**, 2369–75.

Olsson, O. and Brown, J.S. (2006). The foraging benefits of information and the penalty of ignorance. *Oikos* **112**, 260–73.

Olsson, O. and Brown, J.S. (2010). Smart, smarter, smartest: foraging information states and coexistence. *Oikos* **119**, 292–303.

Olsson, O. and Holmgren, N.M.A. (1998). The survival-rate-maximizing policy for Bayesian foragers: wait for good news. *Behavioral Ecology* **9**, 345–53.

Olsson, O., Wiktander, U., Holmgren, N.M.A., and Nilsson, S. (1999). Gaining ecological information about Bayesian foragers through their behaviour. II. A field test with woodpeckers. *Oikos* **87**, 264–76.

Olsson, O., Brown, J.S., and Helf, K.L. (2008). A guide to central place effects in foraging. *Theoretical Population Biology* **74**, 22–33.

Orians, G.H. and Pearson, N.E. (1979). On the theory of central place foraging. In *Analysis of ecological systems*, eds D.J. Horn, R.D. Mitchell, and G.R. Stairs, pp. 154–77. Ohio State University Press, Columbus.

Orr, J.C., Fabry, V.J., Aumont, O., Bopp, L., Doney, S.C., Feely, R.A., Gnanadesikan, A., Gruber, N., Ishida, A., Joos, F., Key, R.M., Lindsay, K., Maier-Reimer, E., Matear, R., Monfray, P., Mouchet, A., Najjar, R.G., Plattner, G-K., Rodgers, K.B., Sabine, C.L., Sarmiento, J.L., Schlitzer, R., Slater, R.D., Totterdell, I.J., Weirig, M-F., Yamanaka, Y., and Yool, A. (2005). Anthropogenic ocean acidification over the twenty-first century and its impact on calcifying organisms. *Nature* **437**, 681–86.

Orr, N.W.M. (1966). The feeding of sledge dogs on Antarctic expeditions. *British Journal of Nutrition* **20**, 1–12.

Osborn, H.F. (1897). The limits of organic selection. *American Naturalist* **31**, 944–51.

Ott, B.D. and Secor, S.M. (2007). Adaptive regulation of digestive performance in the genus *Python*. *Journal of Experimental Biology* **210**, 340–56.

Oufiero, C.E. and Garland, T. Jr (2007). Evaluating performance costs of sexually selected traits. *Functional Ecology* **21**, 676–89.

Owen, M. (1981). Abdominal profile—a condition index for wild geese in the field. *Journal of Wildlife Management* **45**, 227–30.

Padilla, D.K. (2001). Food and environmental cues trigger an inducible offense. *Evolutionary Ecology Research* **3**, 15–25.

Padilla, D.K. and Adolph, S.C. (1996). Plastic inducible morphologies are not always adaptive: the importance of time delays in a stochastic environment. *Evolutionary Ecology* **10**, 105–107.

Painter, R.C., Osmond, C., Gluckman, P., Hanson, M., Phillips, D.I.W., and Roseboom, T.J. (2008). Transgenerational effects of prenatal exposure to the Dutch famine on neonatal adiposity and health in later life. *British Journal of Obstetrics and Gynaecology* **115**, 1243–49.

Parmesan, C. and Yohe, G (2003). A globally coherent fingerprint of climate change impacts across natural systems. *Nature* **421**, 37–42.

Parter, M., Kashtan, N., and Alon, U. (2008). Facilitated variation: how evolution learns from past environments to generalize to new environments. *Public Library of Science—Computational Biology* **4**, e1000206.

Partridge, L. and Gems, D. (2007). Benchmarks for ageing studies. *Nature* **450**, 165–67.

Pauly, D. (2007). The Sea Around Us Project: documenting and communicating global fisheries impacts on marine ecosystems. *Ambio* **36**, 290–95.

Payne, W.M.C., Williams, D.R., and Trudel, G. (2007). Space flight rehabilitation. *American Journal of Physical Medicine & Rehabilitation* **86**, 583–91.

Peaker, M. and Linzell, J.L. (1975). *Salt glands in birds and reptiles*. Cambridge University Press, Cambridge.

Pearl, R. (1922). *The biology of death*. J.B. Lippincott, Philadelphia.

Pearl, R. (1928). *The rate of living*. Albert Knopf, New York.

Pennisi, E. (2005). The inner tube of life—the dynamic gut. *Science* **307**, 1896–99.

Pennycuick, C.J. and Battley, P.F. (2003). Burning the engine: A time-marching computation of fat and protein consumption in a 5420 km non-stop flight by great knots *Calidris tenuirostris*. *Oikos* **103**, 323–32.

Pennycuick, C.J., Einarsson, O., Bradbury, T.A.M., and Owen, M. (1996). Migrating whooper swans *Cygnus cygnus*: satellite tracks and flight performance calculations. *Journal of Avian Biology* **27**, 118–34.

Pennycuick, C.J., Alerstam, T., and Hedenström, A. (1997). A new low turbulence wind tunnel for bird flight experiments at Lund University, Sweden. *Journal of Experimental Biology* **200**, 1441–49.

Perez-Campo, R., López-Torres, M., Cadenas, S., Rojas, C., and Barja, G. (1998). The rate of free radical production as a determinant of the rate of aging: evidence from the comparative approach. *Journal of Comparative Physiology B* **168**, 149–58.

Perhonen, M.A., Franco, F., Lane, L.D., Buckey, J.C., Blomqvist, C.G., Zerwekh, J.E., Peschock, R.M., Weatherall, P.T., and Levine, B.D. (2001). Cardiac atrophy after bed rest and spaceflight. *Journal of Applied Physiology* **91**, 645–53.

Perrigo, G. (1987). Breeding and feeding strategies in deer mice and house mice when females are challenged to work for their food. *Animal Behaviour* **35**, 1298–316.

Perry, G. and Pianka, E.R. (1997). Animal foraging: past, present and future. *Trends in Ecology and Evolution* **12**, 360–64.

Peters, R.H. (1983). *The ecological implications of body size*. Cambridge University Press, Cambridge.

Peterson, C.C., Nagy, K.A., and Diamond, J.M.(1990). Sustained metabolic scope. *Proceedings of the National Academy of Sciences, USA* **87**, 2324–28.

Petrusek, A., Tollrian, R., Schwenk, K., Haas, A., and Laforsch, C. (2009). A 'crown of thorns' is an inducible defense that protects *Daphnia* against an ancient predator. *Proceedings of the National Academy of Sciences, USA* **106**, 2248–52.

Pfennig, D.W. (1992). Polyphenism in spadefoot toad tadpoles as a locally adjusted evolutionarily stable strategy. *Evolution* 46, 1408–20.

Pfennig, D.W. and Frankino, W.A. (1997). Kind-mediated morphogenesis in facultatively cannibalistic tadpoles. *Evolution* 51, 1993–99.

Pierre, J.S. and Green, R.F. (2007). A Bayesian approach to optimal foraging in parasitoids. In *Behavioral ecology of insects parasitoids: From theoretical approaches to field applications*, eds É. Wajnberg, C. Bernstein, and K. van Alphen, pp. 357–83. Wiley–Blackwell, Oxford.

Piersma, T. (1987). Hop, skip or jump? Constraints on migration of arctic waders by feeding, fattening, and flight speed. *Limosa* 60, 185–94.

Piersma, T. (1990). Pre-migratory 'fattening' usually involves more than the deposition of fat alone. *Ringing & Migration* 11, 113–15.

Piersma, T. (1997). Do global patterns of habitat use and migration strategies co-evolve with relative investments in immunocompetence due to spatial variation in parasite pressure? *Oikos* 80, 623–31.

Piersma, T. (1998). Phenotypic flexibility during migration: optimization of organ size contingent on the risks and rewards of fueling and flight? *Journal of Avian Biology*, **29** 511–20.

Piersma, T. (2002). Energetic bottlenecks and other design constraints in avian annual cycles. *Integrative and Comparative Biology* **42**, 51–67.

Piersma, T. (2003). 'Coastal' versus 'inland' shorebird species: Interlinked fundamental dichotomies between their life- and demographic histories? *Wader Study Group Bulletin* **100**, 5–9.

Piersma, T. (2007). Using the power of comparison to explain habitat use and migration strategies of shorebirds worldwide. *Journal of Ornithology* 148 (Suppl. 1), S45–S59.

Piersma, T. (2009). Threats to intertidal soft-sediment ecosystems. In *Water policy in The Netherlands: Integrated management in a densely populated delta*, eds S. Reinhard and H. Folmer, pp. 57–69. Resources for the Future, Washington, DC.

Piersma, T. and Davidson, N.C. (1992). The migrations and annual cycles of five subspecies of knots in perspective. *Wader Study Group Bulletin*, 64 (Suppl.), 187–97.

Piersma, T. and Dietz, M.W. (2007). Twofold seasonal variation in the supposedly constant, species-specific, ratio of upstroke to downstroke flight muscles in red knots *Calidris canutus. Journal of Avian Biology* 38, 536–40.

Piersma, T. and Drent, J. (2003). Phenotypic flexibility and the evolution of organismal design. *Trends in Ecology and Evolution* 18, 228–33.

Piersma, T. and Gill, R.E. Jr (1998). Guts don't fly: small digestive organs in obese bar-tailed godwits. *Auk* **115**, 196–203.

Piersma, T. and Jukema, J. (1990). Budgeting the flight of a long-distance migrant: changes in nutrient reserve levels of bar-tailed godwits at successive spring staging sites. *Ardea* 78, 315–37.

Piersma, T. and Jukema, J. (2002). Contrast in adaptive mass gains: Eurasian golden plovers store fat before midwinter and protein before prebreeding flight. *Proceedings of the Royal Society, London, B* 269, 1101–1105.

Piersma, T., de Goeij, P., and Tulp, I. (1993a). An evaluation of intertidal feeding habitats from a shorebird perspective: towards relevant comparisons between temperate and tropical mudflats. *Netherlands Journal of Sea Research* 31, 503–12.

Piersma, T., Koolhaas, A., and Dekinga, A. (1993b). Interactions between stomach structure and diet choice in shorebirds. *Auk* 110, 552–64.

Piersma, T., Hoekstra, R., Dekinga, A., Koolhaas, A., Wolf, P., Battley, P.F., and Wiersma, P. (1993c). Scale and intensity of intertidal habitat use by knots *Calidris canutus* in the western Wadden Sea in relation to food, friends and foes. *Netherlands Journal of Sea Research* **31**, 331–57.

Piersma, T., Tulp, I., and Schekkerman, H. (1994a). Final countdown of waders during starvation: Terminal use of nutrients in relation to structural size and concurrent energy expenditure. In *Close to the edge: Energetic bottlenecks and the evolution of migratory pathways in knots*, ed. T. Piersma, pp. 220–27. Uitgeverij Het Open Boek, Den Burg.

Piersma, T., Verkuil, Y., and Tulp, I. (1994b). Resources for long-distance migration of knots *Calidris canutus islandica* and *C. c. canutus*: how broad is the temporal exploitation window of benthic prey in the western and the eastern Wadden Sea? *Oikos* 71, 393–407.

Piersma, T., Cadée, N., and Daan, S. (1995a). Seasonality in basal metabolic rate and thermal conductance in a long-distance migrant shorebird, the knot (*Calidris canutus*). *Journal of Comparative Physiology B* 165, 37–45.

Piersma, T., van Gils, J., de Goeij, P., and van der Meer, J. (1995b). Holling's functional response model as a tool to link the food-finding mechanism of a probing shorebird with its spatial distribution. *Journal of Animal Ecology* 64, 493–504.

Piersma, T., Bruinzeel, L., Drent, R., Kersten, M., van der Meer, J., and Wiersma, P. (1996a). Variability in basal metabolic rate of a long-distance migrant shorebird (red knot, *Calidris canutus*) reflects shifts in organ sizes. *Physiological Zoology* 69, 191–217.

Piersma, T., Gudmundsson, G.A., Davidson, N.C., and Morrison, R.I.G. (1996b). Do arctic-breeding Red Knots

(*Calidris canutus*) accumulate skeletal calcium before egg laying? *Canadian Journal of Zoology* **74**, 2257–61.

Piersma, T., van Aelst, R., Kurk, K., Berkhoudt, H., and Maas, L.R.M. (1998). A new pressure sensory mechanism for prey detection in birds: the use of principles of seabed dynamics? *Proceedings of the Royal Society, London, B* **265**, 1377–83.

Piersma, T., Dietz, M.W., Dekinga, A., Nebel, S., van Gils, J., Battley, P.F., and Spaans, B. (1999a). Reversible size-changes in stomachs of shorebirds: when, to what extent, and why? *Acta Ornithologica* **34**, 175–81.

Piersma, T., Gudmundsson, G.A., and Lilliendahl, K. (1999b). Rapid changes in the size of different functional organ and muscle groups during refueling in a long-distance migrating shorebird. *Physiological and Biochemical Zoology* **72**, 405–15.

Piersma, T., Reneerkens, J., and Ramenofsky, M. (2000). Baseline corticosterone peaks in shorebirds with maximal energy stores for migration: a general preparatory mechanism for rapid behavioral and metabolic transitions? *General and Comparative Endocrinology* **120**, 118–26.

Piersma, T., Koolhaas, A., Dekinga, A., Beukema, J.J., Dekker, R., and Essink, K. (2001). Long-term indirect effects of mechanical cockle-dredging on intertidal bivalve stocks in the Wadden Sea. *Journal of Applied Ecology* **38**, 976–90.

Piersma, T., Lindström, Å., Drent, R., Tulp, I., Jukema, J., Morrison, R.I.G., Reneerkens, J., Schekkerman, H., and Visser, G.H. (2003). High daily energy expenditure of incubating shorebirds on High Arctic tundra: A circumpolar study. *Functional Ecology* **17**, 356–62.

Piersma, T., Rogers, D.I., González, P.M., Zwarts, L., Niles, L.J., de Lima Serrano do Nascimento, I., Minton, C.D.T., and Baker, A.J. (2005). Fuel storage rates in red knots worldwide: Facing the severest ecological constraint in tropical intertidal conditions? In *Birds of two worlds: The ecology and evolution of migratory birds*, eds R. Greenberg and P.P. Marra, pp 262–74. Johns Hopkins University Press, Baltimore.

Pigliucci, M. (2001a). *Phenotypic plasticity: Beyond nature and nurture*. Johns Hopkins University Press, Baltimore.

Pigliucci, M. (2001b). Phenotypic plasticity. In *Evolutionary ecology: Concepts and case studies*, eds C.W. Fox, D.A. Roff, and D.J. Fairbairn, pp. 58–69. Oxford University Press, New York.

Pigliucci, M., Murren, C.J., and Schlichting, C.D. (2006). Phenotypic plasticity and evolution by genetic assimilation. *Journal of Experimental Biology* **209**, 2362–67.

Polikansky, D. (1982). Sex change in plants and animals. *Annual Review of Ecology and Systematics* **13**, 471–95.

Popper, K. (1959). *The logic of scientific discovery*. Basic Books, New York.

Porter, W.P. and Gates, D.M. (1969). Thermodynamic equilibria of animals with environment. *Ecological Monographs* **39**, 227–44.

Price, P.W. (1980). *Evolutionary biology of parasites*. Princeton University Press, Princeton, NJ.

Price, T.D., Qvarnström, A., and Irwin, D.E. (2003). The role of phenotypic plasticity in driving genetic evolution. *Proceedings of the Royal Society, London, B* **270**, 1433–40.

Pugh, L.G. (1972). The logistics of the polar journeys of Scott, Shackleton and Amundsen. *Proceedings of the Royal Society of Medicine* **65**, 42–47.

Pulliam, H.R. (1974). On the theory of optimal diets. *American Naturalist* **108**, 59–75.

Quaintenne, G., van Gils, J.A., Bocher, P., Dekinga, A., and Piersma, T. (2010). Diet selection in a molluscivore shorebird across Western Europe: does it show short- or long-term rate-maximization? *Journal of Animal Ecology* **79**, 53–62.

Quammen, D. (2006). *The reluctant Mr. Darwin: An intimate portrait of Charles Darwin and the making of his theory of evolution*. Norton, New York.

Radford, A.N., Hollén, L.I., and Bell, M.B.V. (2009). The higher the better: Sentinel height influences foraging success in a social bird. *Proceeding of the Royal Society, London, B* **276**, 2437–42.

Ramos, M., Irschick, D.J., and Christenson, T.E. (2004). Overcoming an evolutionary conflict: Removal of a reproductive organ greatly increases locomotor performance. *Proceedings of the National Academy of Sciences, USA* **101**, 4883–87.

Ranta, E., Lundberg, P., and Kaitala, V. (2006). *Ecology of populations*. Cambridge University Press, Cambridge.

Rauw, W.M., Kanis, E., Noordhuizen-Stassen, E.N., and Grommers, F.J. (1998). Undesirable side effects of selection for high production efficiency in farm animals: a review. *Lifestock Production Science* **56**, 15–33.

Redfield, A.C. (1934). On the proportions of organic derivations in sea water and their relation to the composition of plankton. In *James Johnstone memorial volume*, ed. R. J. Daniel, pp. 177–92. University Press of Liverpool, Liverpool.

Redfield, A.C. (1958). The biological control of chemical factors in the environment. *American Scientist* **46**, 205–21.

Reid, J.M., Bignal, E.M., Bignal, S., McCracken, D.I., and Monaghan, P. (2003). Environmental variability, life-history covariation and cohort effects in the red-billed chough *Pyrrhocorax pyrrhocorax*. *Journal of Animal Ecology* **72**, 36–46.

Relyea, R.A. (2001a). Morphological and behavioral plasticity of larval anurans in response to different predators. *Ecology* **82**, 523–40.

Relyea, R.A. (2001b). The lasting effects of adaptive plasticity: Predator-induced tadpoles become long-legged frogs. *Ecology* **82**, 1947–55.

Relyea, R.A. (2002a). Costs of phenotypic plasticity. *American Naturalist* **159**, 272–82.

Relyea, R.A. (2002b). Competitor-induced plasticity in tadpoles: Consequences, cues, and connections to predator-induced plasticity. *Ecological Monographs* **72**, 523–40.

Relyea, R.A. (2003a). How prey respond to combined predators: a review and an empirical test. *Ecology* **84**, 1827–39.

Relyea, R.A. (2003b). Predators come and predators go: the reversibility of predator-induced traits. *Ecology* **84**, 1840–48.

Relyea, R.A. (2005). The heritability of inducible defenses in tadpoles. *Journal of Evolutionary Biology* **18**, 856–66.

Relyea, R.A. (2007). Getting out alive: how predators affect the decision to metamorphose. *Oecologia* **152**, 389–400.

Relyea, R.A. and Auld, J.R. (2004). Having the guts to compete: how intestinal plasticity explains the costs of inducible defences. *Ecology Letters* **7**, 869–75.

Relyea, R.A. and Auld, J.R. (2005). Predator- and competitor-induced plasticity: how changes in foraging morphology affect phenotypic trade-offs. *Ecology* **86**, 1723–29.

Renaudeau, D. and Noblet, J. (2001). Effects of exposure to high ambient temperature and dietary protein level on sow milk production and performance of piglets. *Journal of Animal Science* **79**, 1540–48.

Reneerkens, J., Piersma, T., and Sinninghe Damsté, J.S. (2002). Sandpipers (Scolopacidae) switch from mono to diester preen waxes during courtship and incubation, but why? *Proceedings of the Royal Society, London, B* **269**, 2135–39.

Reneerkens, J., Piersma, T., and Sinninghe Damste, J.S. (2007). Expression of annual cycles in preen wax composition in red knots: Constraints on the changing phenotype. *Journal of Experimental Zoology A* **307A,** 127–39.

Ricklefs, R.E. (1969). Preliminary models for growth rates in altricial birds. *Ecology* **50**, 1031–39.

Ricklefs, R.E. (1991). Structures and transformations of life histories. *Functional Ecology* **5**, 174–83.

Ricklefs, R.E. (1998). Evolutionary theories of aging: confirmation of a fundamental prediction, with implications for the genetic basis and evolution of life span. *American Naturalist* **152**, 22–44.

Ricklefs, R.E. (2000). Intrinsic aging-related mortality in birds. *Journal of Avian Biology* **31**, 103–11.

Ricklefs, R.E. (2008). The evolution of senescence from a comparative perspective. *Functional Ecology* **22**, 379–92.

Ricklefs, R.E. and Scheuerlein, A. (2001). Comparison of aging-related mortality among birds and mammals. *Experimental Gerontology* **36**, 845–57.

Ricklefs, R.E. and Wikelski, M. (2002). The physiology–life history nexus. *Trends in Ecology and Evolution* **17**, 462–68.

Ricklefs, R.E., Konarzewski, M., and Daan, S. (1996). The relationship between basal metabolic rate and daily energy expenditure in birds and mammals. *American Naturalist* **147**, 1047–71.

Ridley, M. (1994). *The Red Queen: Sex and the evolution of human nature*. Penguin Books, London.

Riebesell, U., Zondervan, I., Rost, B., Tortell, P.D., Zeebe, R.E., and Morel, F.M.M. (2000). Reduced calcification of marine plankton in response to increased atmospheric CO_2. *Nature* **407**, 364–67.

Riessen, H.P. and Trevett-Smith, J.B. (2009). Turning inducible defenses on and off: Adaptive responses of *Daphnia* to a gape-limited predator. *Ecology* **90**, 3455–69.

Ripple, W.J. and Beschta, R.L. (2004). Wolves and the ecology of fear: can predation risk structure ecosystems? *Bioscience* **54**, 755–66.

Ripple, W.J. and Beschta, R.L. (2005). Linking wolves and plants: Aldo Leopold on trophic cascades. *Bioscience* **55**, 613–21.

Rittweger, J. and Felsenberg, D. (2009). Recovery of muscle atrophy and bone loss from 90 days bed rest: results from a one-year follow-up. *Bone* **44**, 214–24.

Robbins, C.T. (2001). *Wildlife feeding and nutrition*. Academic Press, San Diego.

Robert, V.A. and Casadevall, A. (2009). Vertebrate endothermy restricts most fungi as potential pathogens. *Journal of Infectious Diseases* **200**, 1623–26.

Roberts, W.M. (1996). Hummingbirds' nectar concentration preferences at low volume: the importance of time scale. *Animal Behaviour* **52**, 361–70.

Robin, J-P., Frain, M., Sardet, C., Groscolas, R., and Le Maho, Y. (1988). Protein and lipid utilization during long-term fasting in emperor penguins. *American Journal of Physiology—Regulatory, Integrative and Comparative Physiology* **254**, 61–68.

Robling, A.G., Hinant, F.M., Burr, D.B., and Turner, C.H. (2002). Improved bone structure and strength after long-term mechanical loading is greatest if loading is separated in short bouts. *Journal of Bone and Mineral Research* **17**, 1545–54.

Rogers, D.I., Battley, P.F., Piersma, T., van Gils, J.A., and Rogers, K.G. (2006). High tide habitat choice: Insights from modelling roost selection by shorebirds around a tropical bay. *Animal Behaviour* **72**, 563–75.

Rogowitz, G.L. (1998). Limits to milk flow and energy allocation during lactation of the hispid cotton rat (*Sigmodon hispidus*). *Physiological Zoology* **71**, 312–20.

Roll-Hansen, N. (1979). The genotype theory of Wilhelm Johannsen and its relation to plant breeding and the study of evolution. *Centaurus* **22**, 201–35.

Rollo, C.D. (1995). *Phenotypes: Their epigenetics, ecology and evolution.* Chapman and Hall, London.

Rønning, B., Moe, B., and Bech, C. (2005). Long-term repeatability makes basal metabolic rate a likely heritable trait in the zebra finch *Taeniopygia guttata. Journal of Experimental Biology* **208**, 4663–69.

Roseboom, T., de Rooij, S., and Painter, R. (2006). The Dutch famine and its long-term consequences for adult health. *Early Human Development* **82**, 485–91.

Rowland, J.M. and Emlen, D.J. (2009). Two thresholds, three male forms result in facultative male trimorphism in beetles. *Science* **323**, 773–76.

Royama, T. (1966). Factors governing feeding rate, food requirement and brood size of nestling great tits *Parus major. Ibis* **108**, 313–47.

Rubenstein, D.M. (2007). Stress hormones and sociality: Integrating social and environmental stressors. *Proceedings of the Royal Sciety, London, B* **274**, 967–75.

Rubner, M. (1908). *Das Problem der Lebensdauer und seine Beziehungen zum Wachstum und Ernahrung.* Oldenburg, München.

Ruff, C., Holt, B., and Trinkaus, E. (2006). Who's afraid of the big bad Wolff?: 'Wolff's Law' and bone functional adaptation. *American Journal of Physical Anthropology* **129**, 484–98.

Rutten, A.L., Oosterbeek, K., Ens, B.J., and Verhulst, S. (2006). Optimal foraging on perilous prey: Risk of bill damage reduces optimal prey size in oystercatchers. *Behavioral Ecology* **17**, 297–302.

Ryan, R.T. and Semlitsch, R.D. (1998). Intraspecific heterochrony and life history evolution: Decoupling somatic and sexual development in a facultatively paedomorphic salamander. *Proceedings of the National Academy of Sciences, USA* **95**, 5643–48.

Sadowska, E.T., Baliga-Klimczyk, K., Chrząścik, K.M., and Koteja, P. (2008). Laboratory model of adaptive radiation: a selection experiment in the bank vole. *Physiological and Biochemical Zoology* **81**, 627–40.

Salvador, A. and Savageau, M.A. (2003). Quantitative evolutionary design of glucose 6-phosphate dehydrogenase expression in human erythrocytes. *Proceedings of the National Academy of Sciences, USA* **100**, 14463–68.

Sandoval, C.P. (1994). Plasticity in web design in the spider *Parawixia bistriata*: a response to variable prey type. *Functional Ecology* **8**, 701–707.

Sapir, N., Tsurim, I., Gal, B., and Abramsky, Z. (2004). The effect of water availability on fuel deposition of two staging *Sylvia* warblers. *Journal of Avian Biology* **35**, 25–32.

Sapolsky, R.M. (1997). *The trouble with testosterone, and other essays on the biology of the human predicament.* Simon and Schuster, New York.

Sapolsky, R.M. (2004). *Why zebras don´t get ulcers: The acclaimed guide to stress, stress/related diseases, and coping,* 3rd edition. Owl Books, New York.

Sarkar, S. (2004). From the Reaktionsnorm to the evolution of adaptive plasticity. In *Phenotypic plasticity: Functional and conceptual approaches,* eds T.J. DeWitt and S.M. Scheiner, pp. 10–30. Oxford University Press, New York.

Savage, V.M., Gillooly, J.F., Woodruff, W.H., West, G.B., Allen, A.P., Enquist, B.J., and Brown, J.H. (2004). The predominance of quarter-power scaling in biology. *Functional Ecology* **18**, 257–82.

Schaeffer, P.J., Hokanson, J.F., Wells, D.J., and Lindstedt, S.L. (2001). Cold exposure increases running VO_{2max} and cost of transport in goats. *American Journal of Physiology* **280**, R42–R47.

Schaeffer, P.J., Villarin, J.J., and Lindstedt, S.L. (2003). Chronic cold exposure increases skeletal oxidative structure and function in *Monodelphis domestica*, a marsupial lacking brown adipose tissue. *Physiological and Biochemical Zoology* **76**, 877–87.

Schaeffer, P.J., Villarin, J.J., Pierotti, D.J., Kelly, D.P., and Lindstedt, S.L. (2005). Cost of transport is increased after cold exposure in *Monodelphis domestica*: Training for ineffeiciency. *Journal of Experimental Biology* **208**, 3159–67.

Schekkerman, H., Tulp, I., Piersma, T., and Visser, G.H. (2003). Mechanisms promoting higher growth rates in arctic than temperate shorebirds. *Oecologia* **134**, 332–42.

Schekkerman, H., Teunissen, W., and Oosterveld, E. (2008). Mortality of black-tailed godwit *Limosa limosa* and northern lapwing *Vanellus vanellus* chicks in wet grasslands: influence of predation and agriculture. *Journal of Ornithology* **150**, 133–45.

Schew, W.A. and Ricklefs, R.A. (1998). Developmental plasticity. In *Avian growth and development: evolution within the altricial-precocial spectrum,* eds J.M. Starck and R.E. Ricklefs, pp. 288–304. Oxford University Press, Oxford.

Schlichting, C.D. and Pigliucci, M. (1998). *Phenotypic evolution: A reaction norm perspective.* Sinauer, Sunderland.

Schluter, D. (1981). Does the theory of optimal diets apply in complex environments? *American Naturalist* **118**, 139–47.

Schluter, D. (1993). Adaptive radiation in sticklebacks: size, shape and habitat use efficiency. *Ecology* **74**, 699–709.

Schluter, D. (2000). *The ecology of adaptive radiation.* Oxford University Press, Oxford.

Schmaljohann, H., Liechti, F., and Bruderer, B. (2007). Songbird migration across the Sahara: The non-stop hypothesis rejected! *Proceedings of the Royal Society, London, B* **274**, 735–39.

Schmaljohann, H., Bruderer, B., and Liechti, F. (2008). Sustained bird flights occur at temperatures beyond expected limits of water loss rates. *Animal Behaviour* **76**, 1133–38.

Schmaljohann, H., Liechti, F., and Bruderer, B. (2009). Trans-Sahara migrants select flight altitudes to minimize energy costs rather than water loss. *Behavioral Ecology and Sociobiology* **63**, 1609–19.

Schmid-Hempel, P. and Ebert, D. 2003. On the evolutionary ecology of specific immune defence. *Trends in Ecology and Evolution* **18**, 27–32.

Schmid-Hempel, P., Kacelnik, A., and Houston, A.I. (1985). Honeybees maximize efficiency by not filling their crop. *Behavioral Ecology and Sociobiology* **17**, 61–66.

Schmidt-Nielsen, K. (1984). *Scaling: Why is animal size so important?* Cambridge University Press, Cambridge.

Schmidt-Wellenburg, C.A., Biebach, H., Daan, S., and Visser, G.H. (2006). Energy expenditure and wing beat frequency in relation to body mass in free flying swallows (*Hirundo rustica*). *Journal of Comparative Physiology B* **177**, 327–37.

Schmidt-Wellenburg, C., Engel, S., and Visser, G.H. (2008). Energy expenditure during flight in relation to body mass: Effects of natural increases in mass and artificial load in rose coloured starlings. *Journal of Comparative Physiology B* **178**, 767–77.

Schmitz, O.J., Krivan, V., and Ovadia, O. (2004). Trophic cascades: The primacy of trait-mediated indirect interactions. *Ecology Letters* **7**, 153–63.

Schoeppner, N.M. and Relyea, R.A. (2005). Damage, digestion, and defence: the role of alarm cues and kairomones for inducing prey defences. *Ecology Letters* **8**, 505–12.

Schoeppner, N.M. and Relyea, R.A. (2008). Detecting small environmental differences: Risk-response curves for predator-induced behavior and morphology. *Oecologia* **154**, 743–54.

Scholander, P.F. (1990). *Enjoying a life in science: The autobiography of P.F. Scholander.* University of Alaska Press, Fairbanks.

Scholander, P.F., Hock, R., Walters, V., Johnson, F., and Irving, L. (1950a). Heat regulation in some Arctic and tropical mammals and birds. *Biological Bulletin* **99**, 237–58.

Scholander, P.F., Walters, V., Hock, R., and Irving, L. (1950b). Body insulation of some Arctic and tropical mammals and birds. *Biological Bulletin* **99**, 225–36.

Scholander, P.F., Hock, R., Walters, V., and Irving, L. (1950c). Adaptation to cold in arctic and tropical mammals and birds in relation to body temperature, insulation and basal metabolic rate. *Biological Bulletin* **99**, 259–71.

Schroeder, J., Lourenço, P.M., Hooijmeijer, J.C.E.W., Both, C., and Piersma, T. (2009). A possible case of contemporary selection leading to a decrease in sexual plumage dimorphism in a grassland-breeding shorebird. *Behavioral Ecology* **20**, 797–807.

Schubert, K.A. (2009). Synthesis: environmental quality, energy allocation and the cost of reproduction. In *Breeding on a budget: Fundamental links between energy metabolism and mammalian life history trade-offs*, pp. 120–32. PhD thesis, University of Groningen, The Netherlands.

Schubert, K.A., de Vries, G., Vaanholt, L.M., Meijer, H., Daan, S., and Verhulst, S. (2009). Maternal energy allocation to offspring increases with environmental quality in house mice. *American Naturalist* **173**, 831–40.

Schwilch, R., Piersma, T., Holmgren, N.M.A., and Jenni, L. (2002). Do migratory birds need a nap after a long nonstop flight? *Ardea* **90**, 149–54.

Scott, I. and Evans, P.R. (1992). The metabolic output of avian (*Sturnus vulgaris, Calidris alpina*) adipose tissue, liver and skeletal muscle: Implications for BMR/body mass relationships. *Comparative Biochemistry and Physiology A* **103**, 329–32.

Scott, G.R., Egginton, S., Richards, J.G., and Milsom, W.K. (2009). Evolution of muscle phenotype for extreme high altitude flight in the bar-headed goose. *Proceedings of the Royal Society, London, B* **276**, 3645–53.

Secor, S.M. (2008). Digestive physiology of the Burmese python: broad regulation of integrated performance. *Journal of Experimental Biology* **211**, 3767–74.

Secor, S.M. and Diamond, J.M. (1995). Adaptive responses to feeding in Burmese pythons: pay before pumping. *Journal of Experimental Biology* **198**, 1313–25.

Secor, S.M., Stein, E.D., and Diamond, J.M. (1994). Rapid up-regulation of snake intestine in response to feeding: a new model of intestinal adaptation. *American Journal of Physiology* **266**, G695–G705.

Sedinger, J.S. (1997). Adaptations to and consequences of an herbivorous diet in grouse and waterfowl. *Condor* **99**, 314–26.

Selman, C., McLaren, J.S., Collins, A.R., Duthie, G.G., and Speakman, J.R. (2008). The impact of experimentally elevated energy expenditure on oxidative stress and lifespan in the short-tailed field vole *Microtus agrestis*. *Proceedings of the Royal Society, London, B* **275**, 1907–16.

Selvin, S. (1975). On the Monty Hall problem. *American Statistician* **29**, 134.

Seppänen, J-T., Forsman, J.T., Mönkkönen, M., and Thomson, R.L. (2007). Social information use is a process across time, space, and ecology, reaching heterospecifics. *Ecology* **88**, 1622–33.

Shapiro, A.M. (1976). Seasonal polyphenism. *Evolutionary Biology* **9**, 259–333.

Sheriff, M.J., Krebs, C.J., and Boonstra, R. (2009). The sensitive hare: sublethal effects of predator stress on reproduction in snowshoe hares. *Journal of Animal Ecology* **78**, 1249–58.

Shettleworth, S.J. (1998). *Cognition, evolution and behaviour*. Oxford University Press, Oxford.

Shubin, N. (2008). *Your inner fish: A journey into the 3.5-billion-year history of the human body*. Vintage Books, New York.

Sih, A. and Christensen, B. (2001). Optimal diet theory: when does it work, and when and why does it fail? *Animal Behaviour* **61**, 379–90.

Simmons, L.W. and Emlen, D.J. (2006). Evolutionary trade-off between weapons and testes. *Proceedings of the National Academy of Sciences, USA* **103**, 16346–51.

Sinervo, B. (1999). Mechanistic analysis of natural selection and a refinement of Lack's and Williams's principles. *American Naturalist* **154**, S26–S42.

Sinervo, B. and Basolo, A.L. (1996). Testing adaptation using phenotypic manipulations. In *Adaptation*, eds M.R. Rose and G.V. Lauder, pp. 149–85. Academic Press, San Diego.

Sjodin, A.M., Anderson, A.B., Hogberg, J.M., and Westerterp, K.R. (1994). Energy balance in cross-country skiers: a study using doubly labelled water. *Medicine and Science in Sports and Exercise* **26**, 720–24.

Smith, G.T., Brenowitz, E.A., and Wingfield, F.C. (1997). Roles of photoperiod and testosterone in seasonal plasticity of the avian song control system. *Journal of Neurobiology* **32**, 426–42.

Smith, J.W., Benkman, C.W., and Coffey, K. (1999). The use and misuse of public information by foraging red crossbills. *Behavioral Ecology* **10**, 54–62.

Solomon, S. (2001). *The coldest march: Scott's fatal Antarctic expedition*. Yale University Press, London.

Sonnenfeld, G. (2005). The immune system in space, including earth-based benefits of space-based research. *Current Pharmaceutical Biotechnology* **6**, 343–49.

Spalding, K.L., Bhardwaj, R.D., Buchholz, B.A., Druid, H., and Frisén, J. (2005a). Retrospective birth dating of cells in humans. *Cell* **122**, 133–43.

Spalding, K.L., Buchholz, B.A., Bergman, L-E., Druid, H., and Frisén, J. (2005b). Age written in teeth by nuclear tests. *Nature* **437**, 333–34.

Spalding, K.L., Arner, E., Westermark, P.O., Bernard, S., Buchholz, B.A., Bergmann, O., Blomqvist, L., Hoffstedt, J., Näslund, E., Britton, T., Concha, H., Hassan, M., Rydén, M., Frisén, J., and Arner, P. (2008). Dynamics of fat cell turnover in humans. *Nature* **453**, 783–87.

Speakman, J.R. (1997a). *Doubly labelled water: Theory and practice*. Chapman and Hall, London.

Speakman, J.R. (1997b). Factors influencing the daily energy expenditure of small mammals. *Proceedings of the Nutrition Society* **56**, 1119–36.

Speakman, J.R. (2000). The cost of living: field metabolic rates of small mammals. *Advances in Ecological Research* **30**, 177–297.

Speakman, J.R. (2005). Body size, energy metabolism and lifespan. *Journal of Experimental Biology* **208**, 1717–30.

Speakman, J.R. (2008). The physiological costs of reproduction in small mammals. *Philosophical Transactions of the Royal Society, London, B* **363**, 375–98.

Speakman, J.R. and Johnson, M.S. (2000). Relationships between resting metabolic rate and morphology in lactating mice: what tissues are the major contributors to resting metabolism? In *Living in the cold*, eds G. Heldmaier and M. Klingenspor, Vol 3, pp. 479–86. Springer, Berlin.

Speakman, J.R. and Król, E. (2005). Limits to sustained intake. IX. A review of hypotheses. *Journal of Comparative Physiology B* **175**, 375–94.

Speakman, J.R., Gidney, A., Bett, J., Mitchell, I.P., and Johnson, M.S. (2001). Limits to sustained intake. IV. Effect of variation in food quality on lactating mice *Mus musculus*. *Journal of Experimental Biology* **204**, 1957–65.

Speakman, J.R., Selman, C., McLaren, J., and Harper, E.J. (2002). Living fast, dying when? The link between aging and energetics. *Journal of Nutrition* **132**, 1583–97.

Speakman, J.R., Król, E., and Johnson, M.S. (2004a). The functional significance of individual variation in basal metabolic rate. *Physiological and Biochemical Zoology* **77**, 900–15.

Speakman, J.R., Talbot, D.A., Selman, C., Snart, S., McLaren, J.S., Redman, P., Król, E., Jackson, D.M., Johnson, M.S., and Brand, M.D. (2004b). Uncoupling and surviving: Individual mice with higher metabolism have higher mitochondrial uncoupling and live longer. *Aging Cell* **3**, 87–95.

Staaland, H. (1967). Anatomical and physiological adaptations of nasal glands in Charadriiformes birds. *Comparative Biochemistry and Physiology* **23**, 933–44.

Stamper, C.E., Downie, J.T., Stevens, D.J., and Monaghan, P. (2009). The effects of perceived predation risk on pre- and post-metamorphic phenotypes in the common frog. *Journal of Zoology* **277**, 205–13.

Stanley, S.M. (1970). Relation of shell form to life habits of the Bivalvia (Mollusca). *Geological Society of America emoirs* **125**, 1–296.

Starck, J.M. (2003). Shaping up: how vertebrates adjust their digestive system to changing environmental conditions. *Animal Biology* **53**, 245–57.

Starck, J.M. (2009). Functional morphology and patterns of blood flow in the heart of *Python regius*. *Journal of Morphology* **270**, 673–87.

Starck, J.M. and Beese, K. (2001). Structural flexibility of the intestine of Burmese python in response to feeding. *Journal of Experimental Biology* **201**, 325–35.

Starck, J.M. and Wimmer, C. (2005). Patterns of blood flow during the postprandial response in ball pythons *Python regius*. *Journal of Experimental Biology* **208**, 881–89.

Starck, J.M., Dietz, M.W., and Piersma, T. (2001). The assessment of body composition and other parameters by ultrasound scanning. In *Body composition analysis of animals: A handbook of non-destructive methods*, ed. J.R. Speakman, pp. 188–210. Cambridge University Press, Cambridge.

Starck, J.M., Cruz-Neto, A.P., and Abe, A.S. (2007). Physiological and morphological responses to feeding in broad-nosed caiman (*Caiman latirostris*). *Journal of Experimental Biology* **210**, 2033–45.

Stearns, S.C. (1989). Trade-offs in life-history evolution. *Functional Ecology* **3**, 259–68.

Stearns, S.C. (1992). *The evolution of life histories*. Oxford University Press, New York.

Stein, T.P. and Wade, C.E. (2005). Metabolic consequences of muscle disuse atrophy. *Journal of Nutrition* **135**, 1824S–1828S.

Stephens, D.W. and Krebs, J.R. (1986). *Foraging theory*. Princeton University Press, Princeton, NJ.

Sterner, R.W. and Elser, J.J. (2002). *Ecological stoichiometry: The biology of elements from molecules to the biosphere*. Princeton University Press, Princeton, NJ.

Sterner, R.W. and Hessen, D.O. (1994). Algal nutrient limitation and the nutrition of aquatic herbivores. *Annual Review of Ecology and Systematics* **25**, 1–29.

Stevens, C.E. and Hume, I.D. (1995). *Comparative physiology of the vertebrate digestive system*, 2nd edition. Cambridge University Press, Cambridge.

Stroud, M.A. (1987a). Nutrition and energy balance on the 'footsteps of Scott' expedition 1984–86. *Human Nutrition: Applied Nutrition* **41A**, 426–33.

Stroud, M.A. (1987b). Did Scott simply starve? Lessons from 'The footsteps of the Scott Antarctic expedition'. *Proceedings of the Nutrition Society* **46**, A129.

Stroud, M. (1998). The nutritional demands of very prolonged exercise in man. *Proceedings of the Nutrition Society* **57**, 55–61.

Stroud, M. (2004). *Survival of the fittest: Anatomy of peak physical performance*. Vintage, London.

Stroud, M.A., Coward, W.A., and Sawyer, M.B. (1993). Measurements of energy expenditure using isotope-labelled water ($^2H_2{}^{18}O$) during an Arctic expedition. *European Journal of Applied Physiology* **67**, 375–79.

Stroud, M.A., Ritz, P., Coward, W.A., Sawyer, M.B., Constantin-Teodosiu, D., Greenhaff, P.L. and Macdonald, I.A. (1997). Energy expenditure using isotope-labelled water ($^2H_2{}^{18}O$), exercise performance, skeletal muscle enzyme activities and plasma biochemical parameters in humans during 95 days of endurance exercise with

inadequate energy intake. *European Journal of Applied Physiology* **76**, 243–52.

Stuart-Fox, D., Moussalli, A., and Whiting, M.J. (2008). Predator-specific camouflage in chameleons. *Biology Letters* **4**, 326–29.

Suarez, R.K. (1992). Hummingbird flight: sustaining the highest mass-specific metabolic rates among vertebrates. *Experientia* **48**, 565–70.

Suarez, R.K. (1998). Oxygen and the upper limits to animal design and performance. *Journal of Experimental Biology* **201**, 1065–72.

Suarez, R.K. and Darveau, C-A. (2005). Multi-level regulation and metabolic scaling. *Journal of Experimental Biology* **208**, 1627–34.

Suarez, R.K., Darveau, C-A., and Childress, J.J. (2004). Metabolic scaling: a many-splendoured thing. *Comparative Biochemistry and Physiology B* **139**, 531–41.

Sutherland, W.J. (1987). Why do animals specialize. *Nature* **325**, 483–84.

Sutherland, W.J. and Anderson, C.W. (1993). Predicting the distribution of individuals and the consequences of habitat loss: the role of prey depletion. *Journal of Theoretical Biology* **160**, 223–30.

Sutherland, W.J., Ens, B.J., Goss-Custard, J.D., and Hulscher, J.B. (1996). Specialization. In *The oystercatcher: From individuals to populations*, ed. J.D. Goss-Custard, pp. 56–76. Oxford University Press, Oxford.

Sutton, J.R., Reeves, J.T., Wagner, P.D., Groves, B.M., Cymerman, A., Malconian, M.K., Rock, P.B., Young, P.M., Walter, S.D., and Houston, C.S. (1988). Operation Everest II: Oxygen transport during exercise at extreme simulated altitude. *Journal of Applied Physiology* **64**, 1309–21.

Swanson, D.L. (1993). Cold tolerance and thermogenic capacity in dark-eyed juncos in winter: Geographic variation and comparison with American tree sparrows. *Journal of Thermal Biology* **18**, 275–81.

Swanson, D.L. and Liknes, E.T. (2006). A comparative analysis of thermogenic capacity and cold tolerance in small birds. *Journal of Experimental Biology* **209**, 466–74.

Swart, J.A.A. and van Andel, J. (2008). Rethinking the interface between ecology and society: the case of the cockle controversy in the Dutch Wadden Sea. *Journal of Applied Ecology* **45**, 82–90.

Swennen, C., Leopold, M.F., and de Bruijn, L.L.M. (1989). Time-stressed oystercatchers, *Haematopus ostralegus*, can increase their intake rate. *Animal Behaviour* **38**, 8–22.

Tammaru, T. and Haukioja, E. (1996). Capital breeders and income breeders among Lepidoptera—consequences to population dynamics. *Oikos* **77**, 561–64.

Taylor, C.R. and Weibel, E.R. (1981). Design of the mammalian respiratory system. I. Problem and strategy. *Respiratory Physiology* **44**, 1–10.

Taylor, C.R., Maloiy, G.M.O., Weibel, E.R., Langman, V.A., Kamau, J.M., Seeherman, H.J., and Heglund, N.C. (1981). Design of the mammalian respiratory system. III. Scaling maximum aerobic capacity to body mass: Wild and domestic animals. *Respiratory Physiology* **44**, 25–37.

Taylor, C.R., Weibel, E.R., Weber, J-M., Vock, R., Hoppeler, H., Roberts, T.J., and Bricon, G. (1996). Design of the oxygen and substrate pathways. I. Model and strategy to test symmorphosis in a network structure. *Journal of Experimental Biology* **199**, 1643–49.

Templeton, J.J. and Giraldeau, L-A. (1996). Vicarious sampling: the use of personal and public information by starlings foraging in a simple patchy environment. *Behavioral Ecology and Sociobiology* **38**, 105–14.

Teplitsky, C. and Laurila, A. (2007). Flexible defense strategies: competition modifies investment in behavioral vs. morphological defenses. *Ecology* **88**, 1641–46.

Teplitsky, C., Plénet, S., Léna, J-P., Mermet, N., Malet, E., and Joly, P. (2005a). Escape behaviour and ultimate causes of specfic induced defences in an anuran tadpole. *Journal of Evolutionary Biology* **18**, 180–90.

Teplitsky, C., Plénet, S., and Joly, P. (2005b). Costs and limits of dosage response to predation risk: To what extent can tadpoles invest in anti-predator morphology? *Oecologia* **145**, 364–70.

Teplitsky, C., Mills, J.A., Alho, J.S., Yarrall, J.W., and Merilä, J. (2008). Bergmann's rule and climate change revisited: disentangling environmental and genetic responses in a wild bird population. *Proceedings of the National Academy of Sciences, USA* **105**, 13492–96.

Tesch, P.A., Trieschmann, J.T., and Ekberg, A. (2004). Hypertrophy of chronically unloaded muscle subjected to resistance exercise. *Journal of Applied Physiology* **96**, 1451–58.

Thompson, G.E. (1973). Review of the progress of dairy sciences: climatic physiology of cattle. *Journal of Dairy Research* **40**, 441–73.

Tieleman, B.I. and Williams, J.B. (1999). The role of hyperthermia in the water economy of desert birds. *Physiological and Biochemical Zoology* **72**, 87–100.

Tieleman, B.I. and Williams, J.B. (2000). The adjustment of avian metabolic rates and water fluxes to desert environments. *Physiological and Biochemical Zoology* **73**, 461–79.

Tieleman, B.I., Williams, J.B., and Buschur, M.E. (2002). Physiological adjustments to arid and mesic environments in larks (Alaudidae). *Physiological and Biochemical Zoology* **75**, 305–13.

Tieleman, B.I., Williams, J.B., Ricklefs, R.E., and Klasing, K.C. (2005). Constitutive innate immunity is a component of the pace-of-life syndrome in tropical birds. *Proceedings of the Royal Society, London, B* **272**, 1715–20.

Tinbergen, N. (1963). On the aims and methods of ethology. *Zeitschrift für Tierpsychologie* **20**, 410–33.

Tinbergen, J.M. (1981). Foraging decisions in starlings (*Sturnus vulgaris* L.). *Ardea* **69**, 1–67.

Tinbergen, J.M. and Dietz, M.W. (1994). Parental energy expenditure during brood rearing in the great tit (*Parus major*) in relation to body mass, temperature, food availability and clutch size. *Functional Ecology* **8**, 563–72.

Tinbergen, J.M. and Verhulst, S. (2000). A fixed energetic ceiling to parental effort in the great tit? *Journal of Animal Ecology* **69**, 323–34.

Tinbergen, J.M. and Williams, J.B. (2002). Energetics of incubation. In *Avian incubation: Behaviour, environment and evolution*, ed. D.C. Deeming, pp. 299–313. Oxford University Press, Oxford.

Tinbergen, N., Impekoven, M., and Franck, C.K. (1967). An experiment on spacing-out as a defence against predation. *Behaviour* **28**, 307–20.

Todd, P.A. (2008). Morphological plasticity in scleractinian corals. *Biological Reviews* **83**, 315–37.

Tollrian, R. and Dodson, S.I. (1999). Inducible defenses in Cladocera: Constraints, costs, and multipredator environments. In *The ecology and evolution of inducible defenses*, eds R. Tollrian and C.D. Harvell, pp. 177–202. Princeton University Press, Princeton, NJ.

Tollrian, R. and Harvell, C.D. eds (1999). *The ecology and evolution of inducible defenses*. Princeton University Press, Princeton, NJ.

Toloza, E.M., Lam, M., and Diamond, J.M. (1991). Nutrient extraction by cold-exposed mice: A test of digestive safety margins. *American Journal of Physiology* **261**, G608–20.

Tomkovich, P.S. (2001). A new subspecies of red knot *Calidris canutus* from the New Siberian Islands. *Bulletin British Ornithologists' Club* **121**, 257–63.

Tomkovich, P.S. and Soloviev, M.Y. (1996). Distribution, migrations and biometrics of knots *Calidris canutus canutus* on Taimyr, Siberia. *Ardea* **84**, 85–98.

Touchon, J.C. and Warkentin, K.M. (2008). Fish and dragonfly nymph predators induce opposite shifts in color and morphology of tadpoles. *Oikos* **117**, 634–40.

Tramontin, A.D., Hartman, V.N., and Brenowitz, E.A. (2000). Breeding conditions induce rapid and sequential growth in adult avian song control circuits: a model of seasonal plasticity in the brain. *Journal of Neuroscience* **20**, 854–61.

Trappe, S., Costill, D., Gallagher, P., Creer, A., Peters, J.R., Evans, H., Riley, D.A., and Fitts, R.H. (2009). Exercise in space: human skeletal muscle after 6 months aboard the International Space Station. *Journal of Applied Physiology* **106**, 1159–68.

Trussell, G.C. and Smith, L.D. (2000). Induced defenses in response to an invading crab predator: An explanation of historical and geographic phenotypic change. *Proceedings of the National Academy of Sciences, USA* **97**, 2123–27.

Tsurim, I., Sapir, N., Belmaker, J., Shanni, I., Izhaki, I., Wojciechowski, M.S., Karasov, W.H., and Pinshow, B. (2008). Drinking water boosts food intake rate, body mass increase and fat accumulation in migratory blackcaps (*Sylvia atricapilla*). *Oecologia* **156**, 21–30.

Tulp, I., McChesney, S., and de Goeij, P. (1994). Migratory departures of waders from north-western Australia: behaviour, timing and possible migration routes. *Ardea* **82**, 201–21.

Tulp, I., Schekkerman, H., Bruinzeel, L.W., Jukema, J., Visser, G.H., and Piersma, T. (2009). Energetic demands during incubation and chick-rearing in a uniparental and a biparental arctic breeding shorebird. *Auk* **126**, 155–64.

Turner, J.S. (2007). *The tinkerer's accomplice: How design emerges from life itself.* Harvard University Press, Cambridge, Mass.

Underwood, E.J. (1977). *Trace elements in human and animal nutrition.* Academic Press, New York.

Urabe, J., Clasen, J., and Sterner, R.W. (1997). Phosphorus limitation of *Daphnia* growth: Is it real? *Limnology and Oceanography* **42**, 1436–43.

Vaanholt, L.M., Daan, S., Schubert, K.A., and Visser, G.H. (2009). Metabolism and aging: effects of cold exposure on metabolic rate, body composition and longevity in mice. *Physiological and Biochemical Zoology* **82**, 314–24.

Vahl, W.K., van der Meer, J., Weissing, F.J., van Dullemen, D., and Piersma, T. (2005). The mechanisms of interference competition: two experiments on foraging waders. *Behavioral Ecology* **16**, 845–55.

Valencak, T.G. and Ruf, T. (2009). Energy turnover in European hares is centrally limited during early, but not during peak lactation. *Journal of Comparative Physiology B* **179**, 933–43.

Valencak, T.G., Tataruch, F., and Ruf, T. (2009). Peak energy turnover in lactating European hares: the role of fat reserves. *Journal of Experimental Biology* **212**, 231–37.

Valladares, F., Gianoli, E., and Gomez, J.M. (2007). Ecological limits to plant phenotypic plasticity. *New Phytologist* **176**, 749–63.

Valone, T.J. (1993). Patch information and estimation: a cost of group foraging. *Oikos* **68**, 258–66.

Valone, T.J. (2006). Are animals capable of Bayesian updating? An empirical review. *Oikos* **112**, 252–59.

Valone, T.J. (2007). From eavesdropping on performance to copying the behaviour of others: A review of public information use. *Behavioral Ecology and Sociobiology* **62**, 1–14.

Valone, T.J. and Giraldeau, L-A. (1993). Patch estimation by group foragers: what information is used? *Animal Behaviour* **45**, 721–28.

Van Buskirk, J. and Relyea, R.A. (1998). Selection for phenotypic plasticity in *Rana sylvatica* tadpoles. *Biological Journal of the Linnean Society* **65**, 301–28.

van de Kam, J., Ens, B. J., Piersma, T., and Zwarts, L. (2004). *Shorebirds: An illustrated behavioural ecology.* KNNV Publishers, Utrecht.

van de Kam, J., Battley, P.F., McCaffery, B.J., Rogers, D.I., Jae-Sang, Hong, Moores, N., Ju-Yong, Ki, Lewis, J., and Piersma, T. (2008). *Invisible connections: Why migrating shorebirds need the Yellow Sea.* Wetlands International, Wageningen.

van den Hout, P.J., Piersma, T., Dekinga, A., Lubbe, S.K., and Visser, G.H. (2006). Ruddy turnstones *Arenaria interpres* rapidly build pectoral muscle after raptor scares. *Journal of Avian Biology* **37**, 425–30.

van den Hout, P.J., Spaans, B., and Piersma, T. (2008). Differential mortality of wintering shorebirds on the Banc d'Arguin, Mauritania, due to predation by large falcons. *Ibis* **150 (Suppl. 1)**, 219–30.

van den Hout, P.J., Mathot, K.J., Maas, L.R.M., and Piersma, T. (2010). Predator escape tactics in birds: Linking ecology and aerodynamics. *Behavioral Ecology* **21**, 16–25.

van de Pol, M., Ens, B.J., Oosterbeek, K., Brouwer, L., Verhulst, S., Tinbergen, J.M., Rutten, A.L. and de Jong, M. (2009). Oystercatchers' bill shapes as a proxy for diet specialization: More differentiation than meets the eye. *Ardea* **97**, 335–47.

van der Meer, J. and Ens, B.J. (1997). Models of interference and their consequences for the spatial distribution of ideal and free predators. *Journal of Animal Ecology* **66**, 846–58.

Vandermeer, J.H. and Goldberg, D.E. (2003). *Population ecology: First principles.* Princeton University Press, Princeton, NJ.

van der Meer, J. and Piersma, T. (1994). Physiologically inspired regression models for estimating and predicting nutrient stores and their composition in birds. *Physiological Zoology* **67**, 305–29.

van der Merwe, M. and Brown, J.S. (2008). Mapping the landscape of fear of the Cape ground squirrel. *Journal of Mammalogy* **89**, 1162–69.

van Gils, J.A. (2010). State-dependent Bayesian foraging on spatially autocorrelated food distributions. *Oikos* **119**, 237–44.

van Gils, J.A. and Piersma, T. (2004). Digestively constrained predators evade the cost of interference competition. *Journal of Animal Ecology* **73**, 386–98.

van Gils, J.A. and Tijsen, W. (2007). Short-term foraging costs and long-term fueling rates in central-place foraging swans revealed by giving-up exploitation times. *American Naturalist* **169**, 609–20.

van Gils, J.A., Piersma, T., Dekinga, A., and Dietz, M.W. (2003a). Cost–benefit analysis of mollusc-eating in a shorebird. II. Optimizing gizzard size in the face of seasonal demands. *Journal of Experimental Biology* **206**, 3369–80.

van Gils, J.A., Schenk, I.W., Bos, O., and Piersma, T. (2003b). Incompletely informed shorebirds that face a digestive constraint maximize net energy gain when exploiting patches. *American Naturalist* **161**, 777–93.

van Gils, J.A., Edelaar, P., Escudero, G., and Piersma, T. (2004). Carrying capacity models should not use fixed prey density thresholds: a plea for using more tools of behavioural ecology. *Oikos* **104**, 197–204.

van Gils, J.A., de Rooij, S.R., van Belle, J., van der Meer, J., Dekinga, A., Piersma, T., and Drent, R. (2005a). Digestive bottleneck affects foraging decisions in red knots *Calidris canutus*. I. Prey choice. *Journal of Animal Ecology* **74**, 105–19.

van Gils, J.A., Dekinga, A., Spaans, B., Vahl, W.K., and Piersma, T. (2005b). Digestive bottleneck affects foraging decisions in red knots *Calidris canutus*. II. Patch choice and length of working day. *Journal of Animal Ecology* **74**, 120–30.

van Gils, J.A., Battley, P.F., Piersma, T., and Drent, R. (2005c). Reinterpretation of gizzard sizes of red knots world-wide emphasises overriding importance of prey quality at migratory stopover sites. *Proceedings of the Royal Society, London, B* **272**, 2609–18.

van Gils, J.A., Piersma, T., Dekinga, A., and Battley, P.F. (2006a). Modelling phenotypic flexibility: An optimality analysis of gizzard size in red knots (*Calidris canutus*). *Ardea* **94**, 409–20.

van Gils, J.A., Piersma, T., Dekinga, A., Spaans, B., and Kraan, C. (2006b). Shellfish-dredging pushes a flexible avian top predator out of a protected marine ecosystem. *Public Library of Science—Biology* **4**, 2399–2404.

van Gils, J.A., Spaans, B., Dekinga, A., and Piersma, T. (2006c). Foraging in a tidally structured environment by red knots (*Calidris canutus*): Ideal, but not free. *Ecology* **87**, 1189–1202.

van Gils, J.A., Dekinga, A., van den Hout, P.J., Spaans, B., and Piersma, T. (2007a). Digestive organ size and behavior of red knots (*Calidris canutus*) indicate the quality of their benthic food stocks. *Israel Journal of Ecology and Evolution* **53**, 329–46.

van Gils, J.A., Munster, V.J., Radersma, R., Liefhebber, D., Fouchier, R.A.M., and Klaassen, M. (2007b). Hampered foraging and migratory performance in swans infected with low-pathogenic avian influenza A virus. *Public Library of Science One* **2**, e184.

van Gils, J.A., Beekman, J.H., Coehoorn, P., Corporaal, E., Dekkers, T., Klaassen, M., van Kraaij, R., de Leeuw, R., and de Vries, P.P. (2008). Longer guts and higher food quality increase energy intake in migratory swans. *Journal of Animal Ecology* **77**, 1234–41.

van Gils, J.A., Kraan, C., Dekinga, A., Koolhaas, A., Drent, J., de Goeij, P., and Piersma, T. (2009). Reversed optimality and predictive ecology: burrowing depth forecasts population change in a bivalve. *Biology Letters* **5**, 5–8.

van Roomen, M., van Turnhout, C., van Winden, E., Koks, B., Goedhart, P., Leopold, M., and Smit, C. (2005). Trends in benthivorous waterbirds in the Dutch Wadden Sea 1975-2002: large differences between shellfish-eaters and worm-eaters. *Limosa* **78**, 21–38.

Van Valen, L. (1973). A new evolutionary law. *Evolutionary Theory* **1**, 1–30.

Vermeij, G.J. (1987). *Evolution and escalation: An ecological history of life*. Princeton University Press, Princeton, NJ.

Vermeij, G.J. (1993). *A natural history of shells*. Princeton University Press, Princeton, NJ.

Vézina, F. and Williams, T.D. (2005). The metabolic cost of egg production is repeatable. *Journal of Experimental Biology* **208**, 2533–38.

Vézina, F., Jalvingh, K.M., Dekinga, A., and Piersma, T. (2006). Acclimation to different thermal conditions in a northerly wintering shorebird is driven by body mass-related changes in organ size. *Journal of Experimental Biology* **209**, 3141–54.

Vézina, F., Jalvingh, K.M., Dekinga, A., and Piersma, T. (2007). Thermogenic side-effects to migratory predisposition in shorebirds. *American Journal of Physiology* **292**, R1287–97.

Vézina, F., Dekinga, A., and Piersma, T. (2010). Phenotypic compromise in the face of conflicting ecological demands: an example in red knots *Calidris canutus*. *Journal of Avian Biology* **41**, 88–93.

Via, S. and Lande, R. (1985). Genotype-environment interaction and the evolution of phenotypic plasticity. *Evolution* **39**, 505–22.

Via, S., Gomulkiewicz, R., de Jong, G., Scheiner, S.M., Schlichting, C.D., and van Tienderen, P.H. (1995). Adaptive phenotypic plasticity: concensus and controversy. *Trends in Ecology and Evolution* **10**, 212–17.

Vico, L., Lafage-Proust, M-H., and Alexandre, C. (1998). Effects of gravitational changes on the bone system *in vitro* and *in vivo*. *Bone* **22**, 95S–100S.

Vico, L., Collet, P., Guignandon, A., Lafage-Proust, M-H., Thomas, T., Rehailia, M., and Alexandre, C. (2000). Effects of long-term microgravity exposure on cancellous and cortical weight-bearing bones of cosmonauts. *Lancet* **355**, 1607–11.

Videler, J.J. (2005). *Avian flight*. Oxford University Press, Oxford.

Vince, G. (2006). Your amazing regenerating body/The many ages of man. *New Scientist* **2556**, 15.

Visser, G.H., Dekinga, A., Achterkamp, B., and Piersma, T. (2000). Ingested water equilibrates isotopically with the body water pool of a shorebird with unrivaled water fluxes. *American Journal of Physiology—Regulatory, Integrative and Comparative Physiology* **279**, R1795–1804.

Visser, M.E., Adriaensen, F., van Balen, J.H., Blondel, J., Dhondt, A.A., van Dongen, S., du Feau, C., Ivankina, E.V., Kerimov, A.B., De Laet, J., Matthysen, E., McCleery, R.H., Orell, M., and Thomson, D.L. (2003). Variable responses to large-scale climate change in European *Parus* populations. *Proceedings of the Royal Society, London, B* **270**, 367–72.

Visser, M.E., Both, C., and Lambrechts, M.M. (2004). Global climate change leads to mistimed avian reproduction. *Advances in Ecological Research* **35**, 89–110.

Vitousek, P.M., Mooney, H.A., Lubchenko, J., and Melillo, J.M. (1997). Human domination of earth's ecosystems. *Science* **277**, 494–99.

Vleck, C.M. and Vleck, D. (1979). Metabolic rate in 5 bird species. *Condor* **81**, 89–91.

Vogel, S. (2001). *Prime mover: A natural history of muscle*. Norton, New York.

Vogel, S. (2003). *Comparative biomechanics: Life's physical world*. Princeton University Press, Princeton, NJ.

Vogt, G., Huber, M., Thiemann, M., van den Boogaart, G., Schmitz, O.J., and Schubart, C.D. (2008). Production of different phenotypes from the same genotype in the same environment by developmental variation. *Journal of Experimental Biology* **211**, 510–23.

Vollrath, F. (1998). Dwarf males. *Trends in Ecology and Evolution* **13**, 159–63.

Waage, J.K. (1979). Foraging for patchily distributed hosts by the parasitoid, *Nemeritis canescens*. *Journal of Animal Ecology* **48**, 353–71.

Waddington, C.H. (1942). Canalization of development and the inheritance of acquired characters. *Nature* **150**, 563–65.

Wagner, P.D. (1993). Algebraic analysis of the determinants of VO$_2$max. *Respiratory Physiology* **93**, 221–37.

Wajnberg, É. (2006). Time allocation strategies in insect parasitoids: from ultimate predictions to proximate behavioural mechanisms. *Behavioral Ecology and Sociobiology* **60**, 589–611.

Walsberg, G.E. (1993). Thermal consequences of diurnal microhabitat selection in a small bird. *Ornis Scandinavica* **24**, 174–82.

Walsberg, G.E. (2000). Small mammals in hot deserts: some generalizations revisited. *Bioscience* **50**, 109–20.

Walther, G.R., Berger, S., and Sykes, M.T. (2005). An ecological 'footprint' of climate change. *Proceedings of the Royal Society, London, B* **272**, 1427–32.

Ward, P. and Zahavi, A. (1973). The importance of certain assemblages of birds as 'information centres' for finding food. *Ibis* **115**, 517–34.

Ward, S., Moller, U., Rayner, J.M.V., Jackson, D.M., Nachtigall, W., and Speakman, J.R. (2004). Metabolic power of European starlings *Sturnus vulgaris* during flight in a wind tunnel, estimated from heat transfer modelling, doubly labelled water and mask respirometry. *Journal of Experimental Biology* **207**, 4291–98.

Waters, W.W., Ziegler, M.G., and Meck, J.V. (2002). Postspaceflight orthostatic hypotension occurs mostly in women and is predicted by low vascular resistance. *Journal of Applied Physiology* **92**, 586–94.

Watling, L. and Norse, E.A. (1998). Disturbance of the seabed by mobile fishing gear: a comparison to forest clear-cutting. *Conservation Biology* **12**, 1180–97.

Watt, W.B. (2000). Avoiding paradigm-based limits to knowledge of evolution. *Evolutionary Biology* **32**, 73–96.

Weathers, W.W. (1979). Climatic adaptation in avian standard metabolic rate. *Oecologia* **42**, 81–89.

Weathers, W.W. (1997). Energetics and thermoregulation by small passerines of the humid, lowland tropics. *Auk* **114**, 341–53.

Weber, T.P., Bergudd, J.J., Hedenström, A., Persson, K., and Sandberg, G. (2005). Resistance of flight feathers to mechanical fatigue covaries with moult strategy in two warbler species. *Biology Letters* **1**, 27–30.

Weibel, E.R. (1985). Design and performance of muscular systems: an overview. *Journal of Experimental Biology* **115**: 405–12.

Weibel, E.R. (2000). *Symmorphosis: On form and function in shaping life*. Harvard University Press, Cambridge, Mass.

Weibel, E.R. (2002). The pitfalls of power laws. *Nature* **417**, 131–32.

Weibel, E.R. and Bolis, L. (1998). In memory of Charles Richard Taylor (1939–1995). In *Principles of animal design: The optimization and symmorphosis debate*, eds E.R. Weibel, C.R. Taylor, and L. Bolis, pp. xv–xvii. Cambridge University Press, Cambridge.

Weibel, E.R. and Hoppeler, H. (2005). Exercise-induced maximal metabolic rates scales with muscle aerobic capacity. *Journal of Experimental Biology* **208**, 1635–44.

Weibel, E.R., Taylor, C.R., and Hoppeler, H. (1991). The concept of symmorphosis: a testable hypothesis of structure-function relationship. *Proceedings of the National Academy of Sciences, USA*, **88**, 10357–61.

Weibel, E.R., Taylor, C.R., and Hoppeler, H. (1992). Variations in function and design: testing symmorphosis in the respiratory system. *Respiratory Physiology* **87**, 325–48.

Weibel, E.R., Taylor, C.R., Weber, J-M., Vock, R., Roberts, T.J., and Hoppeler, H. (1996). Design of the oxygen and substrate pathways. VII. Different structural limits for oxygen and substrate supply to muscle mitochondria. *Journal of Experimental Biology* **199**, 1699–1709.

Weibel, E.R., Taylor, C.R., and Bolis, L. eds (1998). *Principles of animal design: The optimization and symmorphosis debate.* Cambridge University Press, Cambridge.

Weibel, E.R., Bacigalupe, L.D., Schmitt, B., and Hoppeler, H. (2004). Allometric scaling of maximal metabolic rate in mammals: muscle aerobic capacity as determinant factor. *Respiratory Physiology and Neurobiology* **140**, 115–52.

Weiner, J. (1992). Physiological limits to sustainable energy budgets in birds and mammals: Ecological implications. *Trends in Ecology and Evolution* **7**, 384–88.

Weiner, J. (1994). *The beak of the finch: A story of evolution in our time.* Jonathan Cape, London.

Weishaupt, M.A., Staernpfli, H., Billeter, R., and Straub, R. (1996). Temperature changes during strenuous exercise in different body compartments of the horse. *Pferdeheilkunde* **12**, 450–54.

Weiss, S.L., Lee, E.A., and Diamond, J.M. (1998). Evolutionary matches of enzyme and transporter capacities to dietary substrate loads in the intestinal brush border. *Proceedings of the National Academy of Sciences, USA* **95**, 2117–21.

Welham, C.V.J. and Ydenberg, R.C. (1993). Efficiency-maximizing flight speeds in parent black terns. *Ecology* **74**, 1893–1901.

Werner, E.E. and Hall, D.J. (1974). Optimal foraging and the size selection of prey by the bluegill sunfish (*Lepomis macrochirus*). *Ecology* **55**, 1042–52.

Werner, E.E. and Peacor, S.D. (2003). A review of trait-mediated indirect interactions in ecological communities. *Ecology* **84**, 1083–1100.

West, G.B., Savage, V.M., Gillooly, J., Enquist, B.J., Woodruff, W.H., and Brown, J.H. (2003). Why does metabolic rate scale with body size? *Nature* **421**, 713.

West-Eberhard, M.J. (1992). Adaptation: current usages. In *Keywords in evolutionary biology*, eds E.F. Keller and E.A. Lloyd, pp. 13–18. Harvard University Press, Cambridge, Mass.

West-Eberhard, M.J. (2003). *Developmental plasticity and evolution.* Oxford University Press, New York.

West-Eberhard, M.J. (2005). Developmental plasticity and the origin of species differences. *Proceedings of the National Academy of Sciences, USA* **102**, 6543–49.

Westerterp, K.R., Saris, W.H.M., van Es, M., and ten Hoor, F. (1986). Use of the doubly labeled water technique in humans during heavy sustained exercise. *Journal of Applied Physiology* **61**, 2162–67.

Whedon, G.D. and Rambaut, P.C. (2006). Effects of long-duration space flight on calcium metabolism: review of human studies from Skylab to the present. *Acta Astronautica* **58**, 59–81.

White, C.R. and Seymour, R.S. (2004). Does basal metabolic rate contain a useful signal? Mammalian BMR allometry and correlations with a selection of physiological, ecological, and life-history variables. *Physiological and Biochemical Zoology* **77**, 929–41.

White, C.R. and Seymour, R.S. (2005). Allometric scaling of mammalian metabolism. *Journal of Experimental Biology* **208**, 1611–19.

White, C.R., Blackburn, T.M., Martin, G.R., and Butler, P.J. (2007). Basal metabolic rate of birds is associated with habitat temperature and precipitation, not primary productivity. *Proceedings of the Royal Society, London, B* **274**, 287–93.

White, C.R., Terblanche, J.S., Kabat, A.P., Blackburn, T.M., Chown, S.L., and Butler, P.J. (2008). Allometric scaling of maximum metabolic rate: the influence of temperature. *Functional Ecology* **22**, 616–23.

White, T.C.R. (1993). *The inadequate environment: Nitrogen and the abundance of animals.* Springer-Verlag, Berlin.

Wiersma, P. and Piersma, T. (1994). Effects of microhabitat, flocking, climate and migratory goal on energy expenditure in the annual cycle of red knots. *Condor* **96**, 257–79.

Wiersma, P. and Piersma, T. (1995). Scoring abdominal profiles to characterize migratory cohorts of shorebirds: an example with red knots. *Journal of Field Ornithology* **66**, 88–98.

Wiersma, P., Bruinzeel, L., and Piersma, T. (1993). Energy savings in waders: studies on the insulation of knots. *Limosa* **66**, 41–52.

Wiersma, P., Selman, C., Speakman, J.R., and Verhulst, S. (2004). Birds sacrifice oxidative protection for reproduction. *Proceedings of the Royal Society, London, B* **271**, S360–63.

Wiersma, P., Muñoz-Garcia, A., Walker, A., and Williams, J.B. (2007). Tropical birds have a slow pace of life. *Proceedings of the National Academy of Sciences, USA* **104**, 9340–45.

Wijnandts, H. (1984). Ecological energetics of the long-eared owl (*Asio otus*). *Ardea* **72**, 1–92.

Wilbur, H.M. (1990). Coping with chaos: toads in ephemeral ponds. *Trends in Ecology and Evolution* **5**, 37.

Wilbur, H.M. and Collins, J.P. (1973). Ecological aspects of amphibian metamorphosis. *Science* **182**, 1305–14.

Williams, G.C. (1966a). Natural selection, the costs of reproduction, and a refinement of Lack's principle. *American Naturalist* **100**, 687–90.

Williams, G.C. (1966b). *Adaptation and natural selection.* Princeton University Press, Princeton, NJ.

Williams, J.B. and Tieleman, B.I. (2000). Flexibility in basal metabolic rate and evaporative water loss among hoopoe larks exposed to different environmental temperatures. *Journal of Experimental Biology* **203**, 3153–59.

Williams, J.B. and Tieleman, B.I. (2005). Physiological adaptation in desert birds. *BioScience* **55**, 416–25.

Willmer, P., Stone, G., and Johnston, I. (2000). *Environmental physiology of animals*. Blackwell Science, Oxford.

Wilson, D.S. (1976). Deducing the energy available in the environment: an application of optimal foraging theory. *Biotropica* **8**, 96–103.

Wilson, J.S., Kraft, P.G., and van Damme, R. (2005). Predator-specfic changes in the morphology and swimming performance of larval *Rana lessonae*. *Functional Ecology* **19**, 238–44.

Windig, J.J., Brakefield, P.M., Reitsma, N., and Wilson, J.G.M. (1994). Seasonal polyphenism in the wild: Survey of wing patterns in five species of *Bicyclus* butterflies in Malawi. *Ecological Entomology* **19**, 285–98.

Wingfield, J.C. (2005). The concept of allostasis: Coping with a capricious environment. *Journal of Mammalogy* **86**, 248–54.

Wingfield, J.C. (2008). Organziation of vertebrate annual cycles: implications for control mechanisms. *Philosophical Transactions of the Royal Society, London, B* **363**, 425–41.

Wingfield, J.C. and Kenagy, G.J. (1991). Natural regulation of reproductive cycles. In *Vertebrate endocrinology: Fundamentals and biomedical implications*, eds M. Schreibman and R.E. Jones, Vol. 4B, pp. 181–241. Academic Press, San Diego.

Winslow, C.E.A. and Herrington, L.P. (1949). *Temperature and human life*. Princeton University Press, Princeton, NJ.

Wolf, T.J. and Schmid-Hempel, P. (1989). Extra loads and foraging life span in honeybee workers. *Journal of Animal Ecology* **58**, 943–54.

Wolff, J. (1892). *Das Gesetz der Transformation der Knochen*. A. Hirchwild, Berlin (published in translation in 1986 as *The law of bone remodelling*. Springer-Verlag, Berlin).

Wolff, W.J. and Smit, C.J. (1990). The Banc d'Arguin, Mauritania, as an environment for coastal birds. *Ardea* **78**, 17–38.

Woltereck, R. (1909). Weitere experimentelle Untersuchungen über Artveränderung, speziell über das Wesen quantitativer Artunterschiede bei Daphniden. *Versuche Deutsche Zoologisches Gesellschaft* **19**, 110–72.

Wouters-Adriaans, M.P.E. and Westerterp, K.R. (2006). Basal metabolic rate as a proxy for overnight energy expenditure: the effect of age. *British Journal of Nutrition* **95**, 1166–70.

Wright, J., Stone, R.E., and Brown, N. (2003). Communal roosts as structured information centres in the raven, *Corvus corax*. *Journal of Animal Ecology* **72**, 1003–14.

Wund, M.A., Baker, J.A., Clancy, B., Golub, J.L., and Foster, S.A. (2008). A test of the 'flexible stem' model of evolution: ancestral plasticity, genetic accommodation, and morphological divergence in the threespine stickleback radiation. *American Naturalist* **172**, 449–62.

Yan, L.J. and Sohal, R.S.(2000). Prevention of flight activity prolongs the life span of the housefly, *Musca domestica*, and attenuates the age-associated oxidative damage to specific mitochondrial proteins. *Free Radical Biology and Medicine* **29**, 1143–50.

Yang, H-Y., Chen, B., and Zhang, Z-W. (2008). Seasonal changes in numbers and species composition of migratory shorebirds in northern Bohai Bay, China. *Wader Study Group Bulletin* **115**, 133–39.

Ydenberg, R.C. and Hurd, P. (1998). Simple models of feeding with time and energy constraints. *Behavioral Ecology* **9**, 49–53.

Ydenberg, R.C., Welham, C.V.J., Schmid-Hempel, R., Schmid-Hempel, P., and Beauchamp, G. (1994). Time and energy constraints and the relationships between currencies in foraging theory. *Behavioral Ecology* **5**, 28–34.

Ydenberg, R.C., Butler, R.W., Lank, D.B., Smith, B.D., and Ireland, J. (2004). Western sandpipers have altered migration tactics as peregrine falcon populations have recovered. *Proceedings of the Royal Society, London, B* **271**, 1263–69.

Ydenberg, R.C., Butler, R.W., and Lank, D.B. (2007). Effects of predator landscapes on the evolutionary ecology of routing, timing and molt by long-distance migrants. *Journal of Avian Biology* **38**, 523–29.

Yoder, J.A., McClain, C.R., Feldman, G.C., and Esaias, W.E. (1993). Annual cycles of phytoplankton chlorophyll concentrations in the global ocean: a satellite view. *Global Biochemical Cycles* **7**, 181–93.

Zuk, M. (2007). *Riddled with life: Friendly worms, ladybug sex, and the parasites that make us who we are*. Harcourt, Orlando.

Zwarts, L. and Wanink, J.H. (1991). The macrobenthos fraction accessible to waders may represent marginal prey. *Oecologia* **87**, 581–87.

Zwarts, L., Blomert, A-M., and Wanink, J.H. (1992). Annual and seasonal variation in the food supply harvestable by knot *Calidris canutus* staging in the Wadden Sea in late summer. *Marine Ecology Progress Series* **83**, 129–39.

Zwarts, L., Ens, B.J., Goss-Custard, J.D., Hulscher, J.B., and Kersten, M. (1996). Why oystercatchers *Haematopus ostralegus* cannot meet their daily energy requirements in a single low water period. *Ardea* **84A**, 269–90.

Zwarts, L., Bijlsma, R.G., van der Kamp, J., and Wymenga, E. (2009). *Living on the edge: Wetlands and birds in a changing Sahel*. KNNV Publishing, Zeist.

Name index

Subject index

In the hope that this index will be as flexible as the subject matter on which it throws light, the page numbers **in bold** *represent chapters or major sub-sections of text. Readers can therefore concentrate on these, for a speedy appraisal of a topic, or delve deeper into the sub-headings for further enlightenment. The extensive cross-referencing should guide you to any related ideas—phenotypes are variable, and so is the terminology used to describe them!*